IET ELECTI
Series I
T0258387
Professor J.R. Wait

Geometrical Theory of Diffraction for Electromagnetic Waves

Third Edition

Other volumes in this series:

Volume 1 **Geometrical theory of diffraction for electromagnetic waves, 3rd edition** G.L. James
Volume 10 **Aperture antennas and diffraction theory** E.V. Jull
Volume 11 **Adaptive array principles** J.E. Hudson
Volume 12 **Microstrip antenna theory and design** J.R. James, P.S. Hall and C. Wood
Volume 15 **The handbook of antenna design, volume 1** A.W. Rudge, K. Milne, A.D. Oliver and P. Knight (Editors)
Volume 16 **The handbook of antenna design, volume 2** A.W. Rudge, K. Milne, A.D. Oliver and P. Knight (Editors)
Volume 18 **Corrugated horns for microwave antennas** P.J.B. Clarricoats and A.D. Oliver
Volume 19 **Microwave antenna theory and design** S. Silver (Editor)
Volume 21 **Waveguide handbook** N. Marcuvitz
Volume 23 **Ferrites at microwave frequencies** A.J. Baden Fuller
Volume 24 **Propagation of short radio waves** D.E. Kerr (Editor)
Volume 25 **Principles of microwave circuits** C.G. Montgomery, R.H. Dicke and E.M. Purcell (Editors)
Volume 26 **Spherical near-field antenna measurements** J.E. Hansen (Editor)
Volume 28 **Handbook of microstrip antennas, 2 volumes** J.R. James and P.S. Hall (Editors)
Volume 31 **Ionospheric radio** K. Davies
Volume 32 **Electromagnetic waveguides: theory and applications** S.F. Mahmoud
Volume 33 **Radio direction finding and superresolution, 2nd edition** P.J.D. Gething
Volume 34 **Electrodynamic theory of superconductors** S.A. Zhou
Volume 35 **VHF and UHF antennas** R.A. Burberry
Volume 36 **Propagation, scattering and diffraction of electromagnetic waves** A.S. Ilyinski, G. Ya.Slepyan and A. Ya.Slepyan
Volume 37 **Geometrical theory of diffraction** V.A. Borovikov and B.Ye. Kinber
Volume 38 **Analysis of metallic antenna and scatterers** B.D. Popovic and B.M. Kolundzija
Volume 39 **Microwave horns and feeds** A.D. Olver, P.J.B. Clarricoats, A.A. Kishk and L. Shafai
Volume 41 **Approximate boundary conditions in electromagnetics** T.B.A. Senior and J.L. Volakis
Volume 42 **Spectral theory and excitation of open structures** V.P. Shestopalov and Y. Shestopalov
Volume 43 **Open electromagnetic waveguides** T. Rozzi and M. Mongiardo
Volume 44 **Theory of nonuniform waveguides: the cross-section method** B.Z. Katsenelenbaum, L. Mercader Del Rio, M. Pereyaslavets, M. Sorella Ayza and M.K.A. Thumm
Volume 45 **Parabolic equation methods for electromagnetic wave propagation** M. Levy
Volume 46 **Advanced electromagnetic analysis of passive and active planar structures** T. Rozzi and M. Farinai
Volume 47 **Electromagnetic mixing formulae and applications** A. Sihvola
Volume 48 **Theory and design of microwave filters** I.C. Hunter
Volume 49 **Handbook of ridge waveguides and passive components** J. Helszajn
Volume 50 **Channels, propagation and antennas for mobile communications** R. Vaughan and J. Bach-Anderson
Volume 51 **Asymptotic and hybrid methods in electromagnetics** F. Molinet, I. Andronov and D. Bouche
Volume 52 **Thermal microwave radiation: applications for remote sensing** C. Matzler (Editor)
Volume 502 **Propagation of radiowaves, 2nd edition** L.W. Barclay (Editor)

Geometrical Theory of Diffraction for Electromagnetic Waves

Third Edition

Graeme L. James

The Institution of Engineering and Technology

Preface

This is the third edition of a book that first appeared in 1976. Over the past decade there have been significant advances in the Geometrical Theory of Diffraction (GTD) and related topics, and hence a substantial revision of the previous editions was necessary in order to bring the text up to date. To this end new material has been included in most chapters, with Chapter 1 being entirely rewritten. In this introductory chapter a concise survey of GTD and its association with related techniques is given. Chapter 2 gives the basic equations in electromagnetic theory required in later work, the special functions which are to be found in diffraction theory, and a section on the asymptotic evaluation of integrals. In Chapter 3 the formal derivations of the solutions to the canonical problems that have formed the basis of GTD are given. The laws of geometrical optics are developed in Chapter 4 from the appropriate canonical problem, and in Chapter 5 high-frequency diffraction by straight edges and surfaces is considered. Chapter 6 is concerned with the application of GTD to curved edges and surfaces. To conclude, a number of worked examples are given in Chapter 7 to demonstrate the practical application of the GTD techniques developed in the earlier chapters.

The purpose of this book, apart from expounding the GTD method, is to present useful formulations that can be readily applied to solve practical engineering problems. It is not essential, therefore, to understand in detail the material in Chapters 2 and 3, and many readers will want to treat these chapters as a reference only. At the end of Chapters 4 to 6 a summary is provided which gives the main formulas developed in the chapter.

I wish to acknowledge the assistance of colleagues at the CSIRO Division of Radiophysics in the preparation of this third edition, in particular Miss M. Vickery and Dr. T.S. Bird. Further, I am indebted to the following people for their contribution to specific sections: Drs. D.P.S. Malik, G.T. Poulton and G. Tong.

G.L. James
October 1985

Chapter 1

Introduction

The *Geometrical Theory of Diffraction* (GTD), conceived by J.B. Keller in the 1950s and developed continuously since then, is now established as a leading analytical technique in the prediction of high-frequency diffraction phenomena. Basically, GTD is an extension of geometrical optics by the inclusion of additional *diffracted rays* to describe the diffracted field. The concept of diffracted rays was developed by Keller from the asymptotic evaluation of the known exact solution to scattering from simple shapes, referred to as the *canonical problems* for GTD. This rigorous mathematical foundation, and the basic simplicity of ray tracing techniques which permits GTD to treat quite complicated structures, are the main attractions of the method.

In this book we will be concerned with GTD and its applications to electromagnetic wave diffraction. Our main interest, apart from expounding the techniques of GTD, is to develop useful formulations that can be readily applied to solve practical engineering problems. The approach taken is to begin with the solutions to canonical problems, from which we develop the GTD method to treat more complicated structures. In this way the laws of geometrical optics are derived from the canonical problem of plane wave reflection and refraction at an infinite planar dielectric interface. The GTD methods for various diffraction phenomena which follow are then seen to be obtained by a natural extension of this approach to other canonical problems. For example, the half-plane and wedge solutions form the basis of the GTD formulation for edge diffraction. Similarly, the GTD formulation for diffraction around a smooth convex surface is developed from the canonical problem of scattering by a circular cylinder.

In many instances the exact solution to the canonical problem does not exist or is in a form not readily amenable to an asymptotic

formulations to evaluate the field. The most well-known and often used fomulations are the classical methods of aperture field and physical optics approximation. Other more accurate integral equation methods are sometimes used, especially in those circumstances where they can test the accuracy of the various GTD formulations. A number of integral equation methods often used in association with GTD are listed in Table 2. (As with ray tracing methods they are commonly abbreviated as shown in the table).

The first three techniques in Table 2 we have already alluded to and they will be discussed further in the text as indicated. The Physical Theory of Diffraction (PTD) was developed by Ufimtsev (1962), and in concept it has close parallels with the development of GTD. Basically, PTD is an extension of physical optics by an additional current derived from the appropriate canonical problem. Fock (1946) originally suggested this method for diffraction around a smooth convex surface, and Ufimtsev extended the concept to include any shape which deviates from an infinite planar metallic surface. In practice however, Ufimtsev only applied this method to edge diffraction. The major difficulty of PTD, as is the case with all integral equation methods, is that the resultant integrals are not always easily evaluated. Nevertheless, it does find application (as with integral equation methods in general) to those regions where the GTD method fails. A comparison between these two high-frequency methods is to be found in Knott and Senior (1974), Ufimtsev (1975), and Lee (1977).

In some cases (notably the half-plane problem), solutions for the scattered field can be represented in terms of the Fourier transform (or the spectrum) of the induced surface current distribution. This correspondence between the scattered field and the currents flowing on the surface of an obstacle was expressed as a general concept by Mittra *et al.* (1976) as the Spectral Theory of Diffraction (STD). Initially STD was applied to problems involving half-planes and has

Table 2 Integral equation methods

		Major references in the text
ECM	Equivalent Current Method	6.7
–	Aperture Field	2.1.3
PO	Physical Optics	2.1.4, 5.6
PTD	Physical Theory of Diffraction	–
STD	Spectral Theory of Diffraction	–
MM	Moment Method	–

subsequently been applied to plane-wave diffraction at a wedge (Clarkowski *et al.*, 1984). As with PTD the resulting integrals are not always readily evaluated. The main use of STD to date has been to provide accurate and sometimes exact solutions with which to test and compare various approximate asymptotic methods. It has so far been limited to analysing simple geometries, and it would not appear to be readily amenable to treat more complex scattering bodies.

The Moment Method (MM) (Harrington, 1968) is a well-known numerical method for analysing field problems associated with obstacles of small dimensions. In certain applications the combination of MM with asymptotic techniques such as GTD can yield solutions which neither method can achieve effectively alone. For a concise overview of these hybrid methods and their applications, the reader is referred to the paper by Thiele (1982). It is likely that we shall see an increase in the practical application of combined GTD-MM and other hybrid techniques as they continue to be developed.

A number of general reviews of asymptotic methods in diffraction theory are available, in particular those by Kouyoumjian (1965), Borovikov and Kinber (1974), Knott (1985), and Keller (1985). These papers, which are largely non-mathematical, provide additional insight into GTD and associated asymptotic methods.

References

BOROVIKOV, V.A., and KINBER, B.Ye. (1974): 'Some problems in the asymptotic theory of diffraction', *Proc. IEEE*, **62**, pp. 1416–1437.

BURNSIDE, W.D., WANG, N., and PELTON, E.L. (1980): 'Near-field pattern analysis of airborne antennas', *IEEE Trans.*, **AP-28**, pp. 318–327.

CHAMBERLIN, K.A., and LUEBBERS, R.J. (1982): 'An evaluation of Longley-Rice and GTD propagation models', *ibid.*, **AP-30**, pp. 1093–1098.

CLARKOWSKI, A., BOERSMA, J., and MITTRA, R., (1984): 'Plane-wave diffraction by a wedge – a spectral domain approach', *ibid.*, **AP-32**, pp. 20–29.

FOCK, V.A. (1946): 'The distribution of currents induced by a plane wave on the surface of a conductor', *J. Phys., USSR*, **10**, pp. 130–136.

HANSEN, R.C. (Ed.) (1981): 'Geometrical theory of diffraction', (IEEE Press).

HARRINGTON, R.F. (1968): 'Field Computation by Moment Methods', (Macmillan, New York).

JONES, R.M. (1984): 'How edge diffraction couples ground wave modes at a shoreline', *Rad. Sci.*, **19**, pp. 959–965.

KELLER, J.B. (1957a): 'Diffraction by an aperture', *J. Appl. Phys.*, **28**, pp. 426–444.

KELLER, J.B., LEWIS, R.M., and SECKLER, B.D. (1957b): 'Diffraction by an aperture II', *ibid*, **28** pp. 570–579.

KELLER, J.B. (1962): 'Geometrical theory of diffraction', *J. Opt. Soc. Am.*, **52**, pp. 116–130.

KELLER, J.B. (1985): 'One hundred years of diffraction theory', *IEEE Trans.*, **AP-33**, pp. 123–126.
KINBER, B.Ye. (1961): 'Sidelobe radiation of reflector antenna', *Radio Eng. and Electron. Phys.*, **6**, pp. 545–558.
KINBER, B.Ye. (1962a): 'The role of diffraction at the edges of a paraboloid in fringe radiation', *ibid.*, **7**, pp. 79–86.
KINBER, B.Ye. (1962b): 'Diffraction at the open end of a sectoral horn', *ibid.*, **7**, pp. 1620–1623.
KNOTT, E.F., and SENIOR, T.B.A. (1974): 'Comparison of three high-frequency diffraction techniques', *Proc. IEEE*, **62**, pp. 1468–1474.
KNOTT, E.F. (1985): 'A progression of high-frequency RCS prediction techniques', *ibid.*, **73**, pp. 252–264.
KOUYOUMJIAN, R.G. (1965): 'Asymptotic high-frequency methods', *ibid.*, **53**, pp. 864–876.
LEE, S.W. (1977): 'Comparison of uniform asymptotic theory and Ufimtsev's theory of electromagnetic edge diffraction', *IEEE Trans.*, **AP-25**, pp. 162–170.
LEVY, B.R., and KELLER, J.B. (1959): 'Diffraction by a smooth object', *Comm. Pure Appl. Math.*, **12**, pp. 159–209.
LUEBBERS, R.J. (1984): 'Propagation prediction for hilly terrain using GTD wedge diffraction', *IEEE Trans.*, **AP-32**, pp. 951–955.
MITTRA, R., RAHMAT-SAMII, Y., and KO, W.L. (1976): 'Spectral theory of diffraction', *Applied Physics*, **10**, pp. 1–13.
THIELE, G.A. (1982): 'Overview of hybrid methods which combine the moment method and asymptotic techniques', *Proc. SPIE Int. Soc. Opt. Eng.* (*USA*), **358**, pp. 73–79.
UFIMTSEV, P.Ya. (1962): 'The method of fringe waves in the physical theory of diffraction', Sovyetskoye radio, Moscow. Now translated and available from the US Air Force Foreign Technology Division, Wright-Patterson, AFB, Ohio, USA.
UFIMTSEV, P.Ya. (1975): Comments on 'Comparison of three high-frequency diffraction techniques', *IEEE Proc.*, **63**, pp. 1734–1737.

Chapter 2

Electromagnetic fields

In this chapter we begin with the field equations and representations of the electromagnetic field which will be required in later developments. Our main concern is with radiation and scattering from current distributions, and use of potentials, both scalar and vector, to describe the field. This is succeeded by a section on special functions which appear in diffraction theory, namely the Fresnel integral, Airy, Fock and Hankel functions. Fresnel integral functions have an essential role in edge diffraction phenomena while both Airy and Fock functions are used in describing diffraction around a convex surface. The discussion of the Hankel functions is concerned with their asymptotic behaviour for large values of argument. These various functions are not always well tabulated, and in some instances it is necessary to resort to numerical evaluation.

An asymptotic evaluation of the field integrals is given in the last section. The solutions will prove to be useful in extending the methods of GTD and in clarifying associated asymptotic methods such as the physical optics approximation.

2.1 Basic equations

2.1.1 Field equations

Maxwell's equations for time-harmonic electromagnetic fields, with the time dependence exp $(j\omega t)$ suppressed, are

$$\begin{aligned} -\nabla \times \boldsymbol{E} &= j\omega \boldsymbol{B} + \boldsymbol{M} \\ \nabla \times \boldsymbol{H} &= j\omega \boldsymbol{D} + \boldsymbol{J} + \boldsymbol{J}^c \end{aligned} \quad (1)$$

where $\boldsymbol{E}$ is the *electric field intensity*, $\boldsymbol{H}$ is the *magnetic field intensity*, $\boldsymbol{B}$ is the *magnetic flux density*, $\boldsymbol{D}$ is the *electric flux density*, $\boldsymbol{M}$ is the *magnetic source current*, $\boldsymbol{J}$ is the *electric source current*, $\boldsymbol{J}^c$ is the

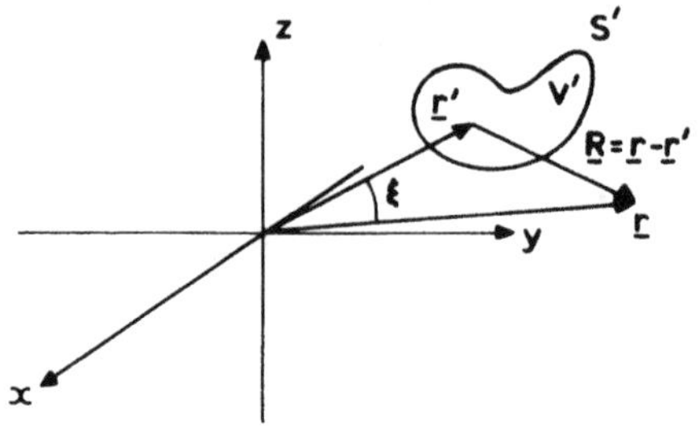

Fig. 2.1 Source distribution contained within a volume V'

$$\nabla G = -\left(jk + \frac{1}{R}\right) G\hat{R}; \qquad R\hat{R} = r - r'$$

$$J \cdot \nabla\nabla G = \left[(J \cdot \hat{R}) R \left\{ -k^2 + \frac{3}{R}\left(jk + \frac{1}{R}\right)\right\} - \frac{1}{R}\left(jk + \frac{1}{R}\right) J \right] G$$

and a similar expression exists for the quantity $M \cdot \nabla\nabla G$. Substitution into eqn. 10 gives

$$E(r) = -jk\int_{V'} \left(M(r') \times \hat{R} + \sqrt{\left(\frac{\mu}{\epsilon}\right)}\, [J(r') - \{J(r') \cdot \hat{R}\}\hat{R}] \right) G(r, r') dV' + 0\left(\frac{1}{R^2}\right)$$

$$H(r) = jk\int_{V'} \left(J(r') \times \hat{R} - \sqrt{\left(\frac{\epsilon}{\mu}\right)}\, [M(r') - \{M(r') \cdot \hat{R}\}\hat{R}] \right) G(r, r') dV' + 0\left(\frac{1}{R^2}\right) \qquad (11)$$

This solution for the electromagnetic field is applicable only in a source-free region (which, as formulated in eqn. 11, is the region external to the volume V' containing the sources). In radiation problems this source-free region is conveniently divided up into three overlapping zones; namely, the near zone, the Fresnel zone, and the far zone. In the *near zone* where R is small, no approximations in the evaluation of eqn. 11 are permissible and we must include all higher order terms in R. For the *Fresnel zone*, R is sufficiently large for the field to be given, to a good approximation, by the leading term in eqn. 11. Beyond the Fresnel zone we have the *far zone* where $r \gg r'_{max}$ and the following approximations are made in the first term of eqn. 11. From Fig. 2.1

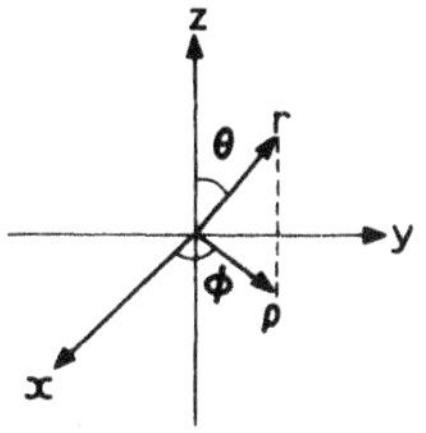

Fig. 2.2 Co-ordinate orientation

$$R = (r^2 + r'^2 - 2rr'\cos\xi)^{\frac{1}{2}}$$

$$R \simeq r - r'\cos\xi \qquad \text{(for phase terms)}$$

$$R \simeq r \qquad \text{(for amplitude terms)}$$

and eqn. 11 for the far zone simplifies to

$$E(r) \sim -jk\,\frac{\exp(-jkr)}{4\pi r}\int_{V'}\left(M(r')\times\hat{r} + \sqrt{\left(\frac{\mu}{\hat{\epsilon}}\right)}\,[J(r') - \{J(r')\cdot\hat{r}\}\hat{r}]\right)$$

$$\exp(jkr'\cos\xi)dV' \qquad (12)$$

$$H(r) \sim \sqrt{\left(\frac{\hat{\epsilon}}{\mu}\right)}\,\hat{r}\times E(r); \qquad r \gg r'_{max}$$

In the far zone we see that the electric and magnetic field components $\boldsymbol{E}$ and $\boldsymbol{H}$ are perpendicular to each other and to the direction of propagation.

In evaluating field quantities we shall use the rectangular (x, y, z), cylindrical (ρ, ϕ, z), and spherical (r, θ, ϕ) co-ordinate systems as defined in Fig. 2.2. For eqn. 12 we require $r'\cos\xi$ in terms of the source co-ordinates. From $\boldsymbol{r}\cdot\boldsymbol{r}' = rr'\cos\xi$ we obtain

$$r'\cos\xi = (x'\cos\phi + y'\sin\phi)\sin\theta + z'\cos\theta \qquad (13a)$$

$$r'\cos\xi = \rho'\cos(\phi-\phi')\sin\theta + z'\cos\theta \qquad (13b)$$

$$r'\cos\xi = r'(\cos(\phi-\phi')\sin\theta\sin\theta' + \cos\theta\cos\theta') \qquad (13c)$$

So far, only 3-dimensional current distributions have been considered. For 2-dimensional distributions the field equations given by eqn. 10 are valid, provided that the correct Green's function is used. With the electromagnetic field independent of the z-direction, it has the form

$$G(\boldsymbol{\rho}, \boldsymbol{\rho}') = \frac{1}{4j}H_0^{(2)}(k\,|\boldsymbol{\rho} - \boldsymbol{\rho}'|) \qquad (14)$$

where $H_0^{(2)}$ is the zero-order Hankel function of the second kind. In solving eqn. 10 for the Fresnel zone we can take the asymptotic expansion of the Hankel function

$$H_\nu^{(2)}(k\alpha) \sim \sqrt{\left(\frac{2j}{\pi k\alpha}\right)}\, j^\nu \exp(-jk\alpha) \qquad \text{for } k\alpha \gg \nu \tag{15}$$

so that the substitution into eqn. 10 yields for the Fresnel zone

$$\begin{aligned} \boldsymbol{E}(\boldsymbol{\rho}) &= -jk\int_{A'}\left\{\boldsymbol{M}(\boldsymbol{\rho}')\times\hat{\boldsymbol{P}} + \sqrt{\left(\frac{\mu}{\hat{\epsilon}}\right)}\,[\boldsymbol{J}(\boldsymbol{\rho}') - \{\boldsymbol{J}(\boldsymbol{\rho}')\cdot\hat{\boldsymbol{P}}\}\hat{\boldsymbol{P}}]\right\} \\ &\qquad \frac{\exp(-jkP)}{\sqrt{(8j\pi kP)}}\, dA' \\ \boldsymbol{H}(\boldsymbol{\rho}) &= jk\int_{A'}\left\{\boldsymbol{J}(\boldsymbol{\rho}')\times\hat{\boldsymbol{P}} - \sqrt{\left(\frac{\hat{\epsilon}}{\mu}\right)}\,[\boldsymbol{M}(\boldsymbol{\rho}') - \{\boldsymbol{M}(\boldsymbol{\rho}')\cdot\hat{\boldsymbol{P}}\}\hat{\boldsymbol{P}}]\right\} \\ &\qquad \frac{\exp(-jkP)}{\sqrt{(8j\pi kP)}}\, dA' \end{aligned} \tag{16}$$

where A' is the area enclosing the sources and

$$P\hat{\boldsymbol{P}} = \boldsymbol{\rho} - \boldsymbol{\rho}'$$

In the far zone where $\rho \gg \rho'_{max}$ this reduces further to

$$\begin{aligned} \boldsymbol{E}(\boldsymbol{\rho}) &\sim -jk\,\frac{\exp(-jk\rho)}{\sqrt{(8j\pi k\rho)}}\int_{A'}\left(\boldsymbol{M}(\boldsymbol{\rho}')\times\hat{\boldsymbol{\rho}} + \sqrt{\left(\frac{\mu}{\hat{\epsilon}}\right)}\,[\boldsymbol{J}(\boldsymbol{\rho}') - \{\boldsymbol{J}(\boldsymbol{\rho}')\cdot\hat{\boldsymbol{\rho}}\}\hat{\boldsymbol{\rho}}]\right) \\ &\qquad \exp\{jk\rho'\cos(\phi-\phi')\}dA' \\ \boldsymbol{H}(\boldsymbol{\rho}) &\sim \sqrt{\left(\frac{\hat{\epsilon}}{\mu}\right)}\,\hat{\boldsymbol{\rho}}\times\boldsymbol{E}(\boldsymbol{\rho}); \qquad \rho \gg \rho'_{max} \end{aligned} \tag{17}$$

*2.1.3 *Equivalent source distributions*

In formulating radiation problems it is often convenient to replace the actual sources of the field with an equivalent source distribution. For example, in Fig. 2.3*a* the volume V' contains the sources of a field $\boldsymbol{E}_1, \boldsymbol{H}_1$ and the evaluation of this field in the source free region external to the volume V' is given by eqn. 10. If we now specify a source-free field $\boldsymbol{E}_2, \boldsymbol{H}_2$ internal to V', and maintain the original field external to V' as in Fig. 2.3*b*, then on the bounding surface S' one finds there must exist surface currents

* The interpretation of $\sqrt{j}$ in the above equations and throughout this book is that it is taken to be equivalent to $\exp(j\pi/4)$.

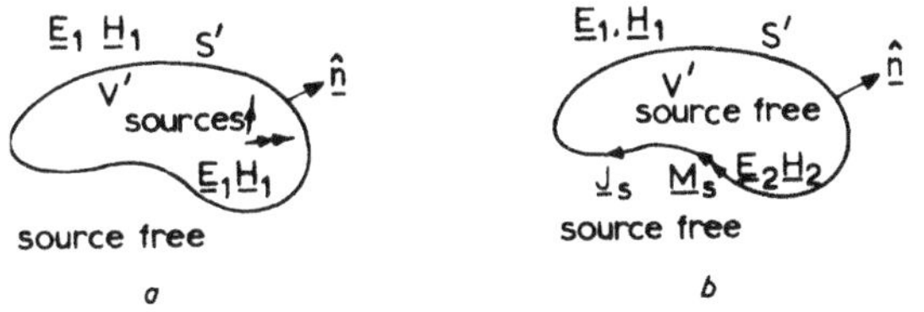

Fig. 2.3 Equivalent source distribution

$$\begin{aligned} \boldsymbol{J}_s &= \hat{\boldsymbol{n}} \times (\boldsymbol{H}_1 - \boldsymbol{H}_2) \\ \boldsymbol{M}_s &= (\boldsymbol{E}_1 - \boldsymbol{E}_2) \times \hat{\boldsymbol{n}} \end{aligned} \qquad (18)$$

to account for the discontinuity in the tangential components of the field. These equations are applicable for a discontinuity in both field and media at the surface S'. When the surface currents $\boldsymbol{J}_s$ and $\boldsymbol{M}_s$ are zero, they state the *boundary conditions* for electromagnetic fields, in that *the tangential components of the electric and magnetic field are continuous across a change in medium provided the conductivity is finite.* For a *perfect electric conductor* $\hat{\epsilon} = \sigma$ is infinite, and a surface conduction current exists while the tangential electric field vanishes. If the region within the volume in Fig. 2.3*b* has infinite conductivity then eqn. 18 becomes

$$0 = \boldsymbol{E}_1 \times \hat{\boldsymbol{n}}; \qquad \boldsymbol{J}_s = \hat{\boldsymbol{n}} \times \boldsymbol{H}_1 \qquad \text{(perfect electric conductor)} \quad (19)$$

Similarly we can mathematically define (although it has no physical meaning) a *perfect magnetic conductor* such that the tangential magnetic field vanishes at its surface. If the region with V' contains this magnetic conductor, eqn. 18 becomes

$$\boldsymbol{M}_s = \boldsymbol{E}_1 \times \hat{\boldsymbol{n}} \qquad 0 = \hat{\boldsymbol{n}} \times \boldsymbol{H}_1 \qquad \text{(perfect magnetic conductor)} \quad (20)$$

Returning to the surface currents of eqn. 18, we can now substitute these sources into eqn. 10 and perform the integration over S' to obtain the field $\boldsymbol{E}_2, \boldsymbol{H}_2$ internal to V', and $\boldsymbol{E}_1, \boldsymbol{H}_1$ external to V'. Thus the equivalent source distribution of eqn. 18 has produced the same field external to the volume as the original sources contained within it. If these sources were external to V', then the resultant field $\boldsymbol{E}_1, \boldsymbol{H}_1$ within the volume would be produced by eqn. 20 where $\hat{n}$ is now the *inward* normal from V', and $\boldsymbol{E}_2, \boldsymbol{H}_2$ is the external field to it. Thus, *having knowledge of the field bounding a source free region allows us to determine the field within that region.* Since the specification of the field $\boldsymbol{E}_2, \boldsymbol{H}_2$ is arbitrary, it is often advantageous to choose a null field for the region containing the actual sources. The surface currents now become

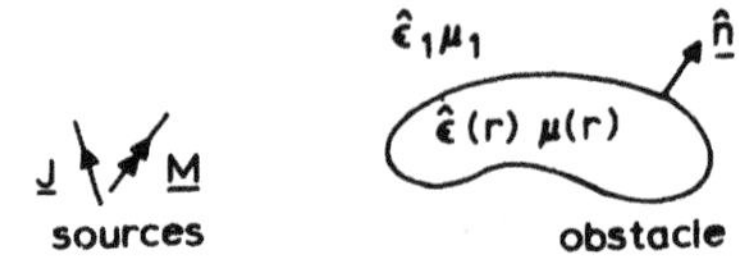

Fig. 2.5 Sources radiating in the presence of an obstacle

integral equations which will determine the field within the obstacle. For a non-magnetic obstacle, $\boldsymbol{M}_{scatt}$ will be zero, and the solution is reduced to solving a single integral equation for the electric field $\boldsymbol{E}$ within the obstacle. Even so, this may lead to formidable computational difficulties if an attempt is made to solve this problem numerically. As a first approximation we may replace $\boldsymbol{E}$, $\boldsymbol{H}$ in eqn. 24 by the known incident field $\boldsymbol{E}^i, \boldsymbol{H}^i$. With $\boldsymbol{E}, \boldsymbol{H}$ calculated by this approximation we may generate an iteration process by substituting these values into eqn. 24 and recalculating *ad lib.*

When the obstacle is perfectly conducting, the magnetic current $\boldsymbol{M}_{scatt}$ will be zero, and the electric current $\boldsymbol{J}_{scatt}$ reduces to a surface current $\boldsymbol{J}_s$ given by $\hat{\boldsymbol{n}} \times \boldsymbol{H}$ as discussed in the previous section. If the obstacle is large and the surface is smoothly varying with large radii of curvature compared to the wavelength, the surface current may be approximated by assuming that each point on the surface behaves locally as if it were part of an infinite ground plane. The tangential component of the magnetic field $\boldsymbol{H}$ at the surface of a ground plane is given by twice the incident tangential field due to the image. Over the illuminated portion of the surface S' we approximate the surface current by

$$\boldsymbol{J}_s = 2\hat{\boldsymbol{n}} \times \boldsymbol{H}^i \tag{25}$$

and the resultant scattered field $\boldsymbol{E}^s, \boldsymbol{H}^s$ is determined from eqn. 21*a*, where the magnetic field $\boldsymbol{H}$ is replaced by $\boldsymbol{H}^i$ and the integration is taken over the illuminated region of S'. This is the basis of the *physical optics approximation* and it is used extensively in electromagnetic scattering problems. Note that, as illustrated in Fig. 2.6, the surface current is assigned a value only over the illuminated region of the surface. In the shadow the surface current is taken to be zero. Thus it is to be expected that this method will be unsuitable for predicting the electromagnetic field in the deep shadow of the obstacle where these neglected currents will be the main producer of the field.

2.1.5 Scalar potentials for source-free regions

In evaluating the field within a homogeneous source-free region using eqn. 4, the potential integral solution of eqn. 8 requires knowledge of

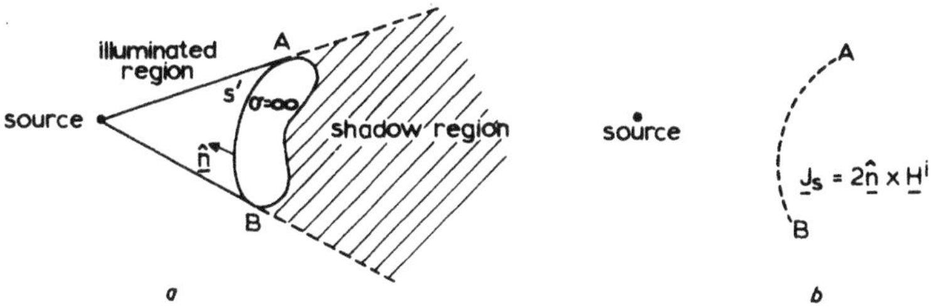

Fig. 2.6 Physical optics approximation for a perfectly conducting obstacle
a Original problem
b Physical optics current distribution

the field bounding the region. An alternative approach is to seek solutions to the Helmholtz equations in eqn. 5 for the potentials within the source-free region. The potentials may then be chosen independently of the actual sources external to the region producing the field. Thus, if we choose $A = \hat{z}u_a$ and $F = \hat{z}u_f$ then the field equations of eqn. 4 become

$$E = \hat{z} \times \nabla u_f - j\omega\mu\hat{z}u_a + \frac{1}{j\omega\hat{\epsilon}}\nabla\left(\frac{\partial u_a}{\partial z}\right)$$
$$H = -\hat{z} \times \nabla u_a - j\omega\hat{\epsilon}\hat{z}u_f + \frac{1}{j\omega\mu}\nabla\left(\frac{\partial u_f}{\partial z}\right) \qquad (26)$$

and the *scalar potentials* u_a, u_f satisfy the scalar Helmholtz equation

$$\nabla^2 u + k^2 u = 0 \qquad (27)$$

For 2-dimensional fields independent of the *z*-direction, the above equations simplify so that

$$u_a = -\frac{1}{j\omega\mu}E_z; \qquad u_f = -\frac{1}{j\omega\hat{\epsilon}}H_z \qquad (28)$$

and the remaining field components are given by

$$-j\omega\hat{\epsilon}E = \hat{z} \times \nabla H_z; \qquad j\omega\mu H = \hat{z} \times \nabla E_z \qquad (29)$$

The simplest electromagnetic field solution to this equation is the plane wave, which has some useful properties that we will exploit later in ray-tracing techniques. As an example, consider an electromagnetic field in a homogeneous medium having *z*-components propagating in the $\hat{n}$-direction, as shown in Fig. 2.7, such that

$$E_z = E_0 \exp\{-jk(x\cos\phi + y\sin\phi)\}; \qquad H_z = 0 \qquad (30a)$$

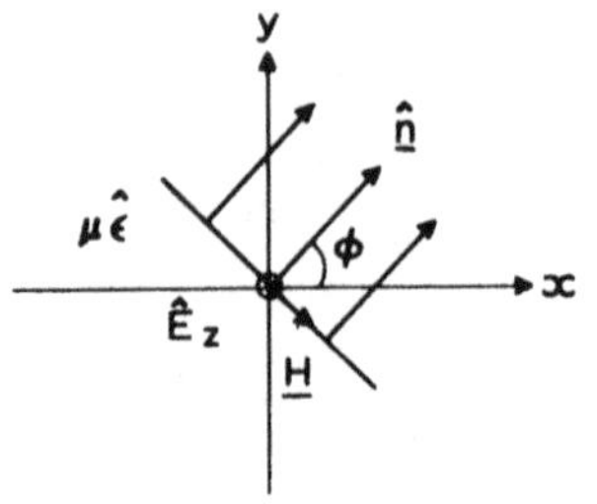

Fig. 2.7 Plane wave propagation in a homogeneous medium

and from eqn. 29 the other field components are

$$H = -\sqrt{\left(\frac{\hat{\epsilon}}{\mu}\right)}\,(\hat{y}\cos\phi - \hat{x}\sin\phi)E_z \tag{30b}$$

From these equations we note a fundamental property of a plane wave in that the electric and magnetic fields are orthogonal to each other and to the direction of propagation. Also, from the Poynting vector defined by eqn. 6, we see that the average flow of energy is in the direction of propagation. This latter statement, however, is true only for isotropic media.

Eqn. 30*a* is an elemental wave function which satisfies the scalar Helmholtz equation of eqn. 27 for 2-dimensional fields. By superposition, a linear sum of elemental wave functions can also represent a solution to eqn. 27, so that we may have

$$u = \sum_n a_n \exp\{-jk(x\cos\phi_n + y\sin\phi_n)\} \tag{31}$$

Alternatively, we can express u in terms of an integration over a range of angle ϕ, such that

$$u = \int_C f(\phi)\exp\{-jk(x\cos\phi + y\sin\phi)\}\,d\phi \tag{32}$$

and the integration path C will, in general, be in the complex ϕ-plane.

For more general solutions to eqn. 27 for 2-dimensional fields independent of the z-direction, we have for u expressed in cylindrical co-ordinates,

$$u = \sum_n a_n \Phi_n(\phi)P_n(\rho) \tag{33}$$

where the functions $\Phi_n(\phi)$ and $P_n(\rho)$ may be of the form

$$\Phi_n(\phi) \quad : \quad \cos n\phi, \quad \sin n\phi, \quad \exp(\pm jn\phi)$$

$$P_n(\rho) \quad : \quad J_n(k\rho), \quad N_n(k\rho), \quad H_n^{(1)}(k\rho), \quad H_n^{(2)}(k\rho)$$

When 3-dimensional field solutions are required, then eqn. 33 becomes a double summation with the addition of the z variable. For different co-ordinate systems we must also use the appropriate functions. Solutions to eqns. 31–33, however, are all that we shall require.

2.2 Special functions

2.2.1 Fresnel integral functions

The *Fresnel integral* is defined for real arguments as

$$F_\pm(x) = \int_x^\infty \exp(\pm jt^2)dt \tag{34}$$

When the argument is zero

$$F_\pm(0) = \frac{\sqrt{(\pi)}}{2} \exp\left(\pm \frac{j\pi}{4}\right) \tag{35}$$

and for large arguments its asymptotic solution is

$$F_\pm(x) \sim \frac{1}{2x} \exp\left\{\pm j\left(x^2 + \frac{\pi}{2}\right)\right\} \sum_{m=0}^{\infty} j^{\mp m} (\tfrac{1}{2})_m x^{-2m} \tag{36}$$

where

$$(\tfrac{1}{2})_0 = 1; \quad (\tfrac{1}{2})_m = \tfrac{1}{2}(\tfrac{1}{2}+1)\ldots(\tfrac{1}{2}+m-1)$$

A useful property is

$$F_\pm(-x) = \sqrt{(\pi)}\exp\left(\pm \frac{j\pi}{4}\right) - F_\pm(x) \tag{37}$$

A function called the *modified Fresnel integral* which we shall use extensively is given by

$$K_\pm(x) = \frac{1}{\sqrt{\pi}} F_\pm(x) \exp\left\{\mp j\left(x^2 + \frac{\pi}{4}\right)\right\} \sim \frac{1}{2x\sqrt{(\pi)}} \exp\left(\pm \frac{j\pi}{4}\right) \sum_{m=0}^{\infty} j^{\mp m} (\tfrac{1}{2})_m x^{-2m} \tag{38a}$$

$$K_\pm(0) = \tfrac{1}{2} \tag{38b}$$

$$K_\pm(-x) = \exp(\mp jx^2) - K_\pm(x) \tag{38c}$$

Another form of $K_-(x)$ is required in the rigorous solution for edge diffraction. Beginning with the well-known result

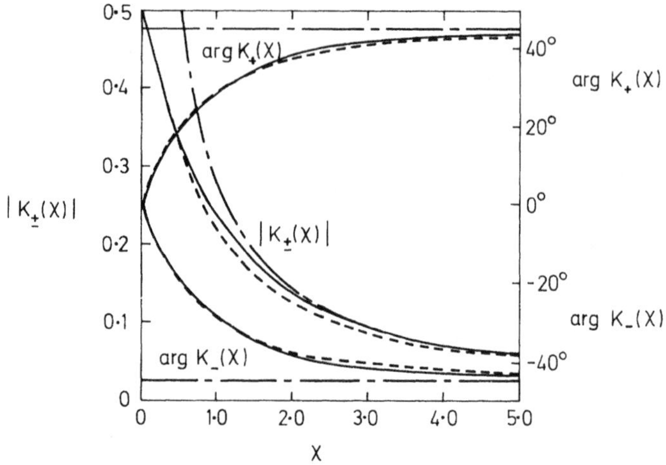

Fig. 2.8 Modified Fresnel integral $K_{\pm}(x)$
——— exact
– – – – approximate
—·—·— asymptotic

2.2.2 *Airy function*

The *Airy function* for complex arguments is given in integral form by

$$\mathrm{Ai}(z) = \frac{1}{2\pi j}\int_C \exp\left(\frac{t^3}{3} - zt\right) dt \tag{50}$$

where C is the contour in the complex t-plane shown in Fig. 2.9*a*. We shall use a different form of the Airy integral to that given in eqn. 50. With the substitution $\eta = t\exp(j\pi/3)$ in eqn. 50, we obtain the relationship

$$\mathrm{Ai}\left(\tau\exp\left(-\frac{j2\pi}{3}\right)\right) = \frac{1}{2\sqrt{(\pi)}}\exp\left(\frac{j\pi}{6}\right)w_1(\tau) \tag{51}$$

where

$$w_1(\tau) = \frac{1}{\sqrt{(\pi)}}\int_\Gamma \exp\left(\tau\eta - \frac{\eta^3}{3}\right) d\eta$$

and the contour Γ is shown in Fig. 2.9*b*.

When the argument is zero

$$\mathrm{Ai}(0) = 0{\cdot}355 \tag{52}$$

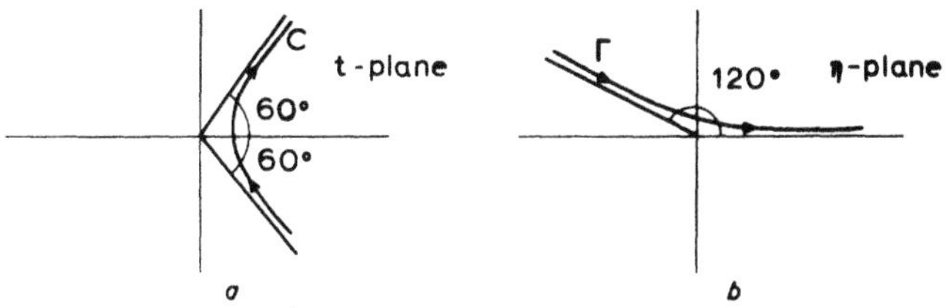

Fig. 2.9 Contours for the Airy integral

and for large arguments its asymptotic solution is

$$\begin{aligned} \mathrm{Ai}(z) &\sim \frac{1}{2\sqrt{(\pi)}} z^{-1/4} \exp(-\tfrac{2}{3} z^{3/2}); \qquad |\arg z| < \pi \\ \mathrm{Ai}(z) &\sim \frac{j}{2\sqrt{(\pi)}} z^{-1/4} \exp(\tfrac{2}{3} z^{3/2}); \qquad \pi < \arg z < 3\pi \end{aligned} \tag{53}$$

For real arguments the contour of eqn. 50 can be deformed into the imaginary axis to give

$$\mathrm{Ai}(x) = \frac{1}{2\pi}\int_{-\infty}^{\infty} \exp\left\{\pm j\left(\frac{t^3}{3} + xt\right)\right\} dt \tag{54}$$

When the argument is negative real, then the Airy function can be rep-represented by the Bessel function relationship

$$\mathrm{Ai}(-x) = \frac{\sqrt{(x)}}{3}\{J_{-1/3}(\tfrac{2}{3}x^{3/2}) + J_{1/3}(\tfrac{2}{3}x^{3/2})\} \tag{55}$$

The zeros and associated values of the Airy function and its first derivative will be required in some applications. A short list is given in Table 2.1 which will suffice for our purposes.

2.2.3 Fock functions

The *Fock functions* are defined by the integrals

$$\begin{aligned} f(x) &= \frac{1}{2\pi}\exp\left(\frac{j\pi}{6}\right)\int_\Gamma \frac{\exp(-jxt)dt}{\mathrm{Ai}\left\{t\exp\left(\frac{-j2\pi}{3}\right)\right\}} = \frac{1}{\sqrt{(\pi)}}\int_\Gamma \frac{\exp(-jxt)dt}{w_1(t)} \\ g(x) &= -\frac{1}{2\pi}\exp\left(-\frac{j\pi}{6}\right)\int_\Gamma \frac{\exp(-jxt)dt}{\mathrm{Ai}'\left\{t\exp\left(\frac{-j2\pi}{3}\right)\right\}} = \frac{1}{\sqrt{(\pi)}}\int_\Gamma \frac{\exp(-jxt)dt}{w_1'(t)} \end{aligned} \tag{56}$$

Table 2.1 Airy function zeros and associated values

	$\mathrm{Ai}(-\alpha_n)=0$		$\mathrm{Ai}'(-\alpha'_n)=0$	
n	α_n	α'_n	$\mathrm{Ai}(-\alpha'_n)$	$\mathrm{Ai}'(-\alpha_n)$
1	2·338	1·019	0·536	0·701
2	4·088	3·248	−0·419	−0·803
3	5·521	4·820	0·380	0·865
4	6·787	6·163	−0·358	−0·911
5	7·944	7·372	0·342	0·947
6	9·023	8·488	−0·330	−0·978
7	10·040	9·535	0·321	1·004
8	11·009	10·528	−0·313	−1·028
9	11·936	11·475	0·307	1·049
10	12·829	12·385	−0·300	−1·068

where the term $w_1(t)$ is the Airy function given in eqn. 51, and Γ is the contour in the complex t-plane as illustrated in Fig. 2.10. The properties of these integrals was originally investigated by Fock (1946) and hence the name. An exhaustive treatment of these and related functions together with tables is to be found in the two reports by Logan (1959).

When the argument is zero

$$f(0) = 0{\cdot}776 \exp\left(\frac{j\pi}{3}\right)$$
$$g(0) = 1{\cdot}399 \qquad (57a)$$

and for positive values of the argument, one may solve the Fock functions in terms of the residues at the poles, where $w_1(t)=0$ and $w_1'(t)=0$. Making use of eqn. 51 we get

$$f(x) = \exp\left(\frac{j\pi}{3}\right)\sum_{n=1}^{\infty}\frac{\exp\left\{\alpha_n x \exp\left(-\frac{j5\pi}{6}\right)\right\}}{\mathrm{Ai}'(-\alpha_n)}$$
$$; \qquad x>0 \quad (57b)$$
$$g(x) = \sum_{n=1}^{\infty}\frac{\exp\left\{\alpha'_n x \exp\left(-\frac{j5\pi}{6}\right)\right\}}{\alpha'_n\,\mathrm{Ai}(-\alpha'_n)}$$

where $-\alpha_n$ and $-\alpha'_n$ are the roots of the Airy function and its derivative, as given in Table 2.1. For large values of argument eqn. 57*b* is seen to reduce exponentially to zero. When the argument is large and negative the asymptotic values of the Fock functions are

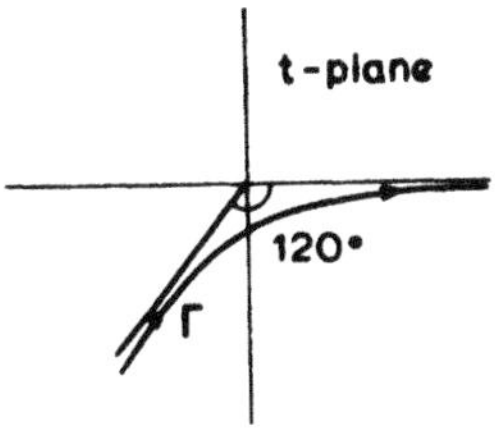

Fig. 2.10 Contour for the Fock functions

$$\left.\begin{aligned} f(x) &\sim -2jx\exp\left(\frac{jx^3}{3}\right) \\ g(x) &\sim 2\exp\left(\frac{jx^3}{3}\right) \end{aligned}\right\} ; \qquad x \to -\infty \qquad (57c)$$

We shall make use of related functions named by Logan (1959) as the *Pekeris carot functions* which are defined for $x > 0$ by the integrals

$$\left.\begin{aligned} \hat{p}(x) &= \frac{1}{\sqrt{(\pi)}}\int_\Gamma \frac{v(t)}{w_1(t)}\exp(-jxt)dt \\ \hat{q}(x) &= \frac{1}{\sqrt{(\pi)}}\int_\Gamma \frac{v'(t)}{w_1'(t)}\exp(-jxt)dt \end{aligned}\right\} ; \qquad v(t) = \sqrt{(\pi)}\,\mathrm{Ai}(t) \quad (58a)$$

To obtain an expression valid for all x we need to change the contour path Γ. In general Γ can be any contour which begins at infinity within the sector $-\pi < \arg t < -\pi/3$, passes between the origin and the first pole of the integral and finally terminates within the sector $-\pi/3 < \arg t < \pi/3$. A more useful representation of the Pekeris carot functions for numerical evaluation and valid for all x is obtained by a change of variable $\zeta = t\exp(-j2\pi/3)$ in eqn. 58*a* and integrating by parts to get

$$\left.\begin{aligned} \hat{p}(x) &= \frac{1}{4\pi\sqrt{(\pi)}jx}\int_L \frac{\exp(\zeta x e^{j\pi/6})d\zeta}{[\mathrm{Ai}(\zeta)]^2} \\ \hat{q}(x) &= \frac{-1}{4\pi\sqrt{(\pi)}jx}\int_L \frac{\exp(\zeta x e^{j\pi/6})\zeta d\zeta}{[\mathrm{Ai}'(\zeta)]^2} \end{aligned}\right\} \qquad (58b)$$

where L is the contour in the complex ζ-plane as illustrated in Fig. 2.11. For positive values of the argument x, the residue series solutions of these functions are

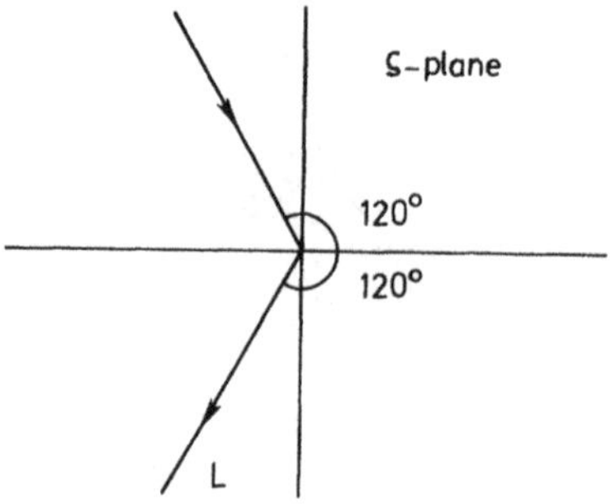

Fig. 2.11 Contour for the Pekeris carot functions

$$\left.\begin{aligned}\hat{p}(x) &= -\frac{\exp\left(\dfrac{j\pi}{6}\right)}{2\sqrt{(\pi)}}\sum_{n=1}^{\infty}\frac{\exp\left\{\alpha_n x\exp\left(-\dfrac{j5\pi}{6}\right)\right\}}{\{\mathrm{Ai}'(-\alpha_n)\}^2}\\ \hat{q}(x) &= -\frac{\exp\left(\dfrac{j\pi}{6}\right)}{2\sqrt{(\pi)}}\sum_{n=1}^{\infty}\frac{\exp\left\{\alpha_n' x\exp\left(-\dfrac{j5\pi}{6}\right)\right\}}{\alpha_n'\{\mathrm{Ai}(-\alpha_n')\}^2}\end{aligned}\right\}; \qquad x>0 \qquad (58c)$$

For a large negative argument the functions are given as

$$\left.\begin{aligned}\hat{p}(x) &\sim \frac{\sqrt{(-x)}}{2}\exp\left\{j\left(\frac{x^3}{12}+\frac{\pi}{4}\right)\right\}\\ \hat{q}(x) &\sim -\frac{\sqrt{(-x)}}{2}\exp\left\{j\left(\frac{x^3}{12}+\frac{\pi}{4}\right)\right\}\end{aligned}\right\}; \qquad x\to-\infty \qquad (58d)$$

Associated functions are the *Pekeris functions* defined by

$$\begin{aligned}p(x) &= \hat{p}(x)+\frac{1}{2\sqrt{(\pi)}x}\\ q(x) &= \hat{q}(x)+\frac{1}{2\sqrt{(\pi)}x}\end{aligned} \qquad (59a)$$

When the argument is zero

$$\begin{aligned}p(0) &= 0{\cdot}354\exp\left(\frac{j\pi}{6}\right)\\ q(0) &= -0{\cdot}307\exp\left(\frac{j\pi}{6}\right)\end{aligned} \qquad (59b)$$

Note that the Pekeris carot functions tend to infinity as the argument tends to zero.

Another Pekeris carot function given by Logan (1969) that we will use is the two-argument function $\hat{V}_2(x, y)$. Here, we relate this function to $\hat{V}(x, y)$ where

$$\hat{V}(x,y) = -\exp(j\pi/4)\hat{V}_2(x,y)$$

$$= \frac{1}{\sqrt{(\pi)}}\int_\Gamma \frac{v'(t)-yv(t)}{w_1'(t)-yw_1(t)}\exp(-jxt)dt; \qquad x>0 \tag{60a}$$

As with eqn. 58*a*, by change of variable and integrating by parts we get an expression valid for all x given by

$$\hat{V}(x,y) = \frac{1}{4\pi\sqrt{(\pi)}jx}$$

$$\times\int_L \frac{(\zeta\exp(j\pi/3)-y^2\exp(-j\pi/3))\exp(\zeta x\exp(j\pi/6))}{(\exp(-j\pi/3)\mathrm{Ai}'(\zeta)-y\exp(j\pi/3)\mathrm{Ai}(\zeta))^2}d\zeta \tag{60b}$$

Evaluating this equation for $x > 0$ by the method of residues yields

$$\hat{V}(x,y) = -\frac{1}{2\sqrt{(\pi)}}e^{j\pi/6}\sum_{n=1}^{\infty}\bar{\beta}_n\exp(\bar{\alpha}_n x\exp(-j5\pi/6)) \tag{60c}$$

where $\bar{\alpha}_n$ are the roots of the equation $\mathrm{Ai}'(-\bar{\alpha}_n)+y\exp(-j\pi/3)\mathrm{Ai}(-\bar{\alpha}_n)=0$, and

$$\bar{\beta}_n = [\bar{\alpha}_n\mathrm{Ai}^2(-\bar{\alpha}_n)+\mathrm{Ai}'^2(-\bar{\alpha}_n)]^{-1}$$

For a large negative argument an asymptotic evaluation of eqn. 60*b* yields

$$\hat{V}(x,y) \sim -\left[\frac{jx-2y}{jx+2y}\right]\frac{\sqrt{-x}}{2}\exp\left(j\left[\frac{x^3}{12}+\frac{\pi}{4}\right]\right); \qquad x\to-\infty \tag{60d}$$

The associated Pekeris function is given by

$$V(x,y) = \hat{V}(x,y)+\frac{1}{2\sqrt{\pi x}} \tag{61}$$

Finally, from the definition of $\hat{V}(x,y)$ we see that $\hat{V}(x,0)=\hat{q}(x)$ and $\hat{V}(x,\infty)=\hat{p}(x)$.

In evaluating the various Fock functions the reports by Logan (1959) are an invaluable reference. However the function $\hat{V}(x,y)$ is not well tabulated for general values of y. Some results are given in James (1980) for real values of y, otherwise it will be necessary to evaluate eqn. 60*b* numerically where x is small and revert to eqns. 60*c* and 60*d* when x is sufficiently large.

2.2.4 *Hankel functions*

The *Hankel functions* can be defined in terms of Bessel functions of the first and second kind as

$$\begin{aligned} H_\nu^{(1)}(z) &= J_\nu(z) + jN_\nu(z) \\ H_\nu^{(2)}(z) &= J_\nu(z) - jN_\nu(z) \end{aligned} \tag{62}$$

which have the Wronskian relation

$$H_\nu^{(1)}(z)H_\nu^{(2)\prime}(z) - H_\nu^{(1)\prime}(z)H_\nu^{(2)}(z) = \frac{-4j}{\pi z} \tag{63}$$

For large values of argument we may replace the functions in eqn. 62 with their asymptotic expansions. We begin with the *uniform asymptotic expansions*, see Jones (1964).

$$\left.\begin{aligned} H_\nu^{(1)}(z) &\sim 2^{7/6}\left(\frac{2}{\nu}\right)^{1/3} \exp\left(-\frac{j\pi}{3}\right)\left\{\frac{\xi\nu^2}{z^2-\nu^2}\right\}^{1/4} \mathrm{Ai}\left(\nu^{2/3}\xi\exp\left(\frac{-j\pi}{3}\right)\right) \\ H_\nu^{(2)}(z) &\sim 2^{7/6}\left(\frac{2}{\nu}\right)^{1/3} \exp\left(\frac{j\pi}{3}\right)\left\{\frac{\xi\nu^2}{z^2-\nu^2}\right\}^{1/4} \mathrm{Ai}\left(\nu^{2/3}\xi\exp\left(\frac{j\pi}{3}\right)\right) \end{aligned}\right\} \tag{64}$$

$$|\arg z| < \pi$$

$$|\arg \nu| \leqslant \frac{\pi}{2}$$

where

$$\begin{aligned} \tfrac{2}{3}\xi^{3/2} &= \left\{\left(\frac{z}{\nu}\right)^2 - 1\right\}^{1/2} - \sec^{-1}\left(\frac{z}{\nu}\right); \quad \left|\frac{z}{\nu}\right| > 1 \\ &= \exp\left(\frac{j3\pi}{2}\right)\left(j\sec^{-1}\left(\frac{z}{\nu}\right) - \left\{1 - \left(\frac{z}{\nu}\right)^2\right\}^{1/2}\right); \quad \left|\frac{z}{\nu}\right| < 1 \end{aligned} \tag{65}$$

For other values of ν and z one must use the continuation formulas

$$H_{-\nu}^{(1)}(z) = \exp(j\pi\nu)H_\nu^{(1)}(z) \tag{66a}$$

$$H_{-\nu}^{(2)}(z) = \exp(-j\pi\nu)H_\nu^{(2)}(z) \tag{66b}$$

$$H_{-\nu}^{(1)}(-z) = -H_\nu^{(2)}(z) \tag{66c}$$

$$H_{-\nu}^{(2)}(-z) = -H_\nu^{(1)}(z) \tag{66d}$$

$$\sec^{-1}(-z) = \pi - \sec^{-1} z \tag{66e}$$

We shall now give the approximations to eqn. 64 for various ranges of ν. Initially we must assume that these equations have restrictions on arg ν and arg z as in eqn. 64. The continuation formulas of eqn. 66 can then be used to determine if they have a wider validity. In particular, we shall require the ranges $\arg z = 0$, $-\frac{\pi}{2} \leqslant \arg \nu < \frac{3\pi}{2}$, so we need only consider the continuation of ν into the remainder of the complex ν-plane where $\frac{\pi}{2} < \arg \nu < \frac{3\pi}{2}$.

For $|\nu| \ll |z|$, eqn. 64 simplifies considerably when the asymptotic value of the Airy function given by eqn. 53 is used. The result is (where z is now replaced by x since $\arg z = 0$)

$$\left.\begin{aligned} H_\nu^{(1)}(x) &\sim \sqrt{\left(\frac{2}{\pi x}\right)} \exp\left\{j\left(x - \frac{\nu\pi}{2} - \frac{\pi}{4}\right)\right\} \\ H_\nu^{(2)}(x) &\sim \sqrt{\left(\frac{2}{\pi x}\right)} \exp\left\{-j\left(x - \frac{\nu\pi}{2} - \frac{\pi}{4}\right)\right\} \end{aligned}\right\} \quad \begin{aligned} &|\nu| \ll x \\ &\text{for all } \arg \nu \end{aligned} \tag{67}$$

When ν approaches x, but is not too close to it, we may still use the Airy function asymptotic value to give

$$\begin{aligned} H_\nu^{(1)}(x) &\sim \left[\frac{2}{\pi(x^2 - \nu^2)^{1/2}}\right]^{1/2} \exp\left[j\left\{(x^2 - \nu^2)^{1/2} - \nu \sec^{-1}\left(\frac{x}{\nu}\right) - \frac{\pi}{4}\right\}\right] \\ H_\nu^{(2)}(x) &\sim \left[\frac{2}{\pi(x^2 - \nu^2)^{1/2}}\right]^{1/2} \exp\left[-j\left\{(x^2 - \nu^2)^{1/2} - \nu \sec^{-1}\left(\frac{x}{\nu}\right) - \frac{\pi}{4}\right\}\right]; \\ &\nu < x \end{aligned} \tag{68}$$

If $\nu \simeq x$ we can simplify the value of ξ given in eqn. 65 by first putting

$$\left(\frac{x}{\nu}\right) = 1 + \delta$$

Expanding the quantities in eqn. 65 for small δ gives

$$\begin{aligned} \sec^{-1}\left(\frac{x}{\nu}\right) &= j\left(\ln\left(\frac{x}{\nu}\right) - \ln\left[1 + \sqrt{\left\{1 - \left(\frac{x}{\nu}\right)^2\right\}}\right]\right) \\ &= \sqrt{2}\left\{\delta^{1/2} - \frac{5}{12}\delta^{3/2} + 0(\delta^2)\right\} \\ \left\{\left(\frac{x}{\nu}\right)^2 - 1\right\}^{1/2} &= \sqrt{2}\{\delta^{1/2} + \tfrac{1}{4}\delta^{3/2} + 0(\delta^2)\} \end{aligned}$$

from which we deduce

$$\xi \simeq 2^{1/3}\delta = 2^{1/3}\left(\frac{x}{\nu}-1\right) \qquad \text{for } z \simeq \nu$$

Inserting this value for ξ into the expressions in eqn. 64 yields

$$\left.\begin{aligned} H_\nu^{(1)}(x) &\sim 2\left(\frac{2}{x}\right)^{1/3} \exp\left(-\frac{j\pi}{3}\right) \mathrm{Ai}\left(\tau \exp\left(\frac{j2\pi}{3}\right)\right) \\ H_\nu^{(2)}(x) &\sim 2\left(\frac{2}{x}\right)^{1/3} \exp\left(\frac{j\pi}{3}\right) \mathrm{Ai}\left\{\tau \exp\left(\frac{-j2\pi}{3}\right)\right\} \\ \tau &= (\nu - x)\left(\frac{2}{x}\right)^{1/3}, \quad \nu \simeq x \end{aligned}\right\} \tag{69}$$

For the condition $|\nu| \gg x$ then the value of ξ is now given approximately as

$$\tfrac{2}{3}\xi^{3/2} \simeq \exp\left(\frac{j3\pi}{2}\right) \ln\left(\frac{2\nu}{ex}\right) \qquad \text{for} \quad |\nu| \gg x$$

and

$$\arg \xi \simeq \pi$$

Substituting into eqn. 64, and using the appropriate Airy function asymptotic expansion in eqn. 53, we get

$$\left.\begin{aligned} H_\nu^{(1)}(x) &\sim -j\sqrt{\left(\frac{2}{\pi\nu}\right)}\left(\frac{2\nu}{ex}\right)^{\nu} \\ H_\nu^{(2)}(x) &\sim j\sqrt{\left(\frac{2}{\pi\nu}\right)}\left(\frac{2\nu}{ex}\right)^{\nu} \end{aligned}\right\} \quad |\arg \nu| \leqslant \frac{\pi}{2}, \quad |\nu| \gg x \tag{70a}$$

and using the continuation formulas in eqn. 66

$$\left.\begin{aligned} H_\nu^{(1)}(x) &\sim \sqrt{\left(\frac{2}{\pi\nu}\right)}\left(\frac{ex}{2\nu}\right)^{\nu} \\ H_\nu^{(2)}(x) &\sim -\sqrt{\left(\frac{2}{\pi\nu}\right)}\left(\frac{ex}{2\nu}\right)^{\nu} \end{aligned}\right\} \quad \frac{\pi}{2} < \arg \nu < \frac{3\pi}{2}, \quad |\nu| \gg x \tag{70b}$$

2.3 Asymptotic evaluation of the field integrals

2.3.1 Method of stationary phase

When the electromagnetic field is determined from an equivalent source distribution over a surface, the resultant integrals may be written in the form

$$I = \iint_S f(x,y)\exp\{jkg(x,y)\}dx\,dy \tag{71}$$

where S is the surface over which the equivalent sources are formulated. Usually the functions $f(x,y)$ and $g(x,y)$ are such as to preclude an analytical evaluation of eqn. 71, and numerical integration has to be used. In many instances, provided that k is large, an asymptotic evaluation of the integral is possible dependent on the behaviour of the phase function $g(x,y)$.

In general, k will be complex, but we will assume in this section that the medium is only slightly lossy so that we have $\arg k \simeq 0$. This assumption simplifies the procedure and will be adequate for most of the applications to be discussed later. We begin by considering the asymptotic behaviour of single integrals.

2.3.1.1 Single integrals

For a 2-dimensional equivalent source distribution, eqn. 71 reduces to

$$I = \int_b^a f(x)\exp\{jkg(x)\}dx \tag{72}$$

where a, b are the limits of the sources. To assist in the asymptotic evaluation we rewrite this equation as

$$\begin{aligned} I &= \int_{-\infty}^{\infty} f(x)\exp\{jkg(x)\}dx - \int_a^{\infty} f(x)\exp\{jkg(x)\}dx \\ &\quad - \int_{-b}^{\infty} f(-x)\exp\{jkg(-x)dx \qquad (73) \\ &= I_0 - I_a - I_{-b} \end{aligned}$$

and consider first the term I_0. If k is large then the phase of I_0 will vary rapidly, and provided that $f(x)$ is a smoothly varying function, the value of I_0 will tend to zero as k tends to infinity. If however, the phase function $g(x)$ has points where it is stationary, i.e. $g'(x) = 0$, then the exponential term $\exp\{jkg(x)\}$ will not vary rapidly in the vicinity of these points, and the major contributions to I_0 will be from those regions where $g'(x) = 0$. By expanding the functions in I_0 around these points we obtain an asymptotic evaluation of the integral. Such points are called stationary phase points and the procedure is known as the *method of stationary phase.* A *first order* stationary phase point at $x = x_0$ is defined by $g'(x_0) = 0, g''(x_0) \neq 0$. Expanding the phase term

$g(x)$ in a Taylor series about x_0, and retaining only the first two non-zero quantities, we have

$$g(x) \simeq g(x_0) + g''(x_0)\frac{s^2}{2}; \qquad s = x - x_0$$

Over the region described by this equation it is assumed that $f(x)$ is slowly varying and can be approximated by $f(x_0)$.

The equation for I_0 now becomes

$$I_0 \sim f(x_0)\exp\{jkg(x_0)\}\int_{-\infty}^{\infty} \exp\left\{jkg''(x_0)\frac{s^2}{2}\right\} ds \tag{74}$$

The integral in eqn. 74 can be rewritten as

$$2\sqrt{\left(\frac{2}{k\,|g''(x_0)|}\right)}\int_0^{\infty} \exp\,[jt^2 \operatorname{sgn}\{g''(x_0)\}]\,dt$$

which is in the form of the Fresnel integral with zero argument as given by eqn. 35. Substituting this equation, after reducing it by the terms of eqn. 35, into eqn. 74 gives

$$I_0 \sim \sqrt{\left(\frac{2\pi}{k\,|g''(x_0)|}\right)} f(x_0)\exp\left(j\left[kg(x_0) + \frac{\pi}{4}\operatorname{sgn}\{g''(x_0)\}\right]\right) \tag{75}$$

If $f(x_0) = 0$, then the next higher order term in the asymptotic expansion, obtained by expressing $f(x)$ as a Taylor series about x_0, is proportional to $k^{-3/2} f''(x_0)$. In this case any end-point contribution, to be discussed below, will be the leading term in the asymptotic evaluation of eqn. 73.

If $g''(x_0) \to 0$, then eqn. 75 will fail, and it is necessary to consider an additional term in the expansion of the phase function $g(x)$

$$g(x) \simeq g(x_0) + g''(x_0)\frac{s^2}{2} + \tfrac{1}{2}g'''(x_0)\frac{s^3}{3}$$

so that eqn. 74 now becomes

$$I_0 \sim f(x_0)\exp\{jkg(x_0)\}\int_{-\infty}^{\infty} \exp\left[jk\left\{g''(x_0)\frac{s^2}{2} + \tfrac{1}{2}g'''(x_0)\frac{s^3}{3}\right\}\right] ds$$

The integral in this equation can be expressed in terms of the Airy function, Ai(x), given earlier (see eqn. 54) for real arguments as

$$\operatorname{Ai}(x) = \frac{1}{2\pi}\int_{-\infty}^{\infty} \exp\left\{\pm j\left(xt + \frac{t^3}{3}\right)\right\} dt$$

from which, by change of variable $t = (\frac{1}{2}k\,|g'''(x_0)|)^{1/3}\left(s + \dfrac{g''(x_0)}{g'''(x_0)}\right)$, we get

$$I_0 \sim 2\pi f \exp(jkg) \exp\left\{j\frac{k}{3}(g'')^3(g''')^{-2}\right\}\left[\frac{2}{k\,|g'''|}\right]^{1/3} \mathrm{Ai}(\zeta)\Bigg|_{x=x_0}$$

where

$$\zeta = \left(\frac{k}{2}\right)^{2/3} |g''|^2\,|g'''|^{-4/3} \exp(\mp j\pi); \qquad g''(x_0) \gtrless 0 \qquad (76)$$

For large values of ζ the asymptotic expression for the Airy function given by the first line of eqn. 53 reduces eqn. 76 to eqn. 75. When $\zeta = 0$, i.e. $g''(x_0) = 0$, we have a *second order* stationary phase point at $x = x_0$. This can be considered as the confluence of two nearby first-order stationary phase points.

If $g'''(x_0) \to 0$ then eqn. 76 becomes invalid. As before we could consider an additional higher order term in the expansion of $g(x)$. The resultant integral, however, cannot be expressed in terms of known functions and in such a situation it is necessary to solve the original integral numerically.

It remains now to evaluate the integrals in eqn. 73 which possess a finite limit in the form

$$I_\alpha = \int_\alpha^\infty f(x) \exp\{jkg(x)\}dx \qquad (77)$$

For stationary phase points removed from the end point at $x = \alpha$, we evaluate their contribution to I_α from either eqn. 75 or eqn. 76, as just discussed. This will give the leading term in the asymptotic evaluation. The next term is given by the contribution from the end point at $x = \alpha$. Writing eqn. 77 as

$$\frac{1}{jk}\int_\alpha^\infty \frac{f(x)}{g'(x)} jkg'(x) \exp\{jkg(x)\}dx$$

and solving by parts, we get for the end-point contribution

$$-\frac{1}{jk}\frac{f(\alpha)}{g'(\alpha)} \exp\{jkg(\alpha)\} + 0(k^{-2}) \qquad (78)$$

The contribution from the upper limit has been removed by tacitly assuming that the medium is slightly lossy, giving k a small imaginary component. Note that this term is of the order $k^{-1/2}$ greater than the stationary phase value for a first order point.

We may proceed in the same way to obtain the next higher order term for the end-point contribution as

$$\frac{1}{(jk)^2}\frac{f'(\alpha)g'(\alpha)-f(\alpha)g''(\alpha)}{\{g'(\alpha)\}^3}\exp\{jkg(\alpha)\}+0(k^{-3}) \qquad (79)$$

and so on. In fact successive evaluation yields the asymptotic series

$$\exp\{jkg(\alpha)\}\sum_{m=1}^{\infty}\left(\frac{j}{k}\right)^m u_m(\alpha)\left(\frac{1}{g'(\alpha)}\right)^{m-1} \qquad (80)$$

where

$$u_m(\alpha) = \frac{\partial}{\partial x}u_{m-1}(x)\bigg|_{x=\alpha} \qquad \text{for} \quad m>1$$

and

$$u_1(x) = \frac{f(x)}{g'(x)}$$

In some instances both $f(x)$ and $g'(x)$ have a first order zero at $x=\alpha$. For this case applying L'Hopital's rule to eqn. 78 gives

$$-\frac{1}{jk}\frac{f'(\alpha)}{g''(\alpha)}\exp\{jkg(\alpha)\} \qquad (81)$$

When a first order stationary phase point approaches the end point at $x=\alpha$, it is necessary to consider the coupling effect between the two points. Thus, when x_0 approaches α, by expanding $g(\alpha)$ in a Taylor series about x_0, we have

$$g(x_0)-g(\alpha)\simeq-\tfrac{1}{2}(\alpha-x_0)^2g''(x_0)$$

or alternatively, expanding $g(x_0)$ about α, we get

$$g(x_0)-g(\alpha)\simeq-(\alpha-x_0)g'(\alpha)$$

so that as $x_0\to\alpha$ with $g''(x_0)\simeq g''(\alpha)$ the following relationship is obtained

$$g'(\alpha)\simeq\tfrac{1}{2}(\alpha-x_0)g''(\alpha) \qquad \text{as} \quad x_0\to\alpha \qquad (82)$$

In order to evaluate the integral of eqn. 77 when $x_0\to\alpha$ we may expand the phase term $g(x)$ either about x_0 or about α. Expansion about x_0 can be found, for example, in Chapter 4 of Felsen and Marcuvitz (1973) but considerably simpler expressions result when the expansion is taken about α. Using this latter approach

$$g(x)\simeq g(\alpha)+sg'(\alpha)+\frac{s^2}{2}g''(\alpha); \qquad s = x-\alpha$$

and on using the sign information in eqn. 82, this may be written as

$$g(x)\simeq g(\alpha)\pm\left\{\epsilon_1 s|g'(\alpha)|+\frac{s^2}{2}|g''(\alpha)|\right\}; \qquad g''(\alpha)\gtrless 0 \qquad (83)$$

where

$$\epsilon_1 = \operatorname{sgn}(\alpha-x_0)$$

Consider first the case where ϵ_1 is positive and the stationary phase point is just outside the integral. Substitution of eqn. 83 into eqn. 77 with the assumption, as before, that $f(x)$ is a slowly varying function, then

$$I_\alpha \simeq f(\alpha) \exp\{jkg(\alpha)\} \int_\alpha^\infty \exp\left[\pm jk \left\{ s\,|g'(\alpha)| + \frac{s^2}{2}\,|g''(\alpha)| \right\}\right] dx;$$

$$g''(\alpha) \gtrless 0$$

By the change of variable

$$t = \sqrt{\left(\frac{k}{2\,|g''(\alpha)|}\right)}\,|g'(\alpha)| + \sqrt{\left(\frac{k}{2}\,|g''(\alpha)|\right)}\,s$$

this equation reduces to

$$I_\alpha \simeq f(\alpha) \exp\{jkg(\alpha) \mp jv^2\} \sqrt{\left(\frac{2}{k\,|g''(\alpha)|}\right)} \int_v^\infty \exp(\pm jt^2)\,dt; \qquad (84)$$

$$\epsilon_1 > 0, \qquad g''(\alpha) \gtrless 0$$

where

$$v = \sqrt{\left(\frac{k}{2\,|g''(\alpha)|}\right)}\,|g'(\alpha)|$$

The integral in eqn. 84 is the Fresnel integral $F_\pm(v)$ and has been defined earlier by eqn. 34. Consider now the case when ϵ_1 is negative and the stationary phase point has moved inside the integral. Eqn. 77 may now be written as

$$I_\alpha = \int_{-\infty}^{\infty} f(x) \exp\{jkg(x)\}dx - \int_{-\infty}^{\alpha} f(x) \exp\{jkg(x)\}dx; \qquad (85)$$

$$\epsilon_1 < 0$$

The first integral will yield the stationary phase contribution, I_0, as in eqn. 75 at $x = x_0$. The second integral is expanded about $x = \alpha$, using eqn. 83 as before, to give

$$f(\alpha) \exp\{jkg(\alpha)\} \int_{-\infty}^{\alpha} \exp\left[\pm jk \left\{ -s\,|g'(\alpha)| + \frac{s^2}{2}\,|g''(\alpha)| \right\}\right] dx;$$

$$g''(\alpha) \gtrless 0$$

By the change of variable

$$t = -\sqrt{\left(\frac{k}{2\,|g''(\alpha)|}\right)}\,|g'(\alpha)| + \sqrt{\left(\frac{k}{2}\,|g''(\alpha)|\right)}\,s$$

this equation becomes

$$f(\alpha) \exp\{jkg(\alpha) \mp jv^2\} \sqrt{\left\{\frac{2}{k\,|g''(\alpha)|}\right\}}\; F_\pm(v)$$

so that the solution for I_α when $\epsilon_1 < 0$ is given by

$$I_\alpha \simeq I_0 - f(\alpha)\exp\{jkg(\alpha) \mp jv^2\}\sqrt{\left\{\frac{2}{k|g''(\alpha)|}\right\}}\, F_\pm(v);$$

$$\epsilon_1 < 0, \quad g''(\alpha) \gtrless 0$$

Combining this result and that of eqn. 84

$$I_\alpha \simeq U(-\epsilon_1)I_0 + \epsilon_1 f(\alpha)\exp\{jkg(\alpha) \mp jv^2\}\sqrt{\left\{\frac{2}{k|g''(\alpha)|}\right\}}\, F_\pm(v);$$

$$g''(\alpha) \gtrless 0 \tag{86a}$$

where

$$\epsilon_1 = \operatorname{sgn}(\alpha - x_0),$$

$$v = \sqrt{\left(\frac{k}{2|g''(\alpha)|}\right)}\,|g'(\alpha)|$$

and $U(x)$ is the unit step function, i.e., $U = 1$ for $x \geqslant 0$ and zero otherwise.

When the Fresnel integral argument v is large, the leading term in the asymptotic expansion of the Fresnel integral as given by eqn. 36 together with the sign information in eqn. 82 reduces eqn. 86*a* to

$$I_\alpha \sim U(-\epsilon_1)I_0 - \frac{1}{jk}\frac{f(\alpha)}{g'(\alpha)}\exp\{jkg(\alpha)\} \tag{86b}$$

which agrees with the result obtained earlier for the leading term in the isolated endpoint contribution. Thus the formulation given in eqn. 86 offers a uniform asymptotic solution as a first order stationary phase point traverses across the endpoint of an integral. Note, however, that this solution has been shown to be uniform by using the sign information in eqn. 82 i.e.,

$$\epsilon_1 = \pm\frac{|g'(\alpha)|}{g'(\alpha)}; \qquad g''(\alpha) \gtrless 0 \tag{86c}$$

which is only true, in general, as $x_0 \to \alpha$. To avoid possible errors in the application of eqn. 86, one should use eqn. 86*a* for $v < 3{\cdot}0$, and then change to eqn. 86*b* since, for $v > 3{\cdot}0$, Fig. 2.8 shows that the asymptotic value for the Fresnel integral is a good approximation. This precaution in the application of eqn. 86 presents no difficulty.

When the stationary phase point is no longer a simple first order point but is evaluated by eqn. 76, then the Fresnel integral is replaced by the incomplete Airy function of the form

$$\int_u^\infty \exp\left\{-j\left(\sigma t + \frac{t^3}{3}\right)\right\} dt \tag{87}$$

Unfortunately such a function is not so easily generated as, for example, the Fresnel integral and complete Airy function. For a discussion on the incomplete Airy function see Levey and Felsen (1969).

2.3.1.2 Double integrals

For 3-dimensional equivalent source distributions, we must use the double integral

$$I = \iint_S f(x,y) \exp\{jkg(x,y)\} dx\, dy \tag{88}$$

where S contains the sources. As before, the major contribution of the integration for large k occurs when $g(x,y)$ is stationary, i.e., $\nabla g(x,y) = 0$. For a stationary phase point at (x_0, y_0) we make the substitutions

$$s_1 = x - x_0, \qquad s_2 = y - y_0$$

and using the notation

$$\frac{\partial^3 g}{\partial x \partial y \partial z} = g'''_{xyz}$$

the expansion of $g(s_1, s_2)$ in a Taylor series about the stationary point at (0, 0) becomes

$$\begin{aligned} g(s_1, s_2) = \{&g + \tfrac{1}{2}(s_1^2 g''_{s_1 s_1} + s_2^2 g''_{s_2 s_2} + 2 s_1 s_2 g''_{s_1 s_2}) \\ &+ \tfrac{1}{6}(s_1^3 g'''_{s_1 s_1 s_1} + s_2^3 g'''_{s_2 s_2 s_2} + 3 s_1^2 s_2 g'''_{s_1 s_1 s_2} + 3 s_1 s_2^2 g'''_{s_1 s_2 s_2}) \\ &+ \ldots\}; \quad \text{where} \quad g = g(0,0) \end{aligned} \tag{89}$$

For a first-order stationary phase point we retain terms up to the quadratic form only. In order to evaluate the resultant integrations, we introduce a change of variables for s_1, s_2 to u_1, u_2 such that $g''_{u_1 u_2} = 0$. This simply involves a co-ordinate rotation and allows us to treat each integration (for a first-order stationary phase point) independently. Using matrix notation

$$s_1^2 g''_{s_1 s_1} + s_2^2 g''_{s_2 s_2} + 2 s_1 s_2 g''_{s_1 s_2} = \boldsymbol{s}^T \mathbf{G}_s \boldsymbol{s}$$

where T denotes the *transpose* of the matrix (i.e., the interchange of rows and columns) and $\boldsymbol{s}$ and $\mathbf{G}_s$ are

$$\boldsymbol{s} = \begin{bmatrix} s_1 \\ s_2 \end{bmatrix} \qquad \mathbf{G}_s = \begin{bmatrix} g''_{s_1 s_1} & g''_{s_1 s_2} \\ g''_{s_1 s_2} & g''_{s_2 s_2} \end{bmatrix}$$

The relationship between $\boldsymbol{s}$ and the new variables $\boldsymbol{u}$ can be written as

$$\boldsymbol{s} = \mathbf{J}\boldsymbol{u}, \quad \mathbf{J} = \begin{bmatrix} \cos\theta & \sin\theta \\ -\sin\theta & \cos\theta \end{bmatrix}, \quad \boldsymbol{u} = \begin{bmatrix} u_1 \\ u_2 \end{bmatrix} \tag{90}$$

Now for u_1, u_2 to be independent of each other $\mathbf{G}_u$ must be a diagonal matrix, i.e.,

$$\mathbf{G}_u = \begin{bmatrix} g''_{u_1u_1} & 0 \\ 0 & g''_{u_2u_2} \end{bmatrix}$$

so that we have

$$s^T \mathbf{G}_s s = u^T \mathbf{G}_u u$$

Substituting for s in this equation and using the reversal rule for a transposed product, i.e.,

$$(\mathbf{AB})^T = \mathbf{B}^T \mathbf{A}^T \tag{91}$$

we arrive at

$$u^T \mathbf{J}^T \mathbf{G}_s \mathbf{J} u = u^T \mathbf{G}_u u \tag{92}$$

Note that det $\mathbf{J} = 1$, so we have

$$\det \mathbf{G}_u = \det \mathbf{G}_s \tag{93a}$$

and since $\dfrac{\partial}{\partial s_1} = \dfrac{\partial}{\partial x}$, $\dfrac{\partial}{\partial s_2} = \dfrac{\partial}{\partial y}$, det $\mathbf{G}_u$ is given as

$$\det \mathbf{G}_u = g''_{xx} g''_{yy} - g''_{xy} g''_{xy} \tag{93b}$$

Evaluating eqn. 92 for θ, the co-ordinate rotation angle, in order to determine u from eqn. 90, gives

$$\theta = \tfrac{1}{2} \tan^{-1} \left[\frac{2g''_{xy}}{g''_{yy} - g''_{xx}} \right] \tag{94}$$

Eqn. 88 may now be written

$$I = \iint_S f(u) \exp\{jkg(u)\} du$$

and if we initially assume that S extends over an infinite surface, then expansion about the stationary phase point at (0, 0) in u gives

$$I \sim I_{00} = f(0,0) \exp\{jkg(0,0)\} P_{u_1}(0,0) P_{u_2}(0,0) \tag{95}$$

where

$$P_{u_n}(u) = \int_{-\infty}^{\infty} \exp\left\{ \frac{jku_n^2}{2} g''_{u_n u_n}(u) \right\} du_n; \qquad n = 1, 2$$

This integral was evaluated earlier in eqns. 74 and 75 as

$$P_{u_n}(u) = \left\{\frac{k}{2\pi}\,|g''_{u_n u_n}(u)|\right\}^{-1/2} \exp\left[\frac{j\pi}{4}\,\mathrm{sgn}\,\{g''_{u_n u_n}(u)\}\right] \tag{96}$$

Combining eqns. 95 and 96, and using eqn. 93, we may write I_{00} in the compact form

$$\begin{aligned} I_{00} &= \frac{2\pi f}{k}\,(|\det \mathbf{G}_u|)^{-1/2} \\ &\exp\left(j\left[kg + \frac{\pi}{4}\,\{\mathrm{sgn}\,(g''_{u_1 u_1}) + \mathrm{sgn}\,(g''_{u_2 u_2})\}\right]\right)\Bigg|_{\substack{u_1=0\\u_2=0}} \end{aligned} \tag{97}$$

There may, of course, be more than one stationary phase point within S, and provided that these points are well separated, the asymptotic evaluation of the integral is given by the sum of the individual contributions as in eqn. 97. A different angle of co-ordinate rotation will, in general, be required for each point.

We now consider the situation where the integration of eqn. 88 has finite limits. It will be found convenient to express the integral in terms of angular variables, ξ, ψ, so that

$$I = \int_0^{2\pi}\int_0^{\psi_1(\xi)} f(\xi, \psi)\exp\{jkg(\xi, \psi)\}\,d\psi\,d\xi \tag{98}$$

where we note that the ξ integration has no endpoint contribution, and the only non-zero endpoint contribution from the ψ integration is at the upper limit $\psi_1(\xi)$. For an asymptotic evaluation we put this equation in the form

$$I = \int_{-\infty}^{\infty}\int_{-\infty}^{\alpha(u_2)} f(u)\exp\{jkg(u)\}\,du \tag{99}$$

which may be rewritten in accordance with eqn. 73 as

$$\begin{aligned} \int_{-\infty}^{\infty}\int_{-\infty}^{\alpha(u_2)} &= \int_{-\infty}^{\infty}\int_{-\infty}^{\infty} - \int_{-\infty}^{\infty}\int_{\alpha(u_2)}^{\infty} \\ &= \quad I_{00} \quad - \quad I_{0\alpha} \end{aligned} \tag{100}$$

Any stationary phase points within $I_{0\alpha}$ will cancel with those in I_{00} to ensure that the total solution will not give a contribution outside the limits of integration. Thus we need consider only the endpoint contribution of $I_{0\alpha}$. On using the solution of eqn. 86*b* for the u_1 integration in $I_{0\alpha}$ when the endpoint is isolated from any stationary phase points

$$I_{0\alpha} \sim \int_{-\infty}^{\infty} \frac{f\{\alpha(u_2), u_2\}}{-jkg'_{u_1}\{\alpha(u_2), u_2\}} \exp[jkg\{\alpha(u_2), u_2\}]\, du_2 \quad (101)$$

This integral can now be evaluated for stationary phase points at $u_2 = u_{e2}$ determined when $g'_{u_2}\{\alpha(u_{e2}), u_{e2}\} = 0$ to give

$$I_{0\alpha} \sim \frac{f\exp(jkg)}{-jkg'_{u_1}} P_{u_2}\Bigg|_{\substack{u_1=\alpha(u_{e2})\\ u_2=u_{e2}}} \quad ; \quad g'_{u_2}\{\alpha(u_{e2}), u_{e2}\} = 0 \quad (102a)$$

Note that, in general, α is a function of u_2.

When the stationary phase point at (0, 0) approaches the endpoint, which alternatively means $|\alpha| \to 0$, we must use the formulation given in eqn. 86 for the endpoint contribution, so that $I_{0\alpha}$ now becomes

$$I_{0\alpha} \sim \pm \frac{|g'_{u_1}|}{g'_{u_1}} f\exp(jkg)\exp(\mp jv^2) \sqrt{\left(\frac{2}{k\,|g''_{u_1u_1}|}\right)} F_{\pm}(v) P_{u_2}\Bigg|_{\substack{u_1=\alpha(u_{e2})\\ u_2=u_{e2}}} ;$$

$$g''_{u_1u_1}\{\alpha(u_{e2}), u_{e2}\} \gtrless 0 \quad (102b)$$

where

$$v = \sqrt{\left(\frac{k}{2\,|g''_{u_1u_1}|}\right)}\,|g'_{u_1}|; \qquad g'_{u_2}\{\alpha(u_{e2}), u_{e2}\} = 0$$

It is important to note that differentiation with respect to the variable u_2 in eqn. 102 must encompass the endpoint function $\alpha(u_2)$. Also, as for eqn. 86, we use the asymptotic solution in eqn. 102*a* when the Fresnel integral argument $v > 3{\cdot}0$.

2.3.2 *Method of steepest descent*

The integrals solved asymptotically in the previous section were for functions involving real variables. A more general integral which we will sometimes be required to solve involves functions of a complex variable. For our purposes it will suffice to consider single integrals of the form

$$I = \int_C f(z)\exp\{jkg(z)\}dz \quad (103)$$

where $f(z)$ and $g(z)$ are regular functions of the complex variable z along the integration path C, which has its endpoints at infinity, and it is still assumed that $\arg k \simeq 0$.

The phase term $g(z)$ in eqn. 103 may be written as

$$g(z) = u(x,y) + jv(x,y), \quad \text{where} \quad z = x + jy, \quad (104)$$

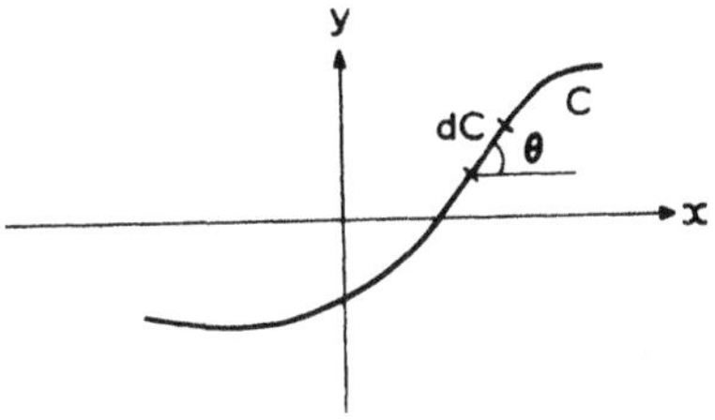

Fig. 2.12 Path *C* in the complex *z*-plane

and u, v are real functions which satisfy the *Cauchy-Riemann* equations

$$\frac{\partial u}{\partial x} = \frac{\partial v}{\partial y}; \qquad \frac{\partial u}{\partial y} = -\frac{\partial v}{\partial x} \tag{105}$$

Substituting eqn. 104 into eqn. 103 gives

$$I = \int_C f(z)\exp(jku)\exp(-kv)dz$$

Clearly, the magnitude of this integral will change most rapidly along the path where $\frac{\partial v}{\partial C}$ is a maximum. Similarly the phase will change most rapidly where $\frac{\partial u}{\partial C}$ is a maximum. From Fig. 2.12 we have

$$\frac{\partial v}{\partial C} = \frac{\partial v}{\partial x}\frac{\partial x}{\partial C} + \frac{\partial v}{\partial y}\frac{\partial y}{\partial C} = \frac{\partial v}{\partial x}\cos\theta + \frac{\partial v}{\partial y}\sin\theta$$

$$\frac{\partial u}{\partial C} = \frac{\partial u}{\partial x}\cos\theta + \frac{\partial u}{\partial y}\sin\theta$$

The values of θ corresponding to the maximum of these functions are determined by

$$\frac{\partial}{\partial\theta}\left(\frac{\partial v}{\partial C}\right) = -\frac{\partial v}{\partial x}\sin\theta + \frac{\partial v}{\partial y}\cos\theta = -\frac{\partial v}{\partial x}\frac{\partial y}{\partial C} + \frac{\partial v}{\partial y}\frac{\partial x}{\partial C} = 0$$

$$\frac{\partial}{\partial\theta}\left(\frac{\partial u}{\partial C}\right) = -\frac{\partial u}{\partial x}\sin\theta + \frac{\partial u}{\partial y}\cos\theta = -\frac{\partial u}{\partial x}\frac{\partial y}{\partial C} + \frac{\partial u}{\partial y}\frac{\partial x}{\partial C} = 0$$

Upon employing the Cauchy-Riemann equations of eqn. 105 we get

$$\frac{\partial u}{\partial C} = 0 \qquad \text{for a maximum change in } v$$

$$\frac{\partial v}{\partial C} = 0 \qquad \text{for a maximum change in } u$$

In other words, along paths of constant phase the amplitude of $\exp\{jkg(z)\}$ is changing most rapidly, and along paths of constant amplitude the phase of $\exp\{jkg(z)\}$ is changing most rapidly.

In line with the method of stationary phase discussed in the previous section, the major contribution to the integral in eqn. 103 for large k will occur in the vicinity of stationary, or saddle points where $g'(z_0) = 0$. The *method of steepest descent* attempts to deform the original contour C in eqn. 103 into a path which, while passing through the saddle point, gives the most rapid *decay* in the magnitude of $\exp\{jkg(z)\}$. Thus the requirement for a steepest descent path through a saddle point at $z = z_0$ is that along the path

$$u(x,y) = \text{constant}, \quad v(x,y) \geqslant v(x_0,y_0) \qquad (106)$$

It is not always possible to determine the complete steepest descent path easily, and it is common practice to expand $g(z)$ in a Taylor series about the saddle points to obtain an asymptotic evaluation of the integral. This procedure is the same as that carried out in the previous section, and for the cases considered there, gives the results in the same form. Note that the method of stationary phase emerges from the above argument as the special case when we choose the path through the saddle point such that $v =$ constant.

We have no need of further discussion of this approach, but for more details the reader is referred to Chapter 4 of Felsen and Marcuvitz (1973).

References

CLEMMOW, P.C., and SENIOR, T.B.A. (1953): 'A note on a generalized Fresnel integral', *Proc. Camb. Phil. Soc.*, **49**, pp. 570–572.

FELSEN, L.B., and MARCUVITZ, N. (1973): 'Radiation and scattering of waves', Prentice-Hall.

FOCK, V.A. (1946): 'The field of a plane wave near the surface of a conducting body', *J. Phys.*, **10**, pp. 399–409. Also see FOCK, V.A. (1965): 'Electromagnetic diffraction and propagation problems', (Pergamon).

JAMES, G.L. (1979): 'An approximation to the Fresnel integral', *Proc. IEEE*, **67**, pp. 677–678.

JAMES, G.L. (1980): 'GTD solution for diffraction by convex corrugated surfaces', *IEE Proc.*, **127**, Pt. H, pp. 257–262.

JONES, D.S. (1964): 'The theory of electromagnetism', (Pergamon), pp. 359–363.

LEVEY, L., and FELSEN, L.B. (1969): 'On incomplete Airy functions and their application to diffraction problems', *Radio Sci.*, **4**, pp. 959–969.

LOGAN, N.A. (1959): 'General research in diffraction theory', Missiles and Space Division, Lockheed Aircraft Corporation, Report LMSD-288087, Vol. 1, and Report LMSD-288088, Vol. 2.

Chapter 3

Canonical problems for GTD

The principle canonical problems which formed the basis of GTD are formally derived in this chapter. All the problems are 2-dimensional and more general formulations are developed from them in later chapters. Thus we develop the methods of geometrical optics from reflection and refraction of a plane wave at an infinite plane dielectric interface. The high frequency behaviour of the half-plane and wedge solution is the starting point for a GTD edge diffraction formulation. Although the half-plane can be considered as a special case of the wedge where the wedge angle is zero, it merits separate consideration since its exact solution is particularly simple and is a classic result in diffraction theory. The example of the circular cylinder forms the basis of diffraction around a smooth convex surface. In all of these cases the structure is assumed to be perfectly conducting which of course limits subsequent diffraction analysis to perfectly conducting bodies. We consider elsewhere structures in which a more general impedance boundary condition applies.

It is not necessary to understand the derivation of the solutions to the canonical problems to make use of the GTD methods given later. Since many readers will want to skip this chapter, a detailed discussion of these solutions is deferred until the appropriate section in Chapters 4–6.

3.1 Reflection and refraction at a plane interface

The canonical problem for geometrical optics is that of reflection and refraction of a plane wave at an infinite plane interface between two differing homogeneous media. This is illustrated in Fig. 3.1 where the x-y plane is chosen as the plane of incidence. We will consider both *electric polarisation*, where the incident electric field is normal to the

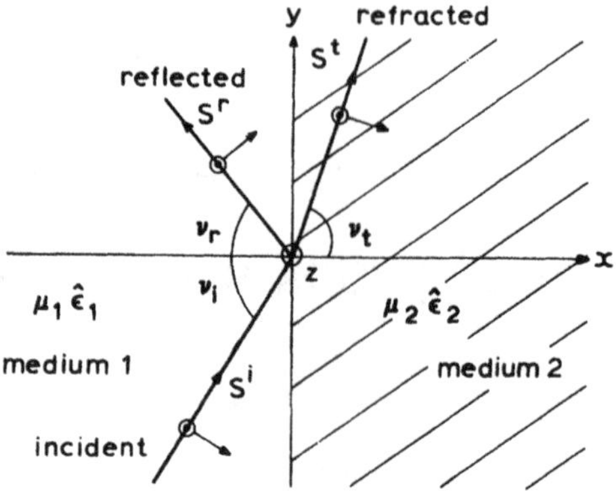

Fig. 3.1 Reflection and refraction at a plane interface

plane of incidence, and *magnetic polarisation*, where the incident magnetic field is normal to the plane of incidence. Since the incident field is independent of z, then the boundary conditions at the interface will ensure that this is true of the total field.

In the representation of the field it will sometimes be convenient to write k in the following forms: if we put $k = k_0 n$ where k_0 is the free space wave-number $\omega\sqrt{(\mu_0 \epsilon_0)}$, then n is the *complex refractive index*. Thus

$$k = k_0 n \quad \text{where} \quad k_0 = \omega\sqrt{(\mu_0 \epsilon_0)}, \qquad n = \sqrt{\left(\frac{\mu\hat{\epsilon}}{\mu_0 \epsilon_0}\right)} \qquad (1a)$$

Writing k in terms of the co-ordinates (x, y) we have

$$k^2 = k_x^2 + k_y^2 \qquad (1b)$$

Finally, since k is generally complex, we put

$$k = \alpha - j\beta \qquad (1c)$$

where α and β are positive real.

3.1.1 Electric polarisation

For electric polarisation the incident electric field E_z^i is given by

$$E_z^i = \exp\{-j(k_x^i x + k_y^i y)\} \qquad (2)$$

where

$$k_x^i = k_1 \cos \nu_i, \qquad k_y^i = k_1 \sin \nu_i$$

We now attempt to obtain a solution in terms of a reflected electric field E_z^r and a refracted, or transmitted, electric field E_z^t of the form

$$\begin{aligned} E_z^r &= R^e \exp\{-j(k_x^r x + k_y^r y)\} \\ E_z^t &= T^e \exp\{-j(k_x^t x + k_y^t y)\} \end{aligned} \tag{3}$$

where

$$k_1^2 = (k_x^i)^2 + (k_y^i)^2 = (k_x^r)^2 + (k_y^r)^2 = \omega^2 \mu_1 \hat{\epsilon}_1 \tag{4}$$
$$k_2^2 = (k_x^t)^2 + (k_y^t)^2 = \omega^2 \mu_2 \hat{\epsilon}_2$$

The continuity of the electric field at $x = 0$ requires $E_z^i + E_z^r = E_z^t$, then we must satisfy

$$1 + R^e = T^e; \qquad k_y^i = k_y^r = k_y^t \tag{5}$$

Since k_y^i, k_y^r are in the same medium, and k_y^i is given by $k_1 \sin \nu_i$, we can write

$$k_y^r = k_1 \sin(\pi - \nu_r) = k_1 \sin \nu_i \tag{6}$$

which expresses *Snell's law of reflection: the incident and reflected waves make equal angles with the normal to the interface.* Similarly, for k_x^r

$$k_x^r = k_1 \cos(\pi - \nu_r) = -k_x^i \tag{7}$$

With these values for k_x^r and k_y^r we can obtain the magnetic reflected field (from eqn. 2.29) to find that the reflected field is also a plane wave.

The transmitted phase term k_x^t can be written as

$$k_x^t = \{k_2^2 - (k_y^i)^2\}^{1/2}$$

As in eqn. 1c, let $k_1 = \alpha_1 - j\beta_1$ and $k_2 = \alpha_2 - j\beta_2$ so that

$$k_x^t = \{(\alpha_2 - j\beta_2)^2 - (\alpha_1 - j\beta_1)^2 \sin^2 \nu_i\}^{1/2} = \tau \exp(-j\gamma) \tag{8}$$

Squaring both sides of this equation and matching real and imaginary parts to obtain values for τ and γ gives

$$\tau = (A^2 + B^2)^{1/4}; \qquad \gamma = \tfrac{1}{2} \tan^{-1} \frac{B}{A} \tag{9}$$

where

$$A = \alpha_2^2 - \beta_2^2 - (\alpha_1^2 - \beta_1^2) \sin^2 \nu_i$$
$$B = 2(\alpha_2 \beta_2 - \alpha_1 \beta_1 \sin^2 \nu_i)$$

By substituting for k_x^t and k_y^t the expression for the transmitted field becomes

$$E_z^t = T^e \exp\{-(\tau \sin \gamma x + \beta_1 \sin \nu_i y)\} \exp\{-j(\tau \cos \gamma x + \alpha_1 \sin \nu_i y)\} \tag{10}$$

We now have surfaces of constant amplitude defined by

$$\tau \sin \gamma x + \beta_1 \sin \nu_i y = \text{const} \tag{11}$$

and surfaces of constant phase defined by

$$\tau \cos \gamma x + \alpha_1 \sin \nu_i y = \text{const} \tag{12}$$

The direction of propagation of a wave is given by the normal to the planes of constant phase. For the refracted wave, the angle of refraction ν_t defined in Fig. 3.1 can now be obtained directly from eqn. 12 such that

$$\cos \nu_t = \frac{\tau \cos \gamma}{(\tau^2 \cos^2 \gamma + \alpha_1^2 \sin^2 \nu_i)^{1/2}}; \qquad \sin \nu_t = \frac{\alpha_1 \sin \nu_i}{(\tau^2 \cos^2 \gamma + \alpha_1^2 \sin^2 \nu_i)^{1/2}} \tag{13}$$

The value of incident angle ν_i giving $\nu_t = \frac{\pi}{2}$ is called the *critical angle* ν_c beyond which a refracted field does not penetrate into the second medium. From eqn. 13 we see that the critical angle is given when

$$\tau(\nu_c) = 0 \tag{14}$$

In particular, if both media are lossless

$$\sin \nu_c = \frac{\alpha_2}{\alpha_1} \tag{15}$$

If $\alpha_1 < \alpha_2$ then there is no critical angle and refraction occurs for all incident angles.

With the electric field given by eqn. 3, the magnetic transmitted field is, by using eqn. 2.29,

$$H^t = \frac{E_z^t}{\omega \mu_2} (\hat{x} k_y^t - \hat{y} k_x^t) \tag{16}$$

with the value of k_x^t given by eqn. 8. This field will not, in general, be transverse to the electric field, and thus the transmitted field is not invariably a plane wave.

The values of the reflection and transmission coefficients R^e and T^e now remain to be calculated. The continuity of H_y at $x = 0$ requires that $H_y^i + H_y^r = H_y^t$ giving

$$\frac{k_x^i}{\mu_1}(1 - R^e) = \frac{k_x^t}{\mu_2} T^e$$

Solving this equation simultaneously with the first equation of eqn. 5

gives the reflection and transmission coefficients

$$R^e = \frac{\mu_2 k_x^i - \mu_1 k_x^t}{\mu_2 k_x^i + \mu_1 k_x^t}; \qquad T^e = \frac{2\mu_2 k_x^i}{\mu_2 k_x^i + \mu_1 k_x^t} \tag{17}$$

These coefficients are in general complex so that both the amplitude and phase of the wave change upon reflection and refraction.

3.1.2 Magnetic polarisation

For magnetic polarisation the incident field is written as

$$H_z^i = \exp\{-j(k_x^i x + k_y^i y)\} \tag{18}$$

and we solve for a reflected magnetic field H_z^r and a transmitted magnetic field H_z^t of the form

$$\begin{aligned} H_z^r &= R^m \exp\{-j(k_x^r x + k_y^r y)\} \\ H_z^t &= T^m \exp\{-j(k_x^t x + k_y^t y)\} \end{aligned} \tag{19}$$

The solution is identical to that for electric polarisation and the values of the reflection and transmission coefficients are given by

$$R^m = \frac{\mu_1 n_2^2 k_x^i - \mu_2 n_1^2 k_x^t}{\mu_1 n_2^2 k_x^i + \mu_2 n_1^2 k_x^t}; \qquad T^m = \frac{2\mu_1 n_2^2 k_x^i}{\mu_1 n_2^2 k_x^i + \mu_2 n_1^2 k_x^t} \tag{20}$$

We have now obtained general formulas for the reflection and refraction of an electromagnetic plane wave at a planar interface between two differing media. The reflected field was always seen to be a plane wave but this was not generally true for the refracted wave. Of particular interest for ray-tracing methods are those cases where the refracted wave either propagates as a plane wave or is highly attenuated, thus we will consider two special cases of the above equations. One case is that when the media are only slightly lossy, so that $\sigma \ll \omega\epsilon$. This is true of most substances, if the frequency is high enough, and of dielectrics in general. The other special case that we will consider is when $\sigma \gg \omega\epsilon$ for one medium and the refracted wave is highly attenuated. This is true of most metals provided that the frequency is not too high. For convenience we will also assume $\mu_1 = \mu_2 = \mu_0$, which is true for many materials. From now on our analysis will be mainly concerned with media which satisfy these conditions.

3.1.3 Slightly lossy media

When both media in Fig. 3.1 are slightly lossy, so that $\sigma_1 \ll \omega\epsilon_1$ and $\sigma_2 \ll \omega\epsilon_2$ we can approximate $k = \omega\sqrt{(\mu_0 \hat{\epsilon})}$ by

$$\omega\sqrt{(\mu\epsilon)}\left(1-\frac{j\sigma}{2\omega\epsilon}\right)$$

to obtain

$$\begin{aligned} \alpha_1 &\simeq \omega\sqrt{(\mu_0\epsilon_1)}; & \beta_1 &\simeq \frac{\sigma_1}{2}\sqrt{\left(\frac{\mu_0}{\epsilon_1}\right)}; & \alpha_1 &\gg \beta_1 \\ \alpha_2 &\simeq \omega\sqrt{(\mu_0\epsilon_2)}; & \beta_2 &\simeq \frac{\sigma_2}{2}\sqrt{\left(\frac{\mu_0}{\epsilon_2}\right)}; & \alpha_2 &\gg \beta_2 \end{aligned} \quad (21)$$

The critical angle is given approximately by eqn. 15, which can be written as

$$\sin\nu_c \simeq \sqrt{\left(\frac{\epsilon_2}{\epsilon_1}\right)} \quad (22)$$

For $\nu_i < \nu_c$, quantities in eqn. 9 simplify such that $|A| \gg |B|$ and the angle γ can be approximated by $\gamma \simeq \dfrac{B}{2A}$. Also, the angle of refraction ν_t defined in eqn. 13 reduces to

$$\sin\nu_t \simeq \sqrt{\left(\frac{\epsilon_1}{\epsilon_2}\right)}\sin\nu_i; \qquad \nu_i < \nu_c \quad (23)$$

which expresses *Snell's law of refraction.*

The transmitted field for electric polarisation given by eqn. 10 becomes, with $\tau \simeq \alpha_2\cos\nu_t$, $\tau\gamma \simeq \beta_2\sec\nu_t - \beta_1\tan\nu_t\sin\nu_i$, and using eqn. 23,

$$\begin{aligned} E_z^t \simeq\ & T^e\exp\{-(\beta_2\sec\nu_t x - \beta_1\tan\nu_t\sin\nu_i x + \beta_1\sin\nu_i y)\} \\ & \exp\{-j\alpha_2(\cos\nu_t x + \sin\nu_t y)\}; \qquad \nu_i < \nu_c \end{aligned}$$

Since $\alpha_1 \gg \beta_1$, $\alpha_2 \gg \beta_2$, the quantities β_1 and β_2 have been retained only in the phase term. Putting $x = s^t\cos\nu_t$ and $y = s^t\sin\nu_t$ in the exponential terms reduces the electric field to

$$E_z^t \simeq T^e\exp(-jk_2 s^t); \qquad \nu_i < \nu_c \quad (24a)$$

and from eqn. 16

$$H^t \simeq \sqrt{\left(\frac{\hat{\epsilon}_2}{\mu_0}\right)}(\hat{x}\sin\nu_t - \hat{y}\cos\nu_t)E_z^t; \qquad \nu_i < \nu_c \quad (24b)$$

Similarly, for magnetic polarisation we get from eqns. 19 and 2.29

$$\left.\begin{aligned} H_z^t &\simeq T^m\exp(-jk_2 s^t) \\ E^t &\simeq \sqrt{\left(\frac{\mu_0}{\hat{\epsilon}_2}\right)}(-\hat{x}\sin\nu_t + \hat{y}\cos\nu_t)H_z^t \end{aligned}\right\}; \qquad \nu_i < \nu_c \quad (25)$$

For slightly lossy media, the transmitted fields as given by eqns. 24 and 25 are seen to propagate as plane waves.

The reflection and transmission coefficients defined by eqns. 17 and 20 become

$$\left.\begin{aligned} R^e &= \frac{\sin(\nu_t-\nu_i)}{\sin(\nu_t+\nu_i)} & T^e &= \frac{2\cos\nu_i\sin\nu_t}{\sin(\nu_t+\nu_i)} \\ R^m &= \frac{\tan(\nu_i-\nu_t)}{\tan(\nu_i+\nu_t)} & T^m &= \frac{2\cos\nu_i\sin\nu_t}{\sin(\nu_i+\nu_t)\cos(\nu_i-\nu_t)} \end{aligned}\right\} ; \quad \nu_i < \nu_c \tag{26}$$

When medium 2 is more dense than medium 1 ($\epsilon_2 > \epsilon_1$) then refraction occurs for all angles of incidence in medium 1. If the reverse is the case then a critical angle exists, as defined by eqn. 22. For $\nu_i > \nu_c$ the angle γ defined in eqn. 9 and the angle ν_t defined in eqn. 13 are both approximately 90°. The electromagnetic field for electric polarisation from eqns. 10 and 16 now becomes

$$\left.\begin{aligned} E_z^t &\simeq T^e \exp(-\tau x)\exp(-j\alpha_1 \sin\nu_i y) \\ H^t &\simeq \frac{1}{\omega\mu_0}(\hat{x}\alpha_1 \sin\nu_i + j\hat{y}\tau)E_z^t \end{aligned}\right\} \quad \nu_i > \nu_c$$

and the reflection and transmission coefficients are complex quantities. For the reflection terms in eqns. 17 and 20 we get

$$R^e = \frac{\alpha_1 \cos\nu_i - j\tau}{\alpha_1 \cos\nu_i + j\tau}; \qquad R^m = \frac{\epsilon_2\alpha_1 \cos\nu_i - j\epsilon_1\tau}{\epsilon_2\alpha_1 \cos\nu_i + j\epsilon_1\tau}; \qquad \nu_i > \nu_c \tag{27}$$

These terms are always seen to have a modulus of one but with a varying phase which is dependent on the incident angle ν_i. This phase dependence is not the same for electric and magnetic polarisation and the reflected field will, in general, be elliptically polarised.

As we discussed earlier, the reflected field is always in the form of a plane wave so that a modulus of one for both R^e and R^m implies that total reflection has occurred. It will be noted that T^e and T^m still remain non-zero. Determining the Poynting vector, given by eqn. 2.6, for the field in medium 2 yields an average energy flow propagating parallel to the interface. For the fields of infinite extent assumed here, this flow of energy in the second medium is practically unimportant in the transfer of energy from the incident to the reflected field.

3.1.4 Highly conducting media

If medium 2 in Fig. 3.1 is highly conducting, then, with the condition that the frequency is sufficiently low for $\sigma_2 \gg \omega\epsilon_2$, we can approximate $k_2 = \omega\sqrt{(\mu_0\hat{\epsilon}_2)}$ by

$$\sqrt{(-j\omega\mu_0\sigma_2)} = \sqrt{(\omega\mu_0\sigma_2)}\exp\left(-\frac{j\pi}{4}\right)$$

to obtain

$$\alpha_2 \simeq \beta_2 \simeq \sqrt{\left(\frac{\omega\mu_0\sigma_2}{2}\right)} \tag{28}$$

This approximation is valid for most metals within the radio frequency spectrum with which we will be chiefly concerned. It must be noted, however, that for sufficiently high frequencies (which for metals occurs at infra-red and higher frequencies) the opposite condition will hold, i.e., $\sigma_2 \ll \omega\epsilon_2$ and the approximations of the previous section are appropriate.

With α_2, β_2 given by eqn. 28, and medium 1 assumed to be slightly lossy, we make a further assumption that $\alpha_2 \gg \alpha_1$, which implies $\sigma_2 \gg 2\omega\epsilon_1$ (where α_1 is given by eqn. 21) so that $|A| \ll |B|$ in eqn. 9 and we get

$$\gamma \simeq \frac{\pi}{4}; \qquad \tau \simeq \sqrt{(\omega\mu_0\sigma_2)}$$

The angle of refraction defined by eqn. 13 now reduces to zero and the transmitted field propagates in the direction normal to the interface for all angles of incidence. From eqn. 10 the transmitted electric field for electric polarisation is

$$E_z^t \simeq T^e \exp(-\alpha_2 x)\exp(-j\alpha_2 x)$$

This field is rapidly attenuated and reduces to $\exp(-1)$ of its surface value at

$$x = \sqrt{\left(\frac{2}{\omega\mu_0\sigma_2}\right)}$$

This value is called the *skin depth* and is a measure of the penetration of the field into the conductor.

When one medium is highly lossy then the reflected field is of primary interest. For high but bounded conductivity the reflection coefficients still remain complex. If we allow the conductivity to approach infinity then

$$R^e \xrightarrow[\sigma_2 \to \infty]{} -1; \qquad R^m \xrightarrow[\sigma_2 \to \infty]{} 1$$

and these values are used when considering reflection from metallic objects at radio frequencies.

3.1.5 Surface impedance of a plane interface

In some applications it is convenient to characterize the plane interface in Fig. 3.1 by a surface impedance. As the transverse field components at $x = 0$ are required to be continuous across the boundary, it follows that the wave impedances normal to the boundary must also be continuous. Consider the field in medium 1 in Fig. 3.1. We can define a surface impedance (looking normally into the interface) from the total field at the surface. Writing $Z^e(Z^m)$ for the surface impedance for the case of electric (magnetic) polarisation we have $Z^e = -E_z/H_y|_{x=0}$, $Z^m = E_y/H_z|_{x=0}$. For two-dimensional fields independent at the $\hat{z}$-direction, the field components are related by eqn. 2.29, and from the definitions for the surface impedances, we can readily deduce the *impedance boundary condition* for the field in medium 1 as

$$\frac{\partial V_z}{\partial n} + \sigma V_z = 0 \tag{29a}$$

where n is taken as the outward normal from the surface. For electric polarization

$$V_z = E_z, \qquad \sigma = -j\omega\mu_1/Z^e \tag{29b}$$

and for magnetic polarization

$$V_z = H_z, \qquad \sigma = -j\omega\hat{\epsilon}_1/Z^m \tag{29c}$$

In the current example of the electric or magnetic polarized plane wave incident from medium 1 on the plane interface we have

$$Z^e = \frac{-(E_z^i + E_z^r)}{(H_y^i + H_y^r)} = \frac{-E_z^t}{H_y^t}, \qquad Z^m = \frac{E_y^t}{H_z^t}$$

which, from the expressions for the transmitted field components, become

$$Z^e = \frac{\omega\mu_2}{k_x^t}, \qquad Z^m = \frac{k_x^t}{\omega\hat{\epsilon}_2}$$

With these values for the surface impedance, the reflection coefficients R^e, R^m for the field in medium 1 given in eqns. 17 and 20 can be rewritten as

$$R^e = \frac{\cos\nu_i - \cos\nu^e}{\cos\nu_i + \cos\nu^e}, \qquad R^m = \frac{\cos\nu_i - \cos\nu^m}{\cos\nu_i + \cos\nu^m} \tag{30a}$$

where $\nu^{e,m}$ are complex angles given by

$$\nu^e = \cos^{-1}(Z_0/Z^e), \qquad \nu^m = \cos^{-1}(Z^m/Z_0) \tag{30b}$$

Here, $Z_0 = \sqrt{\mu_1/\hat{\epsilon}_1}$ is the free space impedance of medium 1. These angles are called the *complex Brewster angles* since they represent the angle of incidence where no reflected wave is present.

3.2 The half-plane

The canonical problem for geometrical optics given in Section 3.1 where a plane wave is incident upon a planar dielectric interface, was seen to yield a rigorous solution in terms of a single elementary wave function u associated with a transmitted and reflected field, where

$$u = \exp\{-j(k_x x + k_y y)\} \tag{31}$$

and is a solution to the scalar Helmholtz equation

$$\nabla^2 u + k^2 u = 0 \tag{32}$$

For the half-plane problem to be discussed in this section we shall also be dealing with 2-dimensional fields, but we will require a more general solution to the Helmholtz equation than that given by eqn. 31. We could of course seek a solution in terms of an infinite series of eigenfunctions as given by eqn. 2.33 and this is the approach we shall use initially for the wedge and circular cylinder to be discussed later. In the case of the half-plane, however, a more tractable solution is obtained if we solve for a continuous spectrum of elemental wave functions as given by eqn. 2.32, viz;

$$u = \int_C f(\xi) \exp\{-jk(x\cos\xi + y\sin\xi)\} d\xi \tag{33}$$

where a suitable path C in the complex ξ-plane is yet to be determined.

The first rigorous solution to diffraction by a perfectly conducting half-plane was given by Sommerfeld (1896) in a classic work in electromagnetic theory. His method involved seeking a solution to the wave equation which had a periodicity of 4π. An account of this approach can be found in Chapter 4 of Baker and Copson (1950). Sommerfeld's result is famous partly because he obtained the exact solution at any point in terms of the well-tabulated Fresnel integral. The half-plane problem has since been solved in a number of ways, most notably using the *Weiner-Hopf technique* (Copson, 1946) which was introduced by Wiener and Hopf (1931) to solve certain types of integral equations. We shall use this technique to derive the exact solution as it involves, for the half-plane problem at least, rather straightforward procedures in contour integration.

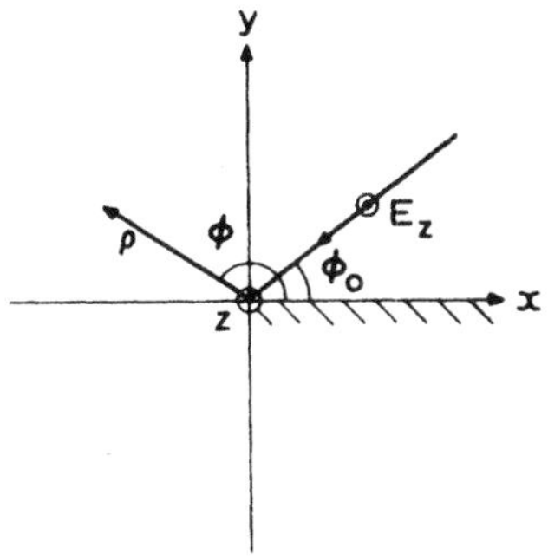

Fig. 3.2 The half-plane

3.2.1 Electric polarisation

For electric polarisation we begin with a normally incident plane wave on the half-plane, as illustrated in Fig. 3.2, of the form

$$E_z^i = \exp\{jk(x\cos\phi_0 + y\sin\phi_0)\} \tag{34}$$

By *normal* incidence we mean that the incoming wavefront is parallel to the edge. For 2-dimensional fields we have from eqn. 2.29 the magnetic field given as

$$j\omega\mu \boldsymbol{H} = \hat{z} \times \nabla E_z \tag{35}$$

so the incident magnetic field is

$$\boldsymbol{H}^i = \sqrt{\left(\frac{\epsilon}{\mu}\right)}(-\hat{x}\sin\phi_0 + \hat{y}\cos\phi_0)\exp\{jk(x\cos\phi_0 + y\sin\phi_0)\} \tag{36}$$

This field incident upon the half-plane will induce electric currents which will radiate a scattered field $\boldsymbol{E}^s$, $\boldsymbol{H}^s$. In the plane $y = 0$, the boundary conditions on the half-plane, which is assumed to be perfectly conducting, require that

$$E_z^i + E_z^s = 0; \qquad y = 0, \quad 0 < x < \infty \tag{37}$$

and the boundary conditions for the remaining half of the $y = 0$ plane where no surface currents exist require

$$H_x^s = H_z^s = 0; \qquad y = 0, \quad -\infty < x < 0 \tag{38}$$

By the symmetry of the problem we can write down immediately, for the non-zero field components that

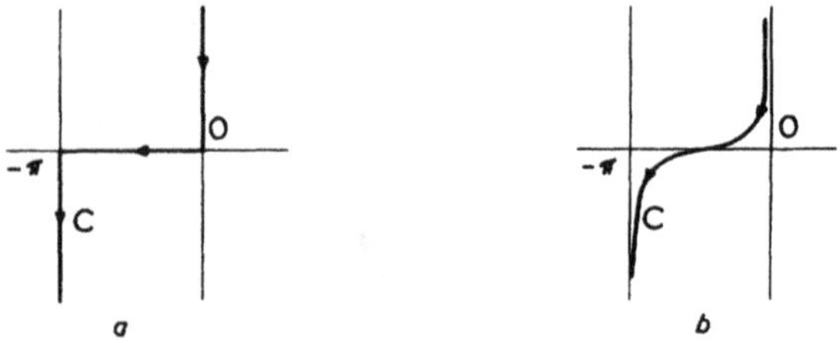

Fig. 3.3 Contours in complex ξ-plane for the integrals in eqns. 40 and 41
a Im $k = 0$
b Im $k \neq 0$

$$
\begin{aligned}
E_z^s(x,y,z) &= E_z^s(x,-y,z)\\
H_x^s(x,y,z) &= -H_x^s(x,-y,z)\\
H_y^s(x,y,z) &= H_y^s(x,-y,z)
\end{aligned}
\tag{39}
$$

We now seek solutions for the scattered field on the $y = 0$ plane in terms of the integral representation of eqn. 33 so that from eqns. 37 and 38

$$E_z^s(x,0) = \int_C f(\xi)\exp(-jkx\cos\xi)d\xi = -E_z^i(x,0); \qquad x > 0 \tag{40}$$

$$H_x^s(x,0) = \int_C \sin\xi f(\xi)\exp(-jkx\cos\xi)d\xi = 0; \qquad x < 0 \tag{41}$$

and we evaluate to obtain the function $f(\xi)$.

In solving these dual integral equations we now construct a Fourier transform of a function, say $g(v)$, to represent the scattered electric field E_z^s on the half-plane so that

$$E_z^s(x,0) = \int_{-\infty}^{\infty} g(v)\exp(-jxv)dv; \qquad x > 0 \tag{42}$$

For this equation to equate with eqn. 40 we must allow

$$v = k\cos\xi, \quad g(v) = \frac{f(\xi)}{k\sin\xi} \tag{43}$$

and the path C in the complex ξ-plane to include all real values of $k\cos\xi$ from ∞ to $-\infty$, as do the paths illustrated in Fig. 3.3. Now the scattered electric field in *all* space can be written from eqn. 33 in terms of the variable v as

$$E_z^s(x,y) = \int_{-\infty}^{\infty} g(v)\exp\left[-j\{xv + y\sqrt{(k^2 - v^2)}\}\right]dv \tag{44}$$

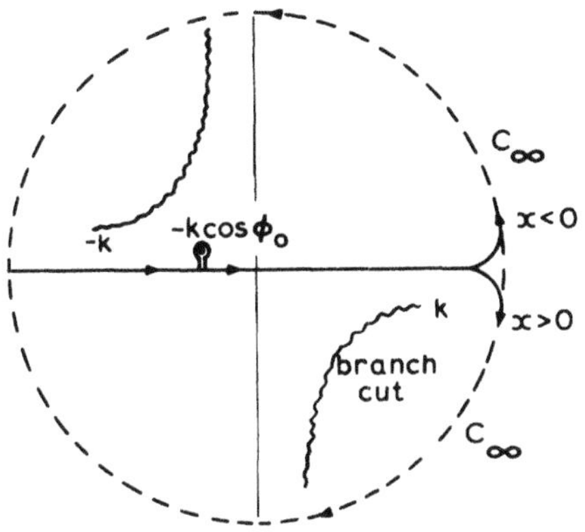

Fig. 3.4 Contours in complex v-plane for the half-plane solution

The square root in the exponential term creates a multivalued function and hence gives rise to branch cuts in the complex v-plane. These branch cuts begin at the branch points where $v = \pm k$. To keep track of any unwanted exponential growth as y increases in eqn. 44, we choose the branch cuts such that the imaginary part of $\sqrt{(k^2 - v^2)}$ is zero on them. The branch cuts are shown in Fig. 3.4, using this criterion, where the medium surrounding the half-plane is assumed to have, in general, some loss so that k will be a complex quantity.

If we now make the substitution $x = \rho\cos\phi$, $y = \rho\sin\phi$, eqn. 44 becomes

$$E_z^s(\rho, \phi) = \int_C f(\xi)\exp\{-jk\rho\cos(\phi \pm \xi)\}d\xi \tag{45}$$

where the combined plus and minus signs $\pm$ arise from the square root in eqn. 44. This ambiguity is resolved by examining the behaviour of the exponential term at the integration limits of $0 + j\infty$ and $-\pi - j\infty$, where

$$\exp\{-jk\rho\cos(\phi \pm \xi)\} \to \exp\{-jk\rho\cos\phi\cos(0, \pi)\text{ch}\,\infty\}$$

$$\exp(\mp k\rho\sin\phi\,\text{sh}\,\infty)$$

It is clear, therefore, that eqn. 45 must be solved as follows:

$$\begin{aligned} E_z^s(\rho, \phi) &= \int_C f(\xi)\exp\{-jk\rho\cos(\phi + \xi)\}d\xi; \qquad 0 < \phi < \pi \\ E_z^s(\rho, \phi) &= \int_C f(\xi)\exp\{-jk\rho\cos(\phi - \xi)\}d\xi; \qquad \pi < \phi < 2\pi \end{aligned} \tag{46}$$

Returning now to eqns. 40 and 41 we have, by the transformation

$v = k\cos\xi$,

$$\int_{-\infty}^{\infty} \frac{f\left(\cos^{-1}\frac{v}{k}\right)}{\sqrt{(k^2 - v^2)}} \exp(-jxv)dv = -\exp(jxv_0); \qquad x > 0 \quad (47)$$

$$\int_{-\infty}^{\infty} f\left(\cos^{-1}\frac{v}{k}\right) \exp(-jxv)dv = 0 \qquad x < 0 \quad (48)$$

where $v_0 = k\cos\phi_0$. These dual integral equations can now be solved for the unknown function $f\left(\cos^{-1}\frac{v}{k}\right)$ by the procedure in contour integration known as the Wiener-Hopf technique.

For the integral in eqn. 48, where x is negative, we may close the path of integration with an infinite semicircle above the real axis in the complex v-plane. No additional contribution will be made to the integral provided that the function $f\left(\cos^{-1}\frac{v}{k}\right)$ is regular (i.e., has no poles or branch cuts) and tends to zero as $|v| \to \infty$ throughout this region. In the integral of eqn. 47, where x is positive, we may close the path of integration with an infinite semicircle below the real axis without changing the value of the integral, provided that $f\left(\cos^{-1}\frac{v}{k}\right)$ $(k^2 - v^2)^{-1/2} \to 0$ as $|v| \to \infty$. The right-hand side, however, is non-zero, and we now equate this to an appropriate pole singularity.

From Cauchy's residue theorem for a simple pole we have

$$f(a) = \pm\frac{1}{2\pi j}\oint \frac{f(z)}{z - a}\,dz \qquad (49)$$

where the integration is around a closed path in the complex z-plane containing the simple pole at $z = a$, and the upper (lower) sign applies for the path taken anticlockwise (clockwise) around the pole. We can now relate the right-hand side of eqn. 47 as the residue of a simple pole at $v = -v_0$, as shown in Fig. 3.4, by writing

$$\frac{f\left(\cos^{-1}\frac{v}{k}\right)}{\sqrt{(k^2 - v^2)}} = \frac{1}{2\pi j}\,\frac{h(v)}{h(-v_0)}\,\frac{1}{v + v_0} \qquad (50)$$

where $h(v)$ is any regular function within the contour of integration, which is deformed as in Fig. 3.4 to capture the pole at $v = -k\cos\phi_0$. This equation can be arranged as

$$\frac{f\left(\cos^{-1}\frac{v}{k}\right)}{\sqrt{(k-v)}}(v+v_0) = \frac{1}{2\pi j}\frac{h(v)}{h(-v_0)}\sqrt{(k+v)} \tag{51}$$

The left-hand side of this equation has no singularities in the upper half-space of the complex v-plane, since the function $f\left(\cos^{-1}\frac{v}{k}\right)$ was assumed to be regular in this region, and the branch cut beginning at $k = v$ is in the lower half-space. Similarly, the right-hand side has no singularities in the lower half-space, since the function $h(v)$ is regular there and the branch cut beginning at $k = -v$ resides in the upper half-space. This rearranging of eqn. 50 into regions of regularity is the essence of the Wiener-Hopf technique, and for the half-plane it is seen to be particularly simple. For other problems, however, it can be a formidable, often intractable, task to obtain an analytic solution.

Returning to eqn. 51, we see that since both sides are to be equal, it implies that either side will be regular throughout the entire complex v-plane. Also with $f\left(\cos^{-1}\frac{v}{k}\right)$ bounded at infinity it is seen that both sides of eqn. 51 must also be bounded at infinity. We can now apply *Liouville's theorem*, which states that a function having no singularities in the entire complex plane and bounded at infinity must be a constant. Thus, applying Liouville's theorem to eqn. 51, we obtain the value of the constant by putting $v = -v_0$ on the right-hand side and immediately derive the expression for $f\left(\cos^{-1}\frac{v}{k}\right)$ as

$$\frac{1}{2\pi j}\frac{\sqrt{(k-v_0)}\sqrt{(k-v)}}{v+v_0}$$

On substituting $v = k\cos\xi$ and $v_0 = k\cos\phi_0$ we have

$$f(\xi) = \frac{1}{\pi j}\frac{\sin\frac{1}{2}\phi_0\sin\frac{1}{2}\xi}{\cos\xi+\cos\phi_0} \tag{52}$$

which can be rewritten as

$$f(\xi) = -\frac{1}{4\pi j}\{\sec\tfrac{1}{2}(\xi-\phi_0)-\sec\tfrac{1}{2}(\xi+\phi_0)\} \tag{53}$$

We now have the exact solution for the function $f(\xi)$, from which we obtain the scattered electric field via eqn. 46 as

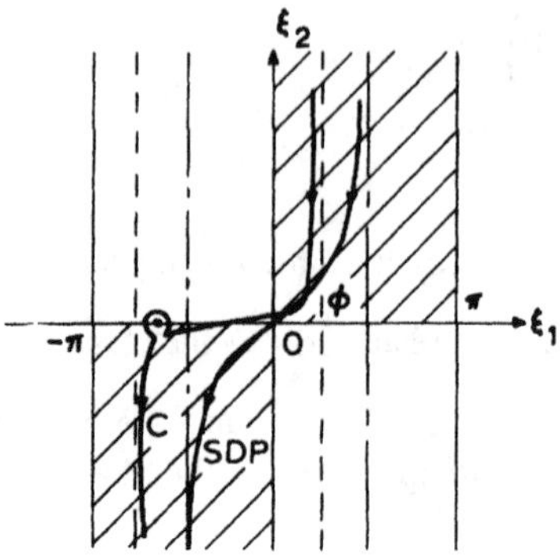

Fig. 3.5 Contours in complex ξ-plane for the integral in eqn. 55

$$E_z^s(\rho,\phi) = -\frac{1}{4\pi j}\int_C \{\sec\tfrac{1}{2}(\xi-\phi_0) - \sec\tfrac{1}{2}(\xi+\phi_0)\} \exp\{-jk\rho\cos(\phi\pm\xi)\}d\xi \qquad (54)$$

where the upper (lower) sign applies for $0<\phi<\pi$ $(\pi<\phi<2\pi)$, and the path C in the complex ξ-plane is illustrated in Fig. 3.3. The path must, if necessary, be indented to include the simple poles (not shown in the figure) at $\xi = \phi_0 - \pi$ and $\pi - \phi_0$ as seen from eqn. 54. It remains now to manipulate eqn. 54 so as to obtain a more useful solution.

Consider first the case when $0<\phi<\pi$. The function $f(\xi)$ as given by eqn. 53 consists of two terms, but it is only necessary to evaluate eqn. 54 with respect to the first term $[\sec\frac{1}{2}(\xi-\phi_0)]$ since the contribution from the second term $[\sec\frac{1}{2}(\xi+\phi_0)]$ will be given simply by changing the sign of the angle ϕ_0. Thus we initially evaluate the integral

$$I = -\frac{1}{4\pi j}\int_C \sec\tfrac{1}{2}(\xi-\phi_0)\exp\{-jk\rho\cos(\phi+\xi)\,d\xi$$

$$\text{for} \quad 0<\phi<\pi$$

By substituting $\xi' = \xi + \phi$ this becomes

$$I = -\frac{1}{4\pi j}\int_C \sec\tfrac{1}{2}(\xi'-\phi-\phi_0)\exp(-jk\rho\cos\xi')\,d\xi'$$

where C is now the path in the complex ξ-plane as shown in Fig. 3.5. This equation can be rewritten as

$$I = -\frac{1}{8\pi j}\int_C \{\sec\tfrac{1}{2}(\xi' - \phi - \phi_0) + \sec\tfrac{1}{2}(\xi' + \phi + \phi_0)\}$$

$$\exp(-jk\rho\cos\xi')\,d\xi'$$

where the second term was obtained by substituting $-\xi'$ for ξ'. We may now rearrange this integral so that

$$I = -\frac{1}{2\pi j}\int_C \frac{\cos\tfrac{1}{2}\xi'\cos\tfrac{1}{2}(\phi+\phi_0)}{\cos\xi' + \cos(\phi+\phi_0)}\exp(-jk\rho\cos\xi')\,d\xi' \quad (55)$$

and attempt to distort the path C into a more useful contour. An obvious initial choice is that contour which traverses the steepest descent path (SDP) which was discussed in Section 2.3.2. By separating the argument of the exponential term in eqn. 55 into real and imaginary components with $\xi' = \xi_1' + j\xi_2'$ and assuming the medium surrounding the half-plane to be, at most, only slightly lossy, then

$$\exp(-jk\rho\cos\xi') = \exp(-jk\rho\cos\xi_1'\,\mathrm{ch}\,\xi_2')\exp(-k\rho\sin\xi_1'\,\mathrm{sh}\,\xi_2')$$

The path of steepest descent from eqn. 2.106 is defined by

$$\cos\xi_1'\,\mathrm{ch}\,\xi_2' = C_0$$

$$\sin\xi_1'\,\mathrm{sh}\,\xi_2' \geqslant 0$$

where C_0 is an arbitrary constant to be determined. (Although this is exact only for lossless media it will give us sufficient accuracy for most materials surrounding the half-plane to be met in practice, in that they will be only slightly lossy). For the second condition to be obeyed, the path must remain within the shaded areas in Fig. 3.5. Thus it intersects the origin, as shown in the figure, and the value of the constant is immediately determined as 1. With $\cos\xi_1'\,\mathrm{ch}\,\xi_2' = 1$ we can readily show that

$$\sin\tfrac{1}{2}\xi_1'\,\mathrm{ch}\,\tfrac{1}{2}\xi_2' = \cos\tfrac{1}{2}\xi_1'\,\mathrm{sh}\,\tfrac{1}{2}\xi_2' = \frac{1}{2}\frac{\mathrm{sh}\,\xi_2'}{\sqrt{(\mathrm{ch}\,\xi_2')}} \quad (56)$$

If we now define a new variable ν as

$$\nu = \sqrt{(2)}\exp\left(-\frac{j\pi}{4}\right)\sin\tfrac{1}{2}\xi' \quad (57)$$

then from eqn. 56 we can see that along the path of steepest descent the variable ν goes through all real values between ∞ and $-\infty$. Transforming eqn. 55 to the steepest descent path, and by change of variable to ν we get

$$I = \frac{1}{2\pi}\exp\left\{-j\left(k\rho - \frac{\pi}{4}\right)\right\}\int_{-\infty}^{\infty}\frac{a\exp(-k\rho\nu^2)d\nu}{\nu^2 + ja^2} + \{\text{pole residue at } \xi' = \phi + \phi_0 - \pi\} \tag{58}$$

where $a = \sqrt{(2)}\cos\frac{1}{2}(\phi + \phi_0)$. The integral in eqn. 58 is in the form of the modified Fresnel integral as seen from eqn. 2.39 with the consequent reduction of I to

$$I = \text{sgn}(a)K_{-}\{|a|\sqrt{(k\rho)}\}\exp(-jk\rho) + \{\text{pole residue}\} \tag{59}$$

To complete the evaluation of eqn. 59 we need to compute the residue of the simple pole of eqn. 55 at $\xi' = \phi + \phi_0 - \pi$. From Fig. 3.5 it is seen that this pole will only be captured when $\phi + \phi_0 - \pi < 0$. By Cauchy's residue theorem, the residue Δ from a simple pole at $z = a$ in the contour integral

$$I = \oint_C g(z)dz$$

is given by

$$\Delta = \pm 2\pi j \lim_{z\to a} g(z)(z - a) \tag{60}$$

where the upper (lower) sign applies to the path C taken anticlockwise (clockwise) about the pole. Applying eqn. 60 to eqn. 55 for the simple pole at $\xi' = \phi + \phi_0 - \pi$, we get for the pole residue of eqn. 59

$$-U(\pi - \phi - \phi_0)\exp\{jk\rho\cos(\phi + \phi_0)\} \tag{61}$$

where U is the unit step function, defined earlier in eqn. 2.86. This term is identified as the reflected wave from the half-plane.

The complete evaluation of eqn. 54 for $0 < \phi < \pi$ simply involves the addition to eqn. 59 of the contribution from the second term in eqn. 54 obtained by changing the sign of ϕ_0. For this latter case the pole contribution is

$$U(\pi - \phi + \phi_0)\exp\{jk\rho\cos(\phi - \phi_0)\} \tag{62}$$

which is associated with the incident field E_z^i.

We now have the scattered field E_z^s for $0 < \phi < \pi$ (i.e., $y > 0$) in a mathematically tractable form. For $\pi < \phi < 2\pi$ (i.e., $y < 0$) the solution is identical, since from eqn. 39

$$E_z^s(x, y, z) = E_z^s(x, -y, z)$$

The addition of the incident field of eqn. 34 and the scattered field of eqn. 54 yields, finally, the total electric field E_z (by expressing eqn. 54 through eqns. 59, 61 and 62), as

$$E_z(\rho,\phi) = U(\epsilon^i)u_0^i(\rho,\phi) - U(\epsilon^r)u_0^r(\rho,\phi) + u_d^i(\rho,\phi) - u_d^r(\rho,\phi) \quad (63)$$

where

$$u_0^{i,r}(\rho,\phi) = \exp\{jk\rho\cos(\phi \mp \phi_0)\};$$

$$\epsilon^{i,r} = \mathrm{sgn}(a^{i,r})$$

$$u_d^{i,r}(\rho,\phi) = -\epsilon^{i,r}K_-(|a^{i,r}|\sqrt{(k\rho)})\exp(-jk\rho)$$

$$a^{i,r} = \sqrt{(2)}\cos\tfrac{1}{2}(\phi \mp \phi_0)$$

This is essentially the result obtained by Sommerfeld in 1896.

The resultant magnetic field components H_ρ, H_ϕ are obtained directly from eqn. 35 as

$$-j\omega\mu H_\rho = \frac{1}{\rho}\frac{\partial E_z}{\partial\phi}; \qquad j\omega\mu H_\phi = \frac{\partial E_z}{\partial\rho} \quad (64)$$

To perform the differentiation on the electric field E_z we write it in the form

$$E_z(\rho,\phi) = u^i(\rho,\phi) - u^r(\rho,\phi) \quad (65a)$$

$$u^{i,r}(\rho,\phi) = U(\epsilon^{i,r})u_0^{i,r} - \epsilon^{i,r}K_-\{|a^{i,r}|\sqrt{(k\rho)}\}\exp(-jk\rho) \quad (65b)$$

Upon invoking the modified Fresnel integral relationship of eqn. 2.38*c* we may rewrite eqn. 65*b* in the more compact form

$$u^{i,r} = K_-\{-\sqrt{(2k\rho)}\cos\tfrac{1}{2}(\phi \mp \phi_0)\}\exp(-jk\rho) \quad (65c)$$

Differentiation of this function yields the magnetic field components as

$$H_\rho(\rho,\phi) = \sqrt{\left(\frac{\hat{\epsilon}}{\mu}\right)}\left\{u^i\sin(\phi-\phi_0) - u^r\sin(\phi+\phi_0) - 4\cos\tfrac{1}{2}\phi\sin\tfrac{1}{2}\phi_0\frac{\exp(-jk\rho)}{\sqrt{(8j\pi k\rho)}}\right\} \quad (66)$$

$$H_\phi(\rho,\phi) = \sqrt{\left(\frac{\hat{\epsilon}}{\mu}\right)}\left\{u^i\cos(\phi-\phi_0) - u^r\cos(\phi+\phi_0) + 4\sin\tfrac{1}{2}\phi\sin\tfrac{1}{2}\phi_0\frac{\exp(-jk\rho)}{\sqrt{(8j\pi k\rho)}}\right\} \quad (67)$$

3.2.2 Magnetic polarisation

For magnetic polarisation we proceed in exactly the same way, with the plane wave at normal incidence assumed to be

$$H_z^i = \exp\{jk(x\cos\phi_0 + y\sin\phi_0)\}$$

$$E^i = \sqrt{\left(\frac{\mu}{\hat{\epsilon}}\right)}\{\hat{x}\sin\phi_0 - \hat{y}\cos\phi_0\}\exp\{jk(x\cos\phi_0 + y\sin\phi_0)\}$$

By satisfying the boundary conditions on the half-plane and representing the scattered magnetic field H_z^s as a Fourier transform of a function $g(v)$ we obtain the following dual integral equations for magnetic polarisation:

$$\int_{-\infty}^{\infty} f\left(\cos^{-1}\frac{v}{k}\right)\exp(-jxv)dv = \sqrt{(k^2 - v_0^2)}\exp(jxv_0); \quad x > 0$$

$$\int_{-\infty}^{\infty} \frac{f\left(\cos^{-1}\frac{v}{K}\right)}{\sqrt{(k^2 - v^2)}}\exp(-jxv)dv = 0; \quad x < 0 \tag{68}$$

As before, these dual integral equations are readily solved using the Wiener-Hopf technique to obtain the function $f(\xi)$ from which the scattered magnetic field is determined from

$$H_z^s(\rho, \phi) = \int_C f(\xi)\exp\{-jk\rho\cos(\phi + \xi)\}d\xi; \quad 0 < \phi < \pi$$

$$H_z^s(\rho, \phi) = -\int_C f(\xi)\exp\{-jk\rho\cos(\phi - \xi)\}d\xi; \quad \pi < \phi < 2\pi \tag{69}$$

The analysis proceeds as before to yield

$$H_z(\rho, \phi) = u^i(\rho, \phi) + u^r(\rho, \phi) \tag{70}$$

and the electric field components are determined (see eqn. 2.29), from

$$j\omega\hat{\epsilon}E_\rho = \frac{1}{\rho}\frac{\partial H_z}{\partial \phi}; \qquad -j\omega\hat{\epsilon}E_\phi = \frac{\partial H_z}{\partial \phi}$$

to give

$$E_\rho(\rho, \phi) = -\sqrt{\left(\frac{\mu}{\hat{\epsilon}}\right)}\left\{u^i\sin(\phi - \phi_0) + u^r\sin(\phi + \phi_0) + 4\sin\tfrac{1}{2}\phi\cos\tfrac{1}{2}\phi_0\frac{\exp(-jk\rho)}{\sqrt{(8j\pi k\rho)}}\right\} \tag{71}$$

$$E_\phi(\rho, \phi) = -\sqrt{\left(\frac{\mu}{\hat{\epsilon}}\right)}\left\{u^i\cos(\phi - \phi_0) + u^r\cos(\phi + \phi_0) + 4\cos\tfrac{1}{2}\phi\cos\tfrac{1}{2}\phi_0\frac{\exp(-jk\rho)}{\sqrt{(8j\pi k\rho)}}\right\} \tag{72}$$

3.2.3 Edge condition

We have now completed the derivation for 2-dimensional plane wave diffraction at a half-plane for both electric and magnetic polarisation. From these results, and those of the wedge to be discussed in Section 3.3, we will develop more general solutions applicable for the methods of GTD. It remains, however, to ensure that the solutions we have are the only possible ones. This uniqueness requirement is established by imposing the *edge condition*, which *requires the electromagnetic energy in a finite neighbourhood of the edge to be finite.* Mathematically we can write this condition from eqn. 2.7 as

$$\oiint E \times H^* \cdot ds \underset{\rho \to 0}{\longrightarrow} 0 \tag{73}$$

where ρ is the radial distance from the edge. The consequences of eqn. 73 for the half-plane as the edge is approached are

$$\begin{aligned} &E_z \to 0; \qquad H_z \to \text{const} \\ &E_\rho, E_\phi, H_\rho, H_\phi \to \rho^{-1/2} \end{aligned} \tag{74}$$

which are clearly satisfied by the solutions above for electric and magnetic polarisation. For a more detailed discussion on the edge condition and uniqueness the reader is referred to pp. 562–569 in Jones (1964).

3.3 The wedge

To treat more general edge diffraction problems, we now consider plane wave diffraction at a perfectly conducting wedge illustrated in Fig. 3.6, where V_z^i is the incident field. One of the earlier contributors to wedge diffraction was Macdonald (1902, 1915) who obtained both an eigenfunction series and integral representations for the field. For a tractable solution at high frequencies it is necessary to take an asymptotic expansion of the integral representation which, unfortunately, leads to an invalid result in the transition regions about the optical boundaries. Such a solution is sometimes referred to as *non-uniform* since it is invalid in some regions. The first *uniform* asymptotic series solution for the wedge (meaning that it is valid in the transition regions) was obtained by Pauli (1938). His solution gave the leading term in the form of a Fresnel integral which reduced to the exact half-plane solution when the wedge angle $\beta = 0$. Another asymptotic expansion was derived later by Oberhettinger (1956). For details of this and other expansions for special cases of wedge angle, the reader is referred to Chapter 6 of Bowman *et al.* (1969).

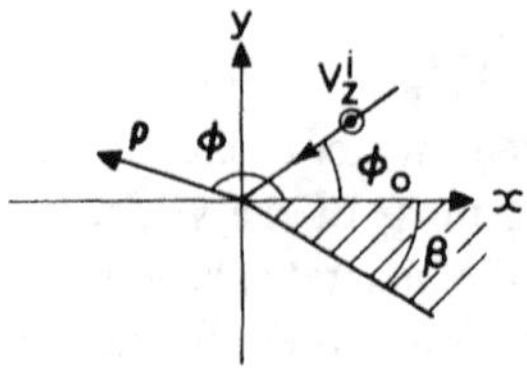

Fig. 3.6 Wedge

More recently, transform methods have been used to solve the wedge diffraction problem. For the half-plane discussed earlier, we made use of the Fourier transform in obtaining our solution. This transform, however, is unsuitable for the wedge but one may use the transform of Kontorowich and Lebedev (1939). [See Jones (1964) pp. 608–612.] The resultant integral representation for the field still requires some form of asymptotic approximation to cast it into a well-tabulated function such as the Fresnel integral. We shall not use the Kontorowich-Lebedev transform directly but will obtain the same integral representation for plane wave incidence from manipulation of the eigenfunction series solution. The resultant integral can then be approximated to yield a uniform asymptotic expansion in terms of the modified Fresnel integral.

In our derivation it will be assumed that the medium surrounding the wedge is only slightly lossy. We shall make use of the representation of the Bessel function given by Morse and Feshback (1953)

$$J_\nu(\sigma) = \frac{1}{2\pi}\int_C \exp\{-j(\sigma \sin\theta - \nu\theta)\}\,d\theta; \qquad \sigma > 0 \qquad (75)$$

where C is a contour in the complex θ-plane as shown in Fig. 3.7. The shaded areas indicate where the integrand in eqn. 75 remains finite as θ goes to $\pm j\infty$, and thus the contour C must begin and end within the shaded sectors as shown.

Consider now the integral

$$\frac{1}{j}\int \frac{\exp\left\{j\alpha\left(\frac{\pi}{2} - \psi_0\right)\right\} f(\alpha)\,d\alpha}{\sin \alpha\psi_0}$$

whose integrand has an infinite series of first order poles along the real α-axis as shown in Fig. 3.8, where $\alpha = \dfrac{m\pi}{\psi_0}$, $m = 0, \pm 1, \pm 2, \ldots$. If we choose a contour Γ which is closed at infinity and traverses the

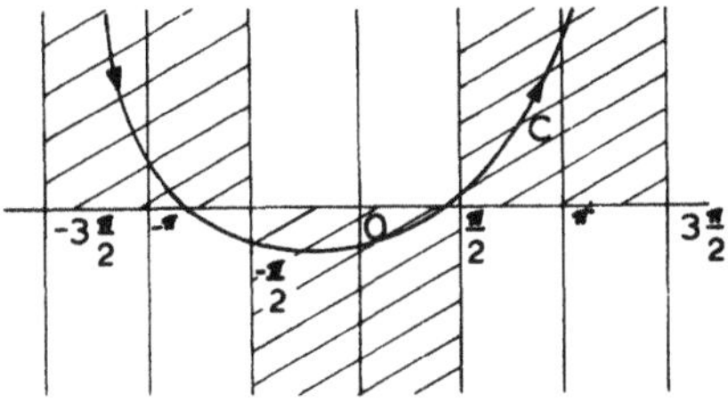

Fig. 3.7 Contour for the Bessel function in complex θ-plane

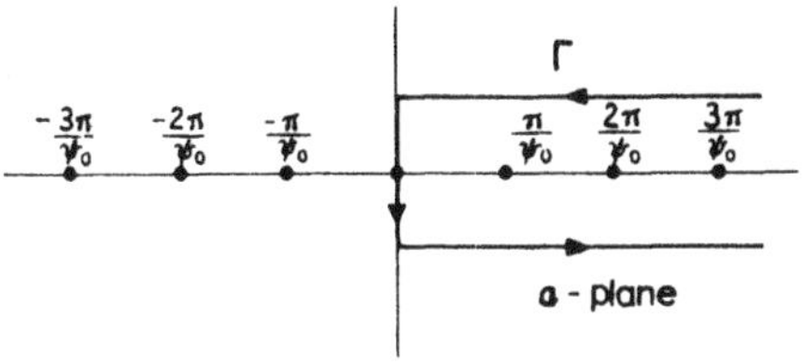

Fig. 3.8 First-order poles along the real axis and the integration path for eqn. 76

complex α-plane as in Fig. 3.8, then, by the sum of the individual first order pole residues given by eqn. 60, we deduce that

$$\frac{1}{j}\oint_{\Gamma} \frac{\exp\left\{j\alpha\left(\frac{\pi}{2}-\psi_0\right)\right\} f(\alpha)d\alpha}{\sin\alpha\psi_0} = \frac{\pi}{\psi_0}\sum_{m=0}^{\infty} \epsilon_\nu j^\nu f(\nu); \quad \nu = \frac{m\pi}{\psi_0} \tag{76}$$

where Neumann's number ϵ_ν is 1 for $\nu = 0$, and 2 for $\nu > 0$. This is introduced since we collect only half the value of the pole at the origin for our chosen contour Γ.

Another series which we will make use of in the next section [see Gradshteyn and Ryzhik (1965)] is

$$1 + 2\sum_{m=1}^{\infty} p^m \cos mx = \frac{p^{-1} - p}{p^{-1} + p - 2\cos x}; \qquad |p| < 1 \tag{77}$$

3.3.1 Electric polarisation

To seek a solution in terms of an infinite series of eigenfunctions we consider initially a line source near the edge at (ρ_0, ϕ_0) as in Fig. 3.9. Plane wave incidence is then the special case when $\rho_0 \to \infty$. A solution is therefore required to the inhomogeneous scalar Helmholtz equation for a line source at (ρ_0, ϕ_0), which for cylindrical co-ordinates independent of the z-direction is

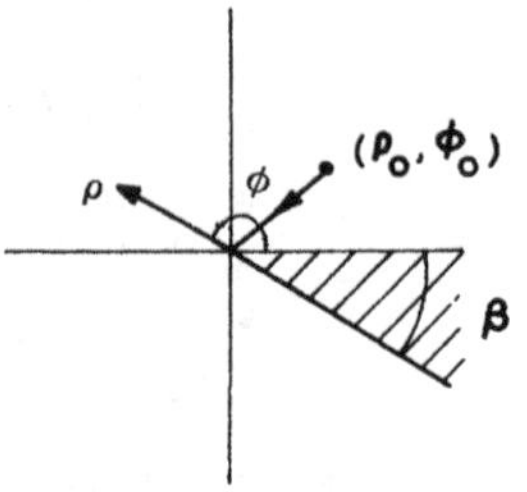

Fig. 3.9 Line source near a wedge

$$\left(\frac{1}{\rho}\frac{\partial}{\partial\rho}\rho\frac{\partial}{\partial\rho}+\frac{1}{\rho^2}\frac{\partial^2}{\partial\phi^2}+k^2\right)E_z = -\frac{\delta(\rho-\rho_0)\delta(\phi-\phi_0)}{\rho} \tag{78}$$

and we seek solutions in the form given by eqn. 2.33:

$$E_z = \sum_n a_n\Phi_n(\phi,\phi_0)P_n(\rho,\rho_0) \tag{79}$$

To satisfy the boundary conditions on the wedge faces at $\phi = 0$ and $2\pi-\beta$, the angular eigenfunctions must be of the form

$$a_n\Phi_n(\phi,\phi_0) = a_n \sin\nu\phi g(\phi_0); \qquad \nu = \frac{n\pi}{2\pi-\beta} \tag{80}$$

Let us now consider $a_n \sin\nu\phi$ in eqn. 80 to be a term in the Fourier sine series of a function $f(\phi)$ over a period $\psi_0 = 2\pi-\beta$ so that

$$\begin{aligned} f(\phi) &= \sum_{n=0}^{\infty} a_n \sin\nu\phi \\ a_n &= \frac{2}{\psi_0}\int_0^{\psi_0} f(\phi')\sin\nu\phi' d\phi' \end{aligned} \tag{81}$$

If we rewrite eqn. 81 as

$$f(\phi) = \int_0^{\psi_0} f(\phi')\left\{\frac{2}{\psi_0}\sum_{n=0}^{\infty}\sin\nu\phi'\sin\nu\phi\right\}d\phi'$$

then the term inside the braces must be equivalent to the delta function $\delta(\phi-\phi')$, i.e.,

$$\delta(\phi-\phi') = \frac{2}{\psi_0}\sum_{n=0}^{\infty}\sin\nu\phi'\sin\nu\phi; \quad \nu = \frac{n\pi}{\psi_0}, \quad \psi_0 = 2\pi-\beta \tag{82}$$

Using eqn. 82 as our choice for the angular eigenfunction, eqn. 79 becomes

$$E_z = \frac{2}{\psi_0} \sum_{n=0}^{\infty} P_n(\rho, \rho_0) \sin \nu\phi \sin \nu\phi_0 \tag{83}$$

and substituting this equation together with eqn. 82 into eqn. 78 gives

$$\left(\frac{1}{\rho} \frac{\partial}{\partial \rho} \rho \frac{\partial}{\partial \rho} - \frac{\nu^2}{\rho^2} + k^2 \right) P_n(\rho, \rho_0) = -\frac{\delta(\rho - \rho_0)}{\rho} \tag{84}$$

For the function $P_n(\rho, \rho_0)$ we must choose

$$\begin{aligned} P_n(\rho, \rho_0) &= b_n J_\nu(k\rho) H_\nu^{(2)}(k\rho_0); & \rho < \rho_0 \\ P_n(\rho, \rho_0) &= b_n J_\nu(k\rho_0) H_\nu^{(2)}(k\rho); & \rho > \rho_0 \end{aligned} \tag{85}$$

to ensure finiteness of the electric field E_z at $\rho = 0, \infty$ and its continuity at $\rho = \rho_0$. Multiplying both sides of eqn. 84 by $\rho \partial \rho$ and integrating from $\rho = \rho_0 + \Delta$ to $\rho_0 - \Delta$, with $\Delta \to 0$ then

$$\frac{\partial}{\partial \rho} P_n(\rho_0^+, \rho_0) - \frac{\partial}{\partial \rho} P_n(\rho_0^-, \rho_0) = -\frac{1}{\rho_0} \tag{86}$$

which yields

$$b_n \{ J_\nu(k\rho_0) H_\nu^{(2)\prime}(k\rho_0) - J_\nu^\prime(k\rho_0) H_\nu^{(2)}(k\rho_0) \} = -\frac{1}{k\rho_0}$$

The expression contained within the braces is simply a Wronskian of Bessel's equation equal to $\dfrac{2}{j\pi k\rho_0}$, see eqn. 2.63, so that $b_n = \dfrac{\pi}{2j}$. Our eigenfunction series solution for cylindrical wave incidence is now complete for electric polarisation, being

$$E_z = \begin{cases} \dfrac{\pi}{2j\psi_0} \displaystyle\sum_{n=0}^{\infty} \epsilon_\nu J_\nu(k\rho) H_\nu^{(2)}(k\rho_0) \sin \nu\phi \sin \nu\phi_0 ; & \rho < \rho_0 \\ \dfrac{\pi}{2j\psi_0} \displaystyle\sum_{n=0}^{\infty} \epsilon_\nu J_\nu(k\rho_0) H_\nu^{(2)}(k\rho) \sin \nu\phi \sin \nu\phi_0 ; & \rho > \rho_0 \end{cases} \tag{87}$$

where the Neumann number ϵ_ν has been introduced to give symmetry with the magnetic polarisation solution to be discussed later.

For plane wave incidence $\rho_0 \to \infty$ for which

$$H_\nu^{(2)}(k\rho_0) \sim \sqrt{\left(\frac{2j}{\pi k\rho_0} \right)} \, j^\nu \exp(-jk\rho_0) \tag{88}$$

Since the field from the unit strength line source at (ρ_0, ϕ_0) is $\dfrac{1}{4j} H_0^{(2)}$ $(k\rho_0)$, see eqn. 2.14, then for $\rho_0 \to \infty$

$$\frac{1}{4j}H_0^{(2)}(k\rho_0) \sim \frac{\exp(-jk\rho_0)}{\sqrt{(8j\pi k\rho_0)}} \tag{89}$$

and one must multiply the first equation in eqn. 87 by the inverse of eqn. 89 to retrieve the solution for plane wave incidence. This yields

$$E_z = \frac{2\pi}{\psi_0}\sum_{n=0}^{\infty} \epsilon_\nu j^\nu J_\nu(k\rho)\sin\nu\phi\sin\nu\phi_0 \tag{90}$$

which may alternatively be written as

$$E_z = \frac{\pi}{\psi_0}\sum_{n=0}^{\infty} \epsilon_\nu j^\nu J_\nu(k\rho)\{\cos\nu(\phi-\phi_0)-\cos\nu(\phi+\phi_0)\} \tag{91}$$

This equation is the required eigenfunction solution for plane wave incidence upon a perfectly conducting wedge. When the total field from eqn. 91 is computed, it is necessary that $\nu \gg k\rho$ in the last few terms for accuracy to be achieved in the solution. Such series, therefore, are poorly convergent for any appreciable value of $k\rho$, and alternative expressions must be sought. In addition, eqn. 91 gives the total field, and is not separated into incident, reflected, and diffracted terms explicitly as for the half-plane solution. We may, although perhaps rather artificially, define a diffracted field by subtracting the known incident and reflected wave components from eqn. 91. This method will be used later in some circumstances, but first we will cast eqn. 91 into a more useful form.

Eqn. 91 consists of two terms of the form

$$I = \frac{\pi}{\psi_0}\sum_{n=0}^{\infty} \epsilon_\nu j^\nu J_\nu(k\rho)\cos\nu\Phi; \qquad \Phi = \phi \mp \phi_0 \tag{92}$$

This series may immediately be expressed as a contour integral by making use of eqn. 76. This gives eqn. 92 as

$$I = \frac{1}{j}\oint_\Gamma \frac{\exp\left\{j\alpha\left(\frac{\pi}{2}-\psi_0\right)\right\} J_\alpha(k\rho)\cos\alpha\Phi}{\sin\alpha\psi_0}\,d\alpha \tag{93}$$

where the contour Γ in the complex α-plane is given in Fig. 3.8. Now using eqn. 75 for the Bessel function yields

$$I = \frac{1}{2\pi j}\oint_\Gamma \frac{\exp\left\{j\alpha\left(\frac{\pi}{2}-\psi_0\right)\right\}\cos\alpha\Phi}{\sin\alpha\psi_0} \int_C \exp\{-j(k\rho\sin\theta-\alpha\theta)\}\,d\theta\,d\alpha \tag{94}$$

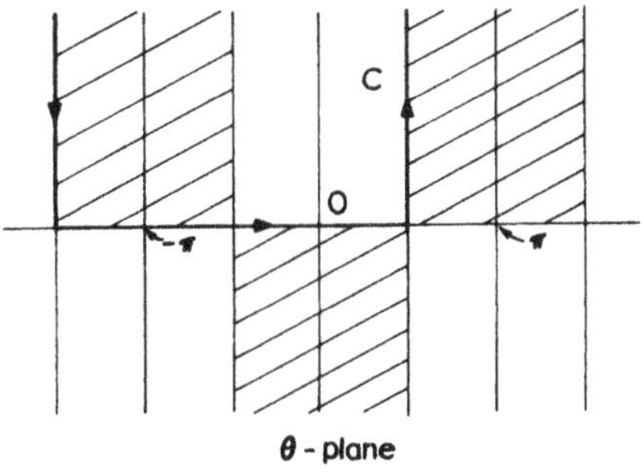

Fig. 3.10 Preferred contour for $J_\nu(k\rho)$

where C is the contour in the complex θ-plane given in Fig. 3.7. Upon rearranging this integral, and ensuring that it still remains bounded, we have

$$I = \frac{1}{2\pi}\int_C \exp(-jk\rho\sin\theta)\,G(\theta)d\theta \tag{95}$$

where

$$G(\theta) = \frac{1}{j}\oint_\Gamma \frac{\exp\left\{j\alpha\left(\frac{\pi}{2}-\psi_0\right)\right\}\cos\alpha\Phi}{\sin\alpha\psi_0}\exp(j\alpha\theta)\,d\alpha \tag{96}$$

For eqn. 96 to remain bounded, the contour C for θ must not go below the real axis, otherwise $\exp(j\alpha\theta)$ will increase exponentially as $\alpha\to\infty$. The contour we choose for C begins at $-\frac{3\pi}{2}+j\infty$ and ends at $\frac{\pi}{2}+j\infty$, as shown in Fig. 3.10. The reason for this particular choice (other than it must remain in the upper half-space) will become clear later.

Eqn. 96 can now be converted back into an infinite series via eqn. 76 to give

$$\begin{aligned} G(\theta) &= \frac{\pi}{\psi_0}\sum_{n=0}^{\infty}\epsilon_\nu j^\nu\cos\nu\Phi\exp(j\nu\theta) \\ &= \frac{\pi}{\psi_0}\left\{1+2\sum_{n=1}^{\infty}\exp(j\nu\xi)\cos\nu\Phi\right\}; \qquad \xi = \theta+\frac{\pi}{2} \end{aligned} \tag{97}$$

We may now use eqn. 77 to simplify this expression by allowing

$$p = \exp\left(\frac{j\pi\xi}{\psi_0}\right); \qquad x = \frac{\pi\Phi}{\psi_0}$$

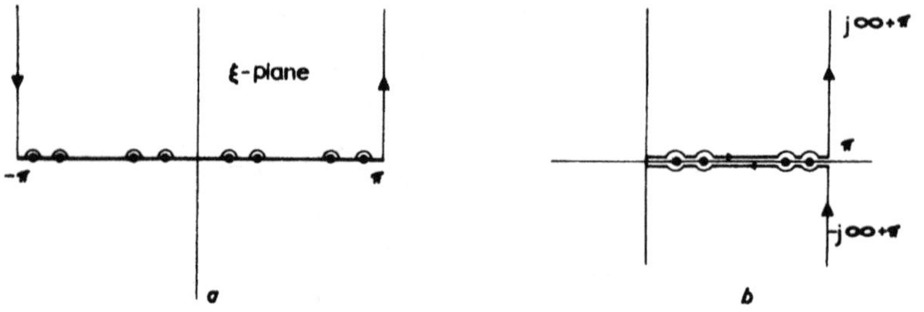

Fig. 3.11 Contours in ξ-plane for eqns. 99 and 101;

so that

$$G(\xi) = \frac{j\pi}{\psi_0}\left[\frac{\sin\dfrac{\pi\xi}{\psi_0}}{\cos\dfrac{\pi\Phi}{\psi_0} - \cos\dfrac{\pi\xi}{\psi_0}}\right] \tag{98}$$

The use of eqn. 77 restricts $|p| < 1$, which in this case requires Im $\xi > 0$. This is satisfied by the contour in Fig. 3.10 which only has a shift of $\frac{\pi}{2}$ to the right for ξ.

Substituting eqn. 98 into eqn. 95, with $\theta = \xi - \frac{\pi}{2}$ and $N = \dfrac{\psi_0}{\pi}$ we get

$$I = \frac{1}{2\pi jN}\int_{j\infty-\pi}^{j\infty+\pi} \frac{\sin\dfrac{\xi}{N}}{\cos\dfrac{\xi}{N} - \cos\dfrac{\Phi}{N}} \exp(jk\rho\cos\xi)d\xi \tag{99}$$

This equation can be seen to have first order poles along the real ξ-axis at

$$\xi = 2n\pi N \pm \Phi; \qquad n = 0, \pm 1, \pm 2, \ldots \tag{100}$$

and it is necessary to deform the contour into the upper half-space around these poles as shown in Fig. 3.11*a*. We now make use of the symmetrical position of these poles with respect to the contour of integration. Splitting the integral of eqn. 99 into

$$\int_0^{j\infty+\pi} + \int_{j\infty-\pi}^{0}$$

and substituting $\xi = -\xi$ in the second integral we get

$$I = \frac{1}{2\pi jN}\int_{-j\infty+\pi}^{j\infty+\pi} \frac{\sin\dfrac{\xi}{N}\exp(jk\rho\cos\xi)d\xi}{\cos\dfrac{\xi}{N} - \cos\dfrac{\Phi}{N}} \tag{101}$$

$$+ \{\text{pole residues in the range } 0 < \xi < \pi\}$$

as seen from the contour in Fig. 3.11*b*. Evaluating the pole residues using eqn. 60 we find they are given by

$$\sum_n \exp\{jk\rho\cos(\Phi + 2n\pi N)\}; \qquad |\Phi + 2n\pi N| < \pi, \quad n = 0, \pm 1, \pm 2, \ldots \tag{102}$$

This clearly gives the incident and reflected wave components, and the integral that remains in eqn. 101 can be associated with the diffracted field. Using the notation for the half-plane solution we have, from eqns. 91, 92, 101 and 102, for an electrically polarised plane wave incident upon a perfectly conducting wedge

$$E_z(\rho, \phi) = u^i(\rho, \phi) - u^r(\rho, \phi) \tag{103}$$

where

$$u^{i,r} = \sum_n U(\epsilon^{i,r})u_0^{i,r} + u_d^{i,r}$$

$$\epsilon^{i,r} = \begin{cases} +1 & \text{for illuminated region} \\ -1 & \text{for shadow region} \end{cases}$$

$$u_0^{i,r} = \exp\{jk\rho\cos(\Phi^{i,r} + 2n\pi N)\}; \qquad \Phi^{i,r} = \phi \mp \phi_0$$

$$n = 0, \pm 1, \pm 2, \ldots$$

The diffracted term given by the integral in eqn. 101 becomes upon substituting $\xi' = \xi - \pi$

$$u_d^{i,r} = \frac{1}{2\pi jN}\int_{-j\infty}^{j\infty} \frac{\sin\left(\dfrac{\xi' + \pi}{N}\right)}{\cos\left(\dfrac{\xi' + \pi}{N}\right) - \cos\dfrac{\Phi^{i,r}}{N}} \exp(-jk\rho\cos\xi')d\xi'; \tag{104}$$

$$N = \frac{\psi_0}{\pi}$$

No further reduction of this integral is possible unless we introduce some approximations. To begin with, let us rearrange it by the identity

$$\cot(x+y)+\cot(x-y) = \frac{2\sin 2x}{\cos 2y - \cos 2x} \tag{105}$$

so that eqn. 104 becomes

$$u_d^{i,r} = f(\Phi^{i,r}) + f(-\Phi^{i,r}) \tag{106}$$

where

$$f(\Phi^{i,r}) = -\frac{1}{4\pi jN}\int_{-j\infty}^{j\infty} \cot\left(\frac{\xi'+\pi+\Phi^{i,r}}{2N}\right)\exp(-jk\rho\cos\xi')\,d\xi' \tag{107}$$

If we now distort the integration path into the steepest descent path, we note that the exponential term in eqn. 107 is identical to that of eqn. 55, and the new contour in the complex ξ'-plane will be as shown in Fig. 3.5 (where we again have assumed that the medium surrounding the wedge is only slightly lossy) where it is denoted by SDP. By using the change of variable ν given by eqn. 57, we include all real values of ν between ∞ and $-\infty$ as we traverse the steepest descent path. Also, by anticipating that eqn. 107 can be converted into a similar integral to that given for the half-plane result in eqn. 58, we let

$$a^2 = 1+\cos(\Phi+2n\pi N) \tag{108}$$

which is analogous to the corresponding value for a in the half-plane solution. Now with the variable $\nu = \sqrt{2}\exp\left(-\frac{j\pi}{4}\right)\sin\frac{1}{2}\xi'$

$$\nu^2 + ja^2 = j\{\cos\xi' + \cos(\Phi+2n\pi N)\} \tag{109}$$

which we will use in eqn. 107. At this stage we are forced to introduce some approximations to convert eqn. 107 into a more useful form. Thus if $k\rho$ is sufficiently large, so that the major contribution to the diffraction term $f(\Phi)$ is in the vicinity of the saddle point at the origin of the complex ξ'-plane, then we may introduce the approximations

$$\nu = \sqrt{2}\exp\left(-\frac{j\pi}{4}\right)\sin\tfrac{1}{2}\xi' \simeq \frac{1}{\sqrt{2}}\exp\left(-\frac{j\pi}{4}\right)\xi'$$

$$\cot\left(\frac{\xi'+\pi+\Phi}{2N}\right) \simeq \cot\left(\frac{\pi+\Phi}{2N}\right)$$

so that substitution into eqn. 107 and using eqns. 108 and 109 gives

$$f(\Phi^{i,r}) \simeq -\frac{a\sqrt{(2j)}}{4\pi N}\cot\left(\frac{\pi+\Phi^{i,r}}{2N}\right)\exp(-jk\rho)\int_{-\infty}^{\infty}\frac{a\exp(-k\rho\nu^2)}{\nu^2+ja^2}\,d\nu$$

The integral in this equation is in the form of the modified Fresnel integral given by eqn. 2.39, so that the function $f(\Phi)$ can be written in a similar form to the diffraction term $u_d^{i,r}$ in eqn. 63 for the half-plane. Thus

$$f(\Phi^{i,r}) \simeq -\operatorname{sgn}(a^{i,r})K_-\{|a^{i,r}|\sqrt{(k\rho)}\}\Lambda^{i,r}\exp(-jk\rho) \quad (110)$$

where

$$a^{i,r} = \sqrt{2}\cos\tfrac{1}{2}(\Phi^{i,r} + 2n\pi N); \qquad \Lambda^{i,r} = \frac{a^{i,r}}{\sqrt{2N}}\cot\left(\frac{\pi + \Phi^{i,r}}{2N}\right) \quad (111)$$

Since the modified Fresnel integral effects a smooth transition across the optical boundaries where $\epsilon^{i,r}$ changes sign, it can be appreciated that the value of n must be such to give, on these boundaries,

$$|\Phi^{i,r} + 2n\pi N| = \pi \quad (112)$$

Then $a^{i,r} = 0$ and the limit of $\Lambda^{i,r}$ in eqn. 110 goes to unity if n is correctly chosen, i.e.,

$$\Lambda^{i,r} \xrightarrow[a^{i,r}\to 0]{} 1 \quad \text{when} \quad |\Phi^{i,r} + 2n\pi N| \to \pi \quad (113a)$$

giving

$$f(\Phi^{i,r}) = -\tfrac{1}{2}\epsilon^{i,r}\exp(-jk\rho) \quad \text{where} \quad \epsilon^{i,r} = \operatorname{sgn}(a^{i,r}) \quad (113b)$$

Note that the complete diffraction term is given by eqn. 106, where for one term the angle Φ is replaced by $-\Phi$ in eqns. 110–112.

Eqn. 110 was developed on the assumption that $k\rho$ was reasonably large. (We shall discuss further in Section 5.2 what is meant by 'reasonably large'.) If this is not true, then we must either evaluate the integral in eqn. 104 numerically, or obtain the diffraction coefficient from eqn. 91 by subtracting the incident and reflected wave components from the total field, i.e.,

$$u_d^{i,r} = \frac{\pi}{\psi_0}\sum_{m=0}^{\infty} \epsilon_\nu j^\nu J_\nu(k\rho)\cos\nu\Phi^{i,r} - \sum_n U(\epsilon^{i,r})u_0^{i,r}; \quad \nu = \frac{m\pi}{\psi_0} \quad (114)$$

This is readily evaluated (requiring up to the first 20 terms in the series) provided that the distance from the edge ρ is not much greater than a wavelength.

The evaluation of the electric field is now completed and the magnetic components may be obtained directly from eqn. 35.

3.3.2 Magnetic polarisation

As for electric polarisation, we initially consider a line source near the edge, and treat plane wave incidence as the special case when it is moved to infinity. We seek an eigenfunction series for the magnetic field of the form

$$H_z = \sum_n a_n \Phi_n(\phi, \phi_0) P_n(\rho, \rho_0)$$

to satisfy the boundary conditions on the wedge faces and the inhomogeneous Helmholtz equation given in eqn. 78. The solution differs from electric polarisation only in the quantity Φ_n, being

$$H_z = \begin{cases} \dfrac{\pi}{2j\psi_0} \sum_{n=0}^{\infty} \epsilon_\nu J_\nu(k\rho) H_\nu^{(2)}(k\rho_0) \cos\nu\phi \cos\nu\phi_0; & \rho < \rho_0 \\ \dfrac{\pi}{2j\psi_0} \sum_{n=0}^{\infty} \epsilon_\nu J_\nu(k\rho_0) H_\nu^{(2)}(k\rho) \cos\nu\phi \cos\nu\phi_0; & \rho > \rho_0 \end{cases} \tag{115}$$

For plane wave incidence when $\rho_0 \to \infty$ we get

$$H_z = \frac{\pi}{\psi_0} \sum_{n=0}^{\infty} \epsilon_\nu j^\nu J_\nu(k\rho)\{\cos\nu(\phi - \phi_0) + \cos\nu(\phi + \phi_0)\} \tag{116}$$

after normalising by eqn. 89. The similarity with eqn. 91 is immediately obvious so we may write

$$H_z = u^i + u^r \tag{117}$$

using $u^{i,r}$ as developed above.

3.4 Circular cylinder

For diffraction by perfectly conducting smooth convex surfaces, we have the canonical problem of plane wave diffraction at an infinitely long circular cylinder. The exact solution was first published by Rayleigh (1881) using the separation of variables technique to give an eigenfunction representation. At high frequencies, as noted earlier in Section 3.3, these series solutions are poorly convergent and we must seek for improved representations of the field. It has only been comparatively recently that suitable high frequency solutions, such as given by Franz (1954) and Goriainov (1958) have been obtained. By transforming the eigenfunction solution into a contour integral, a high frequency solution is obtained from the asymptotic residue series. This procedure shows that within the shadow region the diffracted field decays exponentially away from the shadow boundary. This propagation behaviour was given the name *creeping wave* by Franz.

The asymptotic residue series is only convergent within the shadow region. In the illuminated region it will be seen that the field is dominated by the incident and reflected field. As for the half-plane and

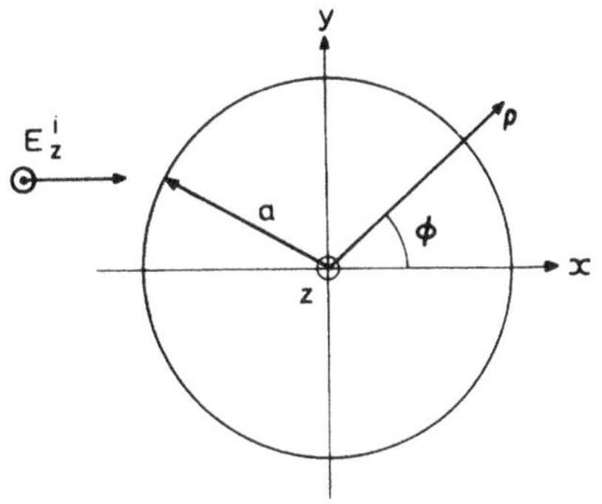

Fig. 3.12 Circular cylinder

wedge, a normally incident (i.e., parallel to the cylinder axis) plane wave is assumed, and the medium surrounding the cylinder is taken to be lossless or only slightly lossy.

The plane wave solution is applicable for sources removed from the surface of the cylinder. To conclude, we shall briefly consider the situation where magnetic line sources lie on the cylindrical surface.

3.4.1 Electric polarisation

Consider an electrically polarised plane wave E_z^i normally incident to a perfectly conducting cylinder as in Fig. 3.12, where

$$E_z^i = \exp(-jkx) = \exp(-jk\rho\cos\phi)$$

This representation may be given alternatively in terms of cylindrical wave functions as

$$E_z^i = \sum_{m=-\infty}^{\infty} j^{-m} J_m(k\rho)\exp(jm\phi) \tag{118}$$

The total electric field is given by the sum of the incident field and a scattered field E_z^s. Since this field can only have outgoing waves at infinity, we may represent it by

$$E_z^s = \sum_{m=-\infty}^{\infty} j^{-m} a_m H_m^{(2)}(k\rho)\exp(jm\phi) \tag{119}$$

so that the total field E_z is

$$E_z = \sum_{m=-\infty}^{\infty} j^{-m}\{J_m(k\rho) + a_m H_m^{(2)}(k\rho)\}\exp(jm\phi)$$

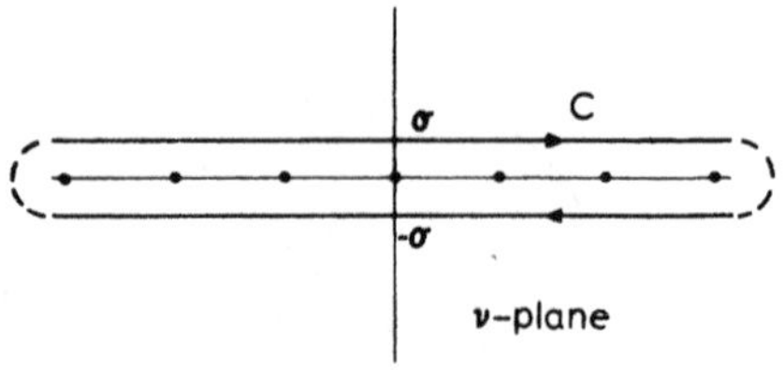

Fig. 3.13 Poles along the real axis and contour for eqn. 121

The values of the coefficients a_m are determined from the boundary condition on the cylinder at $\rho = a$ where the total field must be zero. This is seen to be satisfied if

$$a_m = -\frac{J_m(ka)}{H_m^{(2)}(ka)}$$

so that

$$E_z = \sum_{m=-\infty}^{\infty} \frac{j^{-m}\exp(jm\phi)}{H_m^{(2)}(ka)} \{J_m(k\rho)H_m^{(2)}(ka) - J_m(ka)H_m^{(2)}(k\rho)\} \quad (120)$$

This equation gives the exact eigenfunction solution for electric polarisation. In this form it is of little use for high frequency applications when $ka \gg 1$, due to its poor convergence. We shall now proceed to manipulate eqn. 120 to obtain a suitable expression for large values of ka.

Consider the integral

$$\frac{j}{2}\int \frac{\exp\{j\nu(\phi+\pi)\}}{\sin \nu\pi} f(\nu)\, d\nu$$

which has an infinite series of first order poles along the real ν-axis where $\nu = m$; $m = 0, \pm 1, \pm 2 \ldots$. If we now choose a contour C to enclose all these poles shown in Fig. 3.13, then by summing the first order pole residues, see eqn. 60, we obtain

$$\frac{j}{2}\oint_C \frac{\exp\{j\nu(\phi+\pi)}{\sin \nu\pi} f(\nu) d\nu = \sum_{m=-\infty}^{\infty} \exp(jm\phi) f(m) \quad (121)$$

provided that $f(\nu)$ has no singularities in the neighbourhood of the real axis. It will be noted that a similar integral given by eqn. 76 was used for the wedge solution.

Comparing eqn. 120 with eqn. 121, we get

$$f(\nu) = \frac{j^{-\nu}}{H_\nu^{(2)}(ka)} \left[J_\nu(k\rho)H_\nu^{(2)}(ka) - J_\nu(ka)H_\nu^{(2)}(k\rho)\right] \tag{122}$$

This may be rewritten as

$$f(\nu) = \frac{j^{-\nu}}{2H_\nu^{(2)}(ka)} \left[H_\nu^{(1)}(k\rho)H_\nu^{(2)}(ka) - H_\nu^{(1)}(ka)H_\nu^{(2)}(k\rho)\right] \tag{123}$$

by making use of the relation

$$2J_\nu(z) = H_\nu^{(1)}(z) + H_\nu^{(2)}(z)$$

From the continuation formulas

$$\begin{aligned} H_{-\nu}^{(1)}(z) &= \exp(j\pi\nu)\, H_\nu^{(1)}(z) \\ H_{-\nu}^{(2)}(z) &= \exp(-j\pi\nu)\, H_\nu^{(2)}(z) \end{aligned} \tag{124}$$

we deduce that

$$f(-\nu) = f(\nu)\exp(j2\pi\nu) \tag{125}$$

The total electric field may now be expressed by the integral in eqn. 121 as

$$\begin{aligned} E_z &= \frac{j}{2}\oint_C \frac{\exp\{j\nu(\phi+\pi)\}}{\sin\nu\pi} f(\nu)d\nu \\ &= \frac{j}{2}\left\{\int_{\infty-j\sigma}^{-\infty-j\sigma} + \int_{-\infty+j\sigma}^{\infty+j\sigma}\right\} \\ &= j\int_{\infty-j\sigma}^{-\infty-j\sigma} \frac{j^{2\nu}f(\nu)\cos\nu\phi}{\sin\nu\pi} d\nu \end{aligned} \tag{126}$$

by substituting $-\nu$ for ν in the second integral in the second line and using the relationship given by eqn. 125. The path of integration is seen to run below the real axis and we now attempt to evaluate the integral by closing the contour with an infinite semicircle in the lower half of the complex ν-plane. This requires the location of any poles in $f(\nu)$ in this lower half-space. It can be seen that singularities in the function $f(\nu)$ will occur at the zeros of $H_\nu^{(2)}(ka)$. The values of the complex variable ν where these poles exist are denoted by ν_n so that

$$H_{\nu_n}^{(2)}(ka) = 0 \tag{127}$$

and these values can be determined, on the assumption that ka is large, from the asymptotic formulas for the Hankel function given in Section 2.2.4. For small values of ν, eqn. 127 cannot be satisfied, as seen from eqn. 2.67. As the value of ν approaches ka we deduce from eqn. 2.69

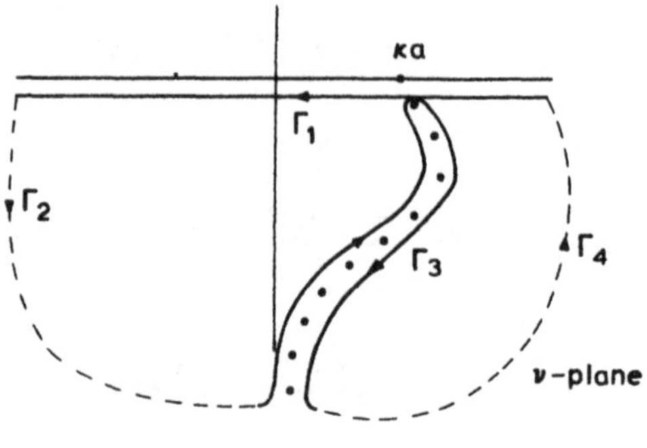

Fig. 3.14 Pole distribution from the zeros of the Hankel function

that

$$\nu_n = ka + \alpha_n \left(\frac{ka}{2}\right)^{1/3} \exp\left(-\frac{j\pi}{3}\right) \tag{128}$$

where $-\alpha_n$ are the zeros of the Airy function as given in Table 2.1. For large ν by setting $\nu = Re^{j\theta}$ we get from eqn. 2.70

$$|H_\nu^{(2)}(x)| \sim \sqrt{\left(\frac{2}{\pi R}\right)} \exp\left[R\left\{\cos\theta \ln\left(\frac{2R}{ex}\right) - \theta \sin\theta\right\}\right]; \qquad |\theta| \leqslant \frac{\pi}{2}$$

$$|H_\nu^{(2)}(x)| \sim \sqrt{\left(\frac{2}{\pi R}\right)} \exp\left[-R\left\{\cos\theta \ln\left(\frac{2R}{ex}\right) + \theta \sin\theta\right\}\right]; \quad \frac{\pi}{2} < \theta < \frac{3\pi}{2}$$

It is evident from these equations that the region $\frac{\pi}{2} < \theta < \frac{3\pi}{2}$ contains no zeros, and in the remaining region zeros occur only in the immediate vicinity of the imaginary axis. Combining this information with the distribution of zeros given by eqn. 128, the poles of the function $f(\nu)$ in the lower half-space will be situated as shown in Fig. 3.14.

With $f(\nu)$ defined by eqn. 123 we see from the previous two equations that

$$f(\nu) \to 0 \quad \text{for} \quad |\nu| \to \infty, \quad |\phi| < \frac{\pi}{2}$$

except for the poles near the imaginary axis. Thus by closing the contour of the integration in eqn. 126 by a semicircle at infinity, as shown in Fig. 3.14, no additional contribution is made to the integral. Its value is now determined from the residue of the poles within the lower half-space.

The contribution from a pole residue is given by eqn. 60, and the application of this equation to each pole in the lower half-space of the complex ν-plane gives, from eqn. 126, the solution to the electric field by the so-called *residue series*

$$E_z = \pi \sum_{n=1}^{\infty} j^{\nu_n} \frac{H_{\nu_n}^{(1)}(ka)\, H_{\nu_n}^{(2)}(k\rho) \cos \nu_n(\phi)}{\left.\frac{\partial}{\partial \nu} H_\nu^{(2)}(ka)\right|_{\nu=\nu_n} \sin \nu_n \pi} \tag{129}$$

For large values of ka it is seen from eqn. 128 and Fig. 3.14 that all values of ν_n have a large negative imaginary component, so that with little loss of accuracy we may let

$$\sin \nu_n \pi \simeq \frac{\exp(j\nu_n \pi)}{2j} \tag{130a}$$

which simplifies the relationship

$$\frac{\cos \nu_n(\phi)}{\sin \nu_n \pi}\, j^{\nu_n} \simeq j\left[\exp\left\{-j\nu_n\left(\frac{\pi}{2}-\phi\right)\right\} + \exp\left\{-j\nu_n\left(\phi+\frac{\pi}{2}\right)\right\}\right] \tag{130b}$$

and the electric field given above now becomes

$$E_z = j\pi \sum_{n=1}^{\infty} \frac{H_{\nu_n}^{(1)}(ka)\, H_{\nu_n}^{(2)}(k\rho)}{\left.\frac{\partial}{\partial \nu} H_\nu^{(2)}(ka)\right|_{\nu=\nu_n}} \left[\exp\left\{-j\nu_n\left(\frac{\pi}{2}-\phi\right)\right\} + \exp\left\{-j\nu_n\left(\phi+\frac{\pi}{2}\right)\right\}\right] \tag{131}$$

Consider now the first few terms in eqn. 131. Using eqn. 2.69 for the Hankel functions when the order and the argument are approximately equal, we derive

$$\begin{aligned} \left.\frac{\partial}{\partial \nu} H_\nu^{(2)}(x)\right|_{\nu=\nu_n} &= -H_{\nu_n}^{(2)\prime}(x) \\ &= 2\left(\frac{2}{x}\right)^{2/3} \mathrm{Ai}'(-\alpha_n) \exp\left(-\frac{j\pi}{3}\right); \qquad x \simeq \nu_n \end{aligned} \tag{132}$$

where $-\alpha_n$ are the zeros of the Airy function Ai as discussed above.

From the Wronskian relation

$$H_\nu^{(1)}(x)\, H_\nu^{(2)\prime}(x) - H_\nu^{(1)\prime}(x)\, H_\nu^{(2)}(x) = -\frac{4j}{\pi x} \tag{133}$$

it is seen that at $\nu = \nu_n$ where the Hankel function $H_{\nu_n}^{(2)}(ka) = 0$

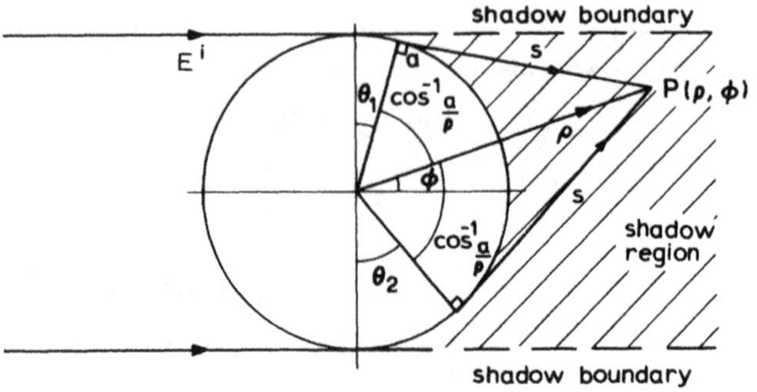

Fig. 3.15 Scattering from the cylinder

$$H_{\nu_n}^{(1)}(ka) = \frac{-4j}{\pi ka H_{\nu_n}^{(2)\prime}(ka)}$$

Substituting this equation, and eqn. 132 into eqn. 131, gives

$$E_z \simeq -\tfrac{1}{2}\left(\frac{ka}{2}\right)^{1/3} \exp\left(\frac{j2\pi}{3}\right) \sum_{n=1}^{N} \frac{H_{\nu_n}^{(2)}(k\rho)}{\{\mathrm{Ai}'(-\alpha_n)\}^2} \left[\exp\left\{-j\nu_n\left(\frac{\pi}{2}-\phi\right)\right\} + \exp\left\{-j\nu_n\left(\phi+\frac{\pi}{2}\right)\right\}\right] \tag{134}$$

Each term in this series is dependent on the validity of eqns. 128 and 132 and therefore limits the upper value of N. For field points removed from the cylinder where $k\rho > \nu_n$, we may use the asymptotic expansion given by eqn. 2.68 for $H_{\nu_n}^{(2)}(k\rho)$, and with the approximations

$$(k\rho)^2 - \nu_n^2 \simeq (ks)^2 \quad \text{where} \quad s = (\rho^2 - a^2)^{1/2}$$

$$\frac{\nu_n}{k\rho} \simeq \frac{a}{\rho}$$

which are valid for the first few terms of the residue series, we obtain

$$E_z \simeq \left(\frac{ka}{2}\right)^{1/3} \exp\left(-\frac{j\pi}{12}\right) \frac{\exp(-jks)}{\sqrt{(2\pi ks)}} \sum_{n=1}^{N} \frac{1}{\{\mathrm{Ai}'(-\alpha_n)\}^2} \left[\exp\left\{-j\nu_n\left(\frac{\pi}{2}-\phi-\cos^{-1}\frac{a}{\rho}\right)\right\} + \exp\left\{-j\nu_n\left(\frac{\pi}{2}+\phi-\cos^{-1}\frac{a}{\rho}\right)\right\}\right] \tag{135}$$

where

$$\nu_n \simeq ka + \alpha_n \left(\frac{ka}{2}\right)^{1/3} \exp\left(-\frac{j\pi}{3}\right)$$

and $-\alpha_n$ are the zeros of the Airy function Ai.

This series will be rapidly convergent within the angular region

$$-\frac{\pi}{2}+\cos^{-1}\frac{a}{\rho}<\phi<\frac{\pi}{2}-\cos^{-1}\frac{a}{\rho}$$

Referring to Fig. 3.15 it is seen that this is the region which corresponds to the geometrical optics shadow. Thus the first few terms of the residue series as given by eqn. 135 will suffice for the evaluation of the field in the shadow region. For the illuminated region, however, we must seek an alternative solution. Making use of the approximation of eqn. 130*a*, the original integral given by eqn. 126 becomes

$$E_z = -\int_{\infty-j\sigma}^{-\infty-j\sigma} f(\nu)\,[\exp(j\nu\phi)+\exp(-j\nu\phi)]\,d\nu \qquad (136)$$

In the region where $0<\phi<\pi$, the first term in the integrand will be dominant, since ν has a large negative imaginary component along the integration path. For the remainder of space where $-\pi<\phi<0$ the second term in the integrand will be dominant. The integral to evaluate, therefore, is

$$E_z = -\int_{\infty-j\sigma}^{-\infty-j\sigma} f(\nu)\exp(j\nu|\phi|)d\nu \qquad 0<|\phi|<\pi$$

On substituting for $f(\nu)$, given by eqn. 123, this becomes

$$E_z = \tfrac{1}{2}\int_{-\infty-j\sigma}^{\infty-j\sigma}\left\{H_\nu^{(1)}(k\rho)-\frac{H_\nu^{(1)}(ka)}{H_\nu^{(2)}(ka)}H_\nu^{(2)}(k\rho)\right\}\exp\left\{j\nu\left(|\phi|-\frac{\pi}{2}\right)\right\}d\nu$$

Since the value of the variable ν is always large along the integration path, we may use the asymptotic expansions of the Hankel functions, so that from Section 2.2.4

$$\frac{H_\nu^{(1)}(ka)}{H_\nu^{(2)}(ka)} \sim -1 \quad \text{for} \quad |\nu|>ka$$

$$\frac{H_\nu^{(1)}(ka)}{H_\nu^{(2)}(ka)} \sim \exp\left(2j\left[\{(ka)^2-\nu^2\}^{1/2}-\nu\cos^{-1}\left(\frac{\nu}{ka}\right)-\frac{\pi}{4}\right]\right)$$

$$\text{for} \quad |\nu|<ka$$

As $k\rho\to\infty$, the terms involving $H_\nu^{(2)}(k\rho)$ will dominate those involving $H_\nu^{(1)}(k\rho)$, since ν has a large negative imaginary component. Thus the integral reduces to

$$E_z = \int_{-\infty-j\sigma}^{\infty-j\sigma} h(\nu)\exp\{jg(\nu)\}d\nu; \qquad k\rho\to\infty \qquad (137)$$

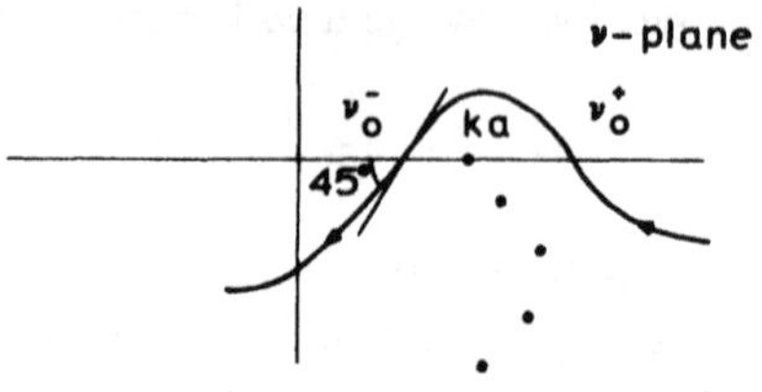

Fig. 3.16 Deforming the contour to pass through the stationary phase points

where upon using the asymptotic expansion of eqn. 2.68 for $H_\nu^{(2)}(k\rho)$

$$\left.\begin{aligned} h(\nu) &= [2\pi\{(k\rho)^2-\nu^2\}^{1/2}]^{-1/2} \\ g(\nu) &= -\{(k\rho)^2-\nu^2\}^{1/2}+\nu\left\{\cos^{-1}\left(\frac{\nu}{k\rho}\right)+|\phi|-\frac{\pi}{2}\right\}+\frac{\pi}{4} \end{aligned}\right\} |\nu|>ka$$

$$\left.\begin{aligned} h(\nu) &= -[2\pi\{(k\rho)^2-\nu^2\}^{1/2}]^{-1/2} \\ g(\nu) &= -\{(k\rho)^2-\nu^2\}^{1/2}+2\{(ka)^2-\nu^2\}^{1/2} \\ &\quad +\nu\left\{\cos^{-1}\left(\frac{\nu}{k\rho}\right)-2\cos^{-1}\left(\frac{\nu}{ka}\right)+|\phi|-\frac{\pi}{2}\right\}-\frac{\pi}{4} \end{aligned}\right\} |\nu|<ka$$

To solve the integral in eqn. 137 by the method of stationary phase requires higher order derivatives of the phase function $g(\nu)$, viz.,

$$\left.\begin{aligned} g'(\nu) &= \cos^{-1}\left(\frac{\nu}{k\rho}\right)+|\phi|-\frac{\pi}{2} \\ g''(\nu) &= -\{(k\rho)^2-\nu^2\}^{-1/2} \end{aligned}\right\} |\nu|>ka$$

$$\left.\begin{aligned} g'(\nu) &= \cos^{-1}\left(\frac{\nu}{k\rho}\right)-2\cos^{-1}\left(\frac{\nu}{ka}\right)+|\phi|-\frac{\pi}{2} \\ g''(\nu) &= -\{(k\rho)^2-\nu^2\}^{-1/2}+2\{(ka)^2-\nu^2\}^{-1/2} \end{aligned}\right\} |\nu|<ka$$

The stationary phase points ν_0 are determined from the condition $g'(\nu_0)=0$ so that from the preceeding equations

$$\nu_0^+ = k\rho\sin|\phi|; \qquad |\nu|>ka$$

$$\nu_0^- = ka\cos\frac{\phi}{2}; \qquad |\nu|<ka,\ k\rho\to\infty$$

Both these points lie on the real axis and the contour is deformed to pass through these points, as in Fig. 3.16, where it is seen to be well removed from any singularities in the complex ν-plane. The formula of eqn. 2.75 for an isolated first order stationary phase point can now be used for each value of ν_0. Relating eqn. 2.75 to the integral in eqn. 137

$$E_z \sim \sum_{\nu_0} \sqrt{\left(\frac{2\pi}{|g''(\nu_0)|}\right)}\, h(\nu_0)\exp\{jg(\nu_0)\}\exp\left(\pm\frac{j\pi}{4}\right); \quad g''(\nu_0) \gtrless 0$$

which is readily solved to yield

$$E_z \sim \exp(-jk\rho\cos\phi) - \sqrt{\left(\frac{a}{2\rho}\sin\frac{\phi}{2}\right)}\exp\left\{-jk\left(\rho - 2a\sin\frac{\phi}{2}\right)\right\}$$

$$k\rho \to \infty, 0 < |\phi| < \pi \tag{138}$$

The first term in this equation is simply the incident field and the second term can be associated with the reflected field from the cylinder.

3.4.2 Magnetic polarisation

Consider the cylinder in Fig. 3.12 to be illuminated by a magnetically polarised plane wave H_z^i where

$$H_z^i = \exp(-jk\rho\cos\phi) = \sum_{m=-\infty}^{\infty} j^{-m} J_m(k\rho)\exp(jm\phi)$$

The scattered field H_z^s can be represented by eqn. 119, as for electric polarisation, and by satisfying the boundary condition on the cylinder, the total magnetic field H_z is given by

$$\begin{aligned} H_z &= \sum_{m=-\infty}^{\infty} \frac{j^{-m}\exp(jm\phi)}{H_m^{(2)\prime}(ka)} \{J_m(k\rho)H_m^{(2)\prime}(ka) - J_m'(ka)H_m^{(2)}(k\rho)\} \\ &= \sum_{m=-\infty}^{\infty} \frac{j^{-m}\exp(jm\phi)}{2H_m^{(2)\prime}(ka)} \{H_m^{(1)}(k\rho)H_m^{(2)\prime}(ka) - H_m^{(1)\prime}(ka)H_m^{(2)}(k\rho)\} \end{aligned} \tag{139}$$

Using the residue series of eqn. 121, the magnetic field may be expressed by the integral

$$H_z = j\int_{\infty - j\sigma}^{-\infty - j\sigma} \frac{\cos\nu(\phi)}{\sin\nu\pi} j^{2\nu} f(\nu)d\nu \tag{140}$$

where

$$f(\nu) = \frac{j^{-\nu}}{2H_\nu^{(2)\prime}(ka)} \{H_\nu^{(1)}(k\rho)H_\nu^{(2)\prime}(ka) - H_\nu^{(1)\prime}(ka)H_\nu^{(2)}(k\rho)\}$$

This function has singularities in the complex ν-plane at $\nu = \nu_n'$ where

$$H_{\nu_n}^{(2)\prime}(ka) = 0$$

When $\nu \simeq x$ we derive from eqn. 2.69

$$\left.\begin{aligned} H_\nu^{(1)'}(x) &\sim -2\left(\frac{2}{x}\right)^{2/3} \text{Ai}'\left\{\tau \exp\left(\frac{j2\pi}{3}\right)\right\} \exp\left(\frac{j\pi}{3}\right) \\ H_\nu^{(2)'}(x) &\sim -2\left(\frac{2}{x}\right)^{2/3} \text{Ai}'\left\{\tau \exp\left(-\frac{j2\pi}{3}\right)\right\} \exp\left(-\frac{j\pi}{3}\right) \\ \tau &= (\nu - x)\left(\frac{2}{x}\right)^{1/3}, \quad \nu \simeq x \end{aligned}\right\} \qquad (141)$$

and the first few zeros of $H_{\nu_n}^{(2)'}(ka)$ are given by

$$\nu_n' = ka + \alpha_n'\left(\frac{ka}{2}\right)^{1/3} \exp\left(-\frac{j\pi}{3}\right) \qquad (142)$$

where $-\alpha_n'$ are the zeros of the derivative of the Airy function. A short list of these zeros is given in Table 2.1.

For large values of ν we derive, from eqn. 2.70

$$\left.\begin{aligned} H_\nu^{(1)'}(x) &\sim -\frac{\nu}{x} H_\nu^{(1)}(x) \\ H_\nu^{(2)'}(x) &\sim -\frac{\nu}{x} H_\nu^{(2)}(x) \end{aligned}\right\} \quad |\arg \nu| \leqslant \frac{\pi}{2} \qquad (143a)$$

$$|\nu| \gg x$$

$$\left.\begin{aligned} H_\nu^{(1)'}(x) &\sim \frac{\nu}{x} H_\nu^{(1)}(x) \\ H_\nu^{(2)'}(x) &\sim \frac{\nu}{x} H_\nu^{(2)}(x) \end{aligned}\right\} \quad \frac{\pi}{2} < \arg \nu < \frac{3\pi}{2} \qquad (143b)$$

The zeros of $H_{\nu_n}^{(2)'}(ka)$ for large ν occur as for $H_{\nu_n}^{(2)}(ka)$, i.e. in the immediate vicinity of the imaginary axis. Combining this information with the distribution of zeros given by eqn. 142, the poles of the function $f(\nu)$ in the lower half-space will be similar to those shown in Fig. 3.14 for electric polarisation. Closing the integration path at infinity and calculating the contribution from the poles, the residue series solution for magnetic polarisation is

$$H_z = \pi \sum_{n=1}^{\infty} j^{\nu_n'} \frac{H_{\nu_n'}^{(1)'}(ka) H_{\nu_n'}^{(2)}(k\rho) \cos \nu_n'(\phi)}{\left.\frac{\partial}{\partial \nu} H_\nu^{(2)'}(ka)\right|_{\nu=\nu_n'} \sin \nu_n' \pi} \qquad (144)$$

which simplifies, by using eqn. 130, to

$$H_z = j\pi \sum_{n=1}^{\infty} \frac{H_{\nu_n'}^{(1)\prime}(ka)\, H_{\nu_n'}^{(2)}(k\rho)}{\left.\dfrac{\partial}{\partial\nu} H_{\nu}^{(2)\prime}(ka)\right|_{\nu=\nu_n'}}$$

$$\left[\exp\left\{-j\nu_n'\left(\frac{\pi}{2}-\phi\right)\right\} + \exp\left\{-j\nu_n'\left(\phi+\frac{\pi}{2}\right)\right\}\right] \tag{145}$$

For the first few terms we may use eqn. 141 to derive

$$\left.\frac{\partial}{\partial\nu} H_{\nu}^{(2)\prime}(ka)\right|_{\nu=\nu_n'} \sim \frac{4}{ka}\,\mathrm{Ai}''(-\alpha_n') = \frac{-4\alpha_n'}{ka}\,\mathrm{Ai}(-\alpha_n'); \quad ka \simeq \nu_n' \tag{146}$$

by making use of the Airy function differential equation

$$\mathrm{Ai}''(x) = x\,\mathrm{Ai}(x)$$

From the Wronskian relation given in eqn. 133 it is seen that at $\nu' = \nu_n'$ where the Hankel function $H_{\nu_n'}^{(2)\prime}(ka) = 0$

$$H_{\nu_n'}^{(1)\prime}(ka) = \frac{4j}{\pi ka H_{\nu_n'}^{(2)}(ka)}$$

and using the asymptotic expansion of eqn. 2.69 for $H_{\nu_n'}^{(2)}(ka)$ this becomes

$$H_{\nu_n'}^{(1)\prime}(ka) \sim \frac{j}{\pi}\left(\frac{2}{ka}\right)^{2/3} \{\mathrm{Ai}(-\alpha_n')\}^{-1} \exp\left(-\frac{j\pi}{3}\right); \quad ka \simeq \nu_n'$$

Substituting this equation and eqn. 146 into eqn. 145 gives

$$H_z = \tfrac{1}{2}\left(\frac{ka}{2}\right)^{1/3} \exp\left(-\frac{j\pi}{3}\right) \sum_{n=1}^{N} \frac{H_{\nu_n'}^{(2)}(k\rho)}{\alpha_n'\{\mathrm{Ai}(-\alpha_n')\}^2}$$

$$\left[\exp\left\{-j\nu_n'\left(\frac{\pi}{2}-\phi\right)\right\} + \exp\left\{-j\nu_n'\left(\phi+\frac{\pi}{2}\right)\right\}\right] \tag{147}$$

Each term is dependent on the assumption that $ka \simeq \nu_n'$ which limits the upper value N of the series. For field points removed from the cylinder where $k\rho > \nu_n'$, we may use the asymptotic expansion given by eqn. 2.68 for $H_{\nu_n'}^{(2)}(k\rho)$ to obtain

$$H_z \sim \left(\frac{ka}{2}\right)^{1/3} \exp\left\{-\left(\frac{j\pi}{12}\right)\right\} \frac{\exp(-jks)}{\sqrt{(2\pi ks)}} \sum_{n=1}^{N} \frac{1}{\alpha_n' \{\mathrm{Ai}(-\alpha_n')\}^2} \left[\exp\left\{-j\nu_n'\left(\frac{\pi}{2} - \phi - \cos^{-1}\frac{a}{\rho}\right)\right\} + \exp\left\{-j\nu_n'\left(\frac{\pi}{2} + \phi - \cos^{-1}\frac{a}{\rho}\right)\right\}\right] \tag{148}$$

where

$$\nu_n' = ka + \alpha_n' \left(\frac{ka}{2}\right)^{1/3} \exp\left(-\frac{j\pi}{3}\right), \quad s = (\rho^2 - a^2)^{1/2}$$

and $-\alpha_n'$ are the zeros of the Airy function derivative Ai$'$. As for electric polarisation this solution is rapidly convergent in the shadow region.

To obtain a useful solution in the illuminated region we proceed as for electric polarisation to yield eqn. 136 but with the function $f(\nu)$ for magnetic polarisation as given in eqn. 140. The dominant term for large $k\rho$ is the integral

$$-\tfrac{1}{2}\int_{-\infty-j\sigma}^{\infty-j\sigma} \frac{H_\nu^{(1)\prime}(ka)}{H_\nu^{(2)\prime}(ka)} H_\nu^{(2)}(k\rho) \exp\left\{j\nu\left(|\phi| - \frac{\pi}{2}\right)\right\} d\nu; \qquad 0 < |\phi| < \pi$$

and the stationary phase evaluation follows very closely the corresponding evaluation of eqn. 137 for electric polarisation which is given in detail above. As before, the incident and reflected field dominates the illuminated region, with the reflected field given by eqn. 138 after a change in sign.

3.4.3 Transition region

In the immediate vicinity of the shadow boundary, the residue series solution is formally correct in the shadow, but requires a large number of terms for convergence. The stationary phase evaluation for the illuminated region, while yielding the reflected field, does not adequately describe the field at the shadow boundary. We shall now give a solution in this transition region for the total field.

For electric polarisation the integral expression for the total field E_z can be written from eqns. 123, 126 and 130 as

$$E_z = \tfrac{1}{2}\int_{-\infty-j\sigma}^{\infty-j\sigma} \left[H_\nu^{(1)}(k\rho) - \frac{H_\nu^{(1)}(ka)}{H_\nu^{(2)}(ka)} H_\nu^{(2)}(k\rho)\right] \left[\exp\left\{-j\nu\left(\frac{\pi}{2} - \phi\right)\right\} + \exp\left\{-j\nu\left(\frac{\pi}{2} + \phi\right)\right\}\right] d\nu \tag{149}$$

From the stationary phase evaluation given earlier, it can be seen that the region giving most contribution to the integral near the shadow boundary for large $k\rho$ is where $\nu \simeq ka$. We now approximate the quantities in eqn. 149 for this condition. Specifically, from eqn. 2.69 we have

$$H_\nu^{(1)}(ka)/H_\nu^{(2)}(ka) \simeq \exp\left(-\frac{j2\pi}{3}\right) \mathrm{Ai}\left\{\tau \exp\left(\frac{j2\pi}{3}\right)\right\} \Big/ \mathrm{Ai}\left\{\tau \exp\left(-\frac{j2\pi}{3}\right)\right\} \tag{150}$$

where $\tau = (\nu - ka)/M, M = (ka/2)^{1/3}$.

For $\rho > a$ and $\nu = M\tau + ka$ we deduce from eqn. 2.68 for small values of τ

$$H_\nu^{(1)}(k\rho) \simeq \sqrt{\left(\frac{2}{\pi ks}\right)} \exp(jX), \qquad H_\nu^{(2)}(k\rho) \sqrt{\left(\frac{2}{\pi ks}\right)} \exp(-jX),$$

$$s = (\rho^2 - a^2)^{1/2} \tag{151}$$

where

$$X \simeq ks + \frac{1}{2ks} M^2\tau^2 - (M\tau + ka)\cos^{-1}\frac{a}{\rho} - \frac{\pi}{4}$$

With the usual stationary phase approach (see Section 2.3.1.1) the integration limits remain at $\pm\infty$ with the substitution of eqns. 150 and 151 into eqn. 149. On doing this we find that the terms in this equation involving $H_\nu^{(1)}(k\rho)$ equate to zero. To ensure convergence of the remaining integral at ∞ we need to use the Airy function relationship

$$\mathrm{Ai}\left\{\tau \exp\left(\frac{j2\pi}{3}\right)\right\} = \exp\left(\frac{j\pi}{3}\right) \mathrm{Ai}(\tau) - \exp\left(\frac{j2\pi}{3}\right) \mathrm{Ai}\left\{\tau \exp\left(-\frac{j2\pi}{3}\right)\right\} \tag{152}$$

Thus substitution of eqns. 150–152 into eqn. 149 and dropping the quadratic terms of τ where possible and changing the integration variable to $\tau = (\nu - ka)/M$ yields

$$E_z = -\sqrt{\left(\frac{2}{jks}\right)} M \exp(-jks)[\exp(-jka\theta_1)\{p(M\theta_1) + I(M\theta_1)\}$$

$$+\exp(-jka\theta_2)\{p(M\theta_2) + I(M\theta_2)\}] \tag{153}$$

where

$$p(x) = \frac{1}{2\sqrt{\pi}} \exp\left(\frac{j\pi}{6}\right) \int_{0-j\sigma/M}^{\infty - j\sigma/M} \frac{\mathrm{Ai}(\tau)}{\mathrm{Ai}\{\tau \exp(-j2\pi/3)\}} \exp(-jx\tau)\, d\tau$$

$$+ \frac{1}{2\sqrt{\pi}} \exp\left(-\frac{j\pi}{6}\right) \int_{-\infty - j\sigma/M}^{0-j\sigma/M} \frac{\mathrm{Ai}\{\tau \exp(j2\pi/3)\}}{\mathrm{Ai}\{\tau \exp(-j2\pi/3)\}} \exp(-jx\tau)\, d\tau \tag{154}$$

and

$$I(x) = \frac{1}{2j\sqrt{\pi}} \int_{0-j\sigma/M}^{\infty-j\sigma/M} \exp\left\{-j\left(x\tau - \frac{M^2\tau^2}{2ks}\right)\right\} d\tau$$

By completing the square in the exponent of $I(x)$ and by a suitable change of variable, we can write $I(x)$ as

$$I(x) = -y\sqrt{j}K_{-}(xy); \quad y = \frac{1}{M}\sqrt{\left(\frac{1}{2}ks\right)} \tag{155}$$

where $K_{-}(x)$ is the modified Fresnel integral described in Section 2.2.1. From eqn. 2.38*c* we can express $I(x)$ as

$$I(x) = -U(-x)y\sqrt{j}\exp\{j(xy)^2\} - y\sqrt{j}\,\mathrm{sgn}(x)K_{-}(|x|y) \tag{156}$$

The function $p(x)$ defined by eqn. 154 is another representation of the Pekeris function discussed in Section 2.2.3. In this form it is valid for all x. By deforming the contour of the second integral in eqn. 154 to the line $\infty \exp(-j\,2\pi/3)$ and using eqn. 152 in the numerator, the representation given by eqns. 2.58*a* and 2.59*a* is retrieved.

The expression given by eqn. 153 yields a uniform solution for the electric field through the transition region. In the deep shadow region when $xy > 3$, $I(x) \sim -1/2\sqrt{\pi x}$ and hence $p(x) + I(x) \sim \hat{p}(x)$, where $\hat{p}(x)$ is the Pekeris carot function. Using the residue series solution of eqn. 2.58*c* for this function, it is easy to show that eqn. 153 reduces to the shadow region formulation given by eqn. 135. In the deep illuminated region where $xy < -3$ then $p(x) + I(x) \sim \hat{p}(x) - y\sqrt{j}\exp\{j(xy)^2\}$. From eqn. 2.58*d*,

$$\hat{p}(x) \sim \frac{\sqrt{(-x)}}{2}\exp\left\{j\left(\frac{x^3}{12} + \frac{\pi}{4}\right)\right\} \quad \text{for} \quad x \to -\infty$$

so that when θ_1, for example, in eqn. 153 goes negative,

$$E_z \sim \exp\left\{-jks\left(1 - \frac{1}{2}|\theta_1|^2\right)\right\} - \sqrt{\frac{(a|\theta_1|)}{4s}}\exp(-jks)$$
$$\exp\left\{jka\left(|\theta_1| - \frac{|\theta_1|^3}{24}\right)\right\} + \text{terms in } \theta_2 \tag{157}$$

Recalling that the value of ka, and therefore of M, is assumed to be large, $M\theta_1$ will increase rapidly to a large negative number as the field point moves across the boundary into the illuminated region (see Fig. 3.15). Making use of the approximations

$$\left.\begin{aligned}\cos\theta_1 &\simeq 1-\tfrac{1}{2}\theta_1^2\\ \sin\frac{|\theta_1|}{2} &\simeq \frac{|\theta_1|}{2}-\frac{|\theta_1|^3}{48}\end{aligned}\right\}\text{(for phase terms)}$$

$$\simeq \tfrac{1}{2}|\theta_1| \qquad \text{(for amplitude terms)}$$

$$|\theta_1| \simeq |\phi|, \quad s \simeq \rho \text{ for } \rho \gg a$$

eqn. 157 retrieves the geometrical optics field as given by eqn. 138. Thus, provided the cylinder is large, the formulation in eqn. 153 runs smoothly into the geometrical optics solution. This can also be shown to be true for any field point near the shadow boundary in the illuminated region, e.g. see eqn. 6.10 as $\phi' \to \pi$.

The same procedure as above can be applied to the magnetic polarisation case where the other Pekeris function $q(x)$ appears. For the uniform solution we define the quantities

$$\tilde{p}_\pm(x,y) = p(x) - y\,\mathrm{sgn}(x)K_\pm(y|x|)\exp\left(\mp\frac{j\pi}{4}\right) \sim \hat{p}(x)$$

$$\tilde{q}_\pm(x,y) = q(x) - y\,\mathrm{sgn}(x)K_\pm(y|x|)\exp\left(\mp\frac{j\pi}{4}\right) \sim \hat{q}(x) \qquad (158)$$

and shall refer to these functions as the *modified Pekeris functions.* The expression for the electric and magnetic fields are now given as

$$\begin{aligned}E_z = -4\sqrt{\pi}M\Big[&\exp\{-jka\theta_1\}\\ &\{\tilde{p}_-(M\theta_1,y) - y\sqrt{j}U(-\theta_1)\exp(j[M\theta_1 y]^2)\}\\ &+\exp\{-jka\theta_2\}\{\tilde{p}_-(M\theta_2,y) - y\sqrt{j}U(-\theta_2)\exp(j[M\theta_2 y]^2)\}\Big]\\ &\frac{\exp(-jks)}{\sqrt{(8j\pi ks)}}\end{aligned}$$

$$\begin{aligned}H_z = -4\sqrt{\pi}M\Big[&\exp\{-jka\theta_1\}\\ &\{\tilde{q}_-(M\theta_1,y) - y\sqrt{j}U(-\theta_1)\exp(j[M\theta_1 y]^2)\}\\ &+\exp\{-jka\theta_2\}\{\tilde{q}_-(M\theta_2,y) - y\sqrt{j}U(-\theta_2)\exp(j[M\theta_2 y]^2)\}\Big]\\ &\frac{\exp(-jks)}{\sqrt{(8j\pi ks)}}\end{aligned} \qquad (159)$$

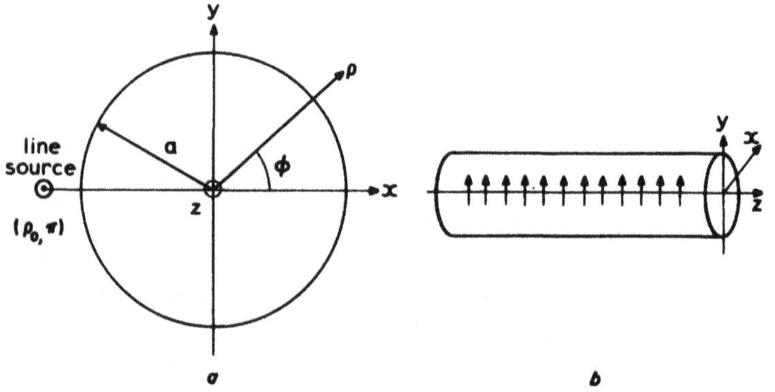

Fig. 3.17 Magnetic line sources near the cylinder
a z-directed elements
b Transverse-directed elements

A solution of this type was first given by Wait and Conda (1959). For field points well removed from the shadow boundary in the illuminated region we must use the geometric optics field directly and not attempt to retrieve it from eqn. 159 as it stands. This will be discussed more fully in Section 6.1.

3.4.4 Sources on the cylinder

The effect of a line source on the surface of the cylinder has a solution where the mathematical details are very similar to those given in Sections 3.4.1 to 3.4.3 and we need only give a brief outline of the solution. Consider first a magnetic line source situated at (ρ_0, π) as shown in Fig. 3.17*a*. The incident field from the source is given by

$$H_z^i = \frac{1}{4j} H_0^{(2)}(k|\boldsymbol{\rho} - \boldsymbol{\rho}_0|) \tag{160}$$

Invoking the *addition theorem* for Hankel functions this becomes

$$H_z^i = \begin{cases} \dfrac{1}{4j}\displaystyle\sum_{m=-\infty}^{\infty} H_m^{(2)}(k\rho_0)J_m(k\rho)\exp\{jm(\phi-\pi)\}; & \rho < \rho_0 \\ \dfrac{1}{4j}\displaystyle\sum_{m=-\infty}^{\infty} H_m^{(2)}(k\rho)J_m(k\rho_0)\exp\{jm(\phi-\pi)\}; & \rho > \rho_0 \end{cases}$$

Constructing the scattered field for $\rho < \rho_0$ as

$$H_z^s = \frac{1}{4j} \sum_{m=-\infty}^{\infty} a_m H_m^{(2)}(k\rho_0)\, H_m^{(2)}(k\rho) \exp\{jm(\phi - \pi)\}; \qquad \rho < \rho_0$$

then by satisfying the boundary conditions on the cylinder, we solve for a_m to give for the total field

$$H_z = \begin{cases} \dfrac{1}{4j} \displaystyle\sum_{m=-\infty}^{\infty} \dfrac{H_m^{(2)}(k\rho_0)}{H_m^{(2)\prime}(ka)} \{H_m^{(2)\prime}(ka)\, J_m(k\rho) - H_m^{(2)}(k\rho)\, J_m'(ka)\} \\ \exp\{jm(\phi - \pi)\}; \qquad \rho < \rho_0 \\ \dfrac{1}{4j} \displaystyle\sum_{m=-\infty}^{\infty} \dfrac{H_m^{(2)}(k\rho)}{H_m^{(2)\prime}(ka)} \{H_m^{(2)\prime}(ka)\, J_m(k\rho_0) - H_m^{(2)}(k\rho_0)\, J_m'(ka)\} \\ \exp\{jm(\phi - \pi)\}; \qquad \rho > \rho_0 \end{cases}$$

When the line source is on the cylinder $(\rho_0 = a)$ then

$$H_z = \frac{1}{4j} \sum_{m=-\infty}^{\infty} \frac{H_m^{(2)}(k\rho)}{H_m^{(2)\prime}(ka)} \{H_m^{(2)\prime}(ka)\, J_m(ka) - H_m^{(2)}(ka)\, J_m'(ka)\}$$

$$\exp\{jm(\phi - \pi)\} \tag{161}$$

$$= -\frac{1}{2\pi ka} \sum_{m=-\infty}^{\infty} \frac{H_m^{(2)}(k\rho)}{H_m^{(2)\prime}(ka)} \exp\{jm(\phi - \pi)\}$$

Making use of eqn. 121 and the approximation of eqn. 130*a* since, as before, σ can be large, eqn. 161 becomes

$$H_z = -\frac{1}{2\pi ka} \int_{-\infty - j\sigma}^{\infty - j\sigma} \frac{H_\nu^{(2)}(k\rho)}{H_\nu^{(2)\prime}(ka)}$$

$$[\exp\{-j\nu(\pi - \phi)\} + \exp\{-j\nu(\pi + \phi)\}] \tag{162}$$

For field points removed from the cylinder and for $\nu \simeq ka$ we have

$$H_\nu^{(2)}(k\rho) \sim \sqrt{\left(\frac{2j}{\pi ks}\right)} \exp(-jks) \exp\left(j\nu \cos^{-1} \frac{a}{\rho}\right);$$

$$s = (\rho^2 - a^2)^{1/2}, \qquad \rho > a$$

$$H_\nu^{(2)\prime}(ka) \sim -2M^{-2}\,\mathrm{Ai}'\left\{\tau\exp\left(-\frac{j2\pi}{3}\right)\right\}\exp\left(-\frac{j\pi}{3}\right),$$

$$\tau = (\nu - ka)M^{-1}, \quad M = \left(\frac{ka}{2}\right)^{1/3}$$

the latter equation following directly from eqn. 2.69. Eqn. 162 can now be written as

$$H_z = \frac{M^2}{\pi ka}\exp\left(\frac{j5\pi}{6}\right)\int_{-\infty-j\sigma}^{\infty-j\sigma}\frac{1}{\mathrm{Ai}'\left\{\tau\exp\left(\frac{-j2\pi}{3}\right)\right\}}$$
$$\{\exp(-j\nu\theta_1)+\exp(-j\nu\theta_2)\}d\nu\,\frac{\exp(-jks)}{\sqrt{(8j\pi ks)}} \qquad (163)$$

where

$$\theta_{1,2} = \pi \mp \phi - \cos^{-1}\frac{a}{\rho}$$

Eqn. 163 is related to the incident field as given by eqn. 160. If we wish to relate the amplitude of H_z to the value of the z, or axially, directed magnetic line source m_a then we multiply eqn. 163 by

$$-jk\sqrt{\left(\frac{\hat{\epsilon}}{\mu}\right)}\,m_a.$$

The other type of line source that may exist on the cylinder is when the elemental current elements are directed transversely to the surface as shown in Fig. 3.17*b*. Proceeding as above the z-component of the field is found to be given by

$$E_z = -\frac{M}{\pi ka}\exp\left(-\frac{j\pi}{3}\right)jkm_t\int_{-\infty-j\sigma}^{\infty-j\sigma}\frac{1}{\mathrm{Ai}\left\{\tau\exp\left(\frac{-j2\pi}{3}\right)\right\}}$$
$$\{\exp(-j\nu\theta_1)+\exp(-j\nu\theta_2)\}d\nu\,\frac{\exp(-jks)}{\sqrt{(8j\pi ks)}}$$

where m_t is the value of the line source with transversely directed elements. With the change of variable $\tau = (\nu - ka)M^{-1}$ and the appropriate deformation of the contour, the integral in this equation and in eqn. 163 can be expressed in terms of the Fock functions given in Section 2.2.3. The result is

$$E_z = -\sqrt{\left(\frac{jk}{8\pi}\right)}\frac{m_t}{jM}\{\exp(-jka\theta_1)f(M\theta_1) + \exp(-jka\theta_2)f(M\theta_2)\}$$

$$\frac{\exp(-jks)}{\sqrt{(s)}} \tag{164}$$

$$H_z = -\sqrt{\left(\frac{\hat{\epsilon}}{\mu}\right)}\sqrt{\left(\frac{jk}{8\pi}\right)}m_a\{\exp(-jka\theta_1)g(M\theta_1)$$

$$+\exp(-jka\theta_2)g(M\theta_2)\}\frac{\exp(-jks)}{\sqrt{(s)}}$$

For field points in the shadow, the Fock function arguments are positive, and we may use the residue series solution of eqn. 2.57*b* to give

$$E_z = -m_t\sqrt{\left(\frac{jk}{8\pi s}\right)}\exp\left(\frac{-j\pi}{6}\right)M^{-1}\sum_{n=1}^{\infty}\frac{1}{\mathrm{Ai}'(-\alpha_n)}$$

$$[\exp\{-j\nu_n\theta_1\} + \exp\{-j\nu_n\theta_2\}]\exp(-jks) \quad \theta_{1,2} > 0$$

$$H_z = -\sqrt{\left(\frac{\hat{\epsilon}}{\mu}\right)}m_a\sqrt{\left(\frac{jk}{8\pi s}\right)}\sum_{n=1}^{\infty}\frac{1}{\alpha_n'\mathrm{Ai}(-\alpha_n')} \tag{165}$$

$$\{\exp(-j\nu_n'\theta_1) + \exp(-j\nu_n'\theta_2)\}\exp(-jks)$$

where

$$\nu_n = ka + \alpha_n M\exp\left(\frac{-j\pi}{3}\right)$$
$$\nu_n' = ka + \alpha_n' M\exp\left(\frac{-j\pi}{3}\right) \tag{166}$$

As for the case of plane wave incidence, eqn. 166 is only true when $\nu \simeq ka$ in the original integrals, and therefore the number of terms that can be taken in the series of eqn. 165 is limited.

In the illuminated region, θ_1 (or θ_2) will be negative, and for large values of ka, and hence M, the Fock functions may be approximated by

$$f(x) \sim -2jx\exp\left(\frac{jx^3}{3}\right); \qquad g(x) \sim 2\exp\left(\frac{jx^3}{3}\right) \quad \text{for} \quad x \to -\infty$$

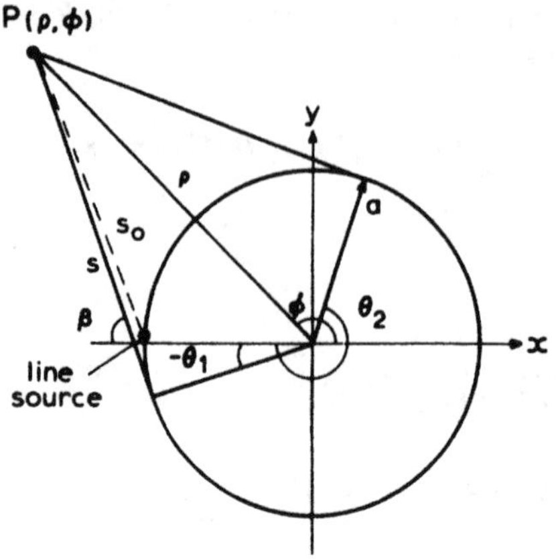

Fig. 3.18 Magnetic line source on the cylinder

so that when θ_1 goes negative as in Fig. 3.18

$$E_z = \left[-jk2m_t|\theta_1|\exp\left\{jka\left(|\theta_1| - \frac{|\theta_1|^3}{6}\right)\right\} + \text{terms in } \theta_2\right] \frac{\exp(-jks)}{\sqrt{(8j\pi ks)}}$$

$$H_z = \left[-jk2m_a\sqrt{\left(\frac{\hat{\epsilon}}{\mu}\right)}\exp\left\{jka\left(|\theta_1| - \frac{|\theta_1|^3}{6}\right)\right\} + \text{terms in } \theta_2\right] \frac{\exp(-jks)}{\sqrt{(8j\pi ks)}} \quad (167)$$

For small values of $|\theta_1|$ we note from Fig. 3.18

$$\cos\beta \simeq \sin|\theta_1| \simeq |\theta_1| - \tfrac{1}{6}|\theta_1|^3 \qquad \text{(for phase terms)}$$

$$\simeq |\theta_1| \qquad \text{(for amplitude terms)}$$

and the first terms in the equations in eqn. 167 are

$$E_z \simeq -jk2m_t\cos\beta\exp(jka\cos\beta)\frac{\exp(-jks)}{\sqrt{(8j\pi ks)}}$$

$$\simeq -jk2m_t\cos\beta\frac{\exp(-jks_0)}{\sqrt{(8j\pi ks_0)}}$$

$$H_z \simeq -jk2m_a\sqrt{\left(\frac{\hat{\epsilon}}{\mu}\right)}\frac{\exp(-jks_0)}{\sqrt{(8j\pi ks_0)}} \quad (168)$$

which is the geometrical optics field radiated by the sources in the presence of the cylinder. Thus, provided that the cylinder is large, the Fock function representation in eqn. 164 passes smoothly into the geometrical optics field in the illuminated region. Eqn. 164 will yield the geometrical optics solution only for field points close to the shadow boundary. For the field removed from this boundary in the illuminated region, we must use the geometrical optics field directly and not attempt to retrieve it from the Fock functions in eqn. 164 as it stands. This point will be discussed more fully in Section 6.3.

References

BAKER, B.B., and COPSON, E.T. (1950): 'The mathematical theory of Huygens' principle' (Oxford University Press, 2nd ed.)

BOWMAN, J.J., SENIOR, T.B.A., and USLENGHI, P.L.E. (1969): 'Electromagnetic and acoustic scattering by simple shapes' (North-Holland Publishing Company)

COPSON, E.T. (1946): 'On an integral equation arising in the theory of diffraction', *Quart. J. Math.*, **17**, pp. 19–34

FRANZ, W. (1954): 'Über die greensche funktionen des zylinders und der kugel', *Z. Naturforsch*, **9a**, pp. 705–716

GORIAINOV, A.S. (1958): 'An asymptotic solution of the problem of diffraction of a plane electromagnetic wave by a conducting cylinder', *Radio Eng. & Electron. Phys.*, **3**, pp. 23–39

GRADSHTEYN, I.S., and RYZHIK, I.M. (1966): 'Table of integrals, sums, series, and products' (Academic Press, 4th edn.) p. 40

JONES, D.S. (1964): 'The theory of electromagnetism' (Pergamon)

KONTOROWICH, M.J., and LEBEDEV, N.N. (1939): 'On a method of solution of some problems of the diffraction theory', *J. Phys. (USSR)*, **1**, pp. 229–241

MACDONALD, H.M. (1902): 'Electric waves' (Cambridge University Press)

MACDONALD, H.M. (1915): 'A class of diffraction problems', *Proceeding of the London Mathematical Society*, 14, pp. 410–427.

MORSE, P.M., and FESHBACH, H. (1953): 'Methods of theoretical physics, Vols. 1 and 2', (McGraw-Hill) p. 623

OBERHETTINGER, F. (1956): 'On asymptotic series for functions occuring in the theory of diffraction of waves by wedges', *J. Math. Phys.*, **34**, pp. 245–255

PAULI, W. (1938): 'On asymptotic series for functions in the theory of diffraction of light', *Phys. Rev.*, **54**, pp. 924–931

RAYLEIGH (1881): 'On the electromagnetic theory of light', *Phil. Mag.*, **12**, pp. 81–101

SOMMERFELD, A. (1896): 'Mathematische theorie der diffraction', *Math. Ann.*, **47**, pp. 317–374

WAIT, J.R., and CONDA, A.M. (1959): 'Diffraction of electromagnetic waves by smooth obstacles for grazing angles', *J. Res. Natl. Bur. Stand.*, **63D**, pp. 181–197

WEINER, N., and HOPF, E. (1931): 'Über eine klasse singulärer integralgleichungen', *Sitzber. Preuss. Akad. Wiss.*, pp. 696–706

Chapter 4

Geometrical optics

4.1 Geometrical optics method

When a plane wave is incident upon a planar interface between two slightly lossy homogeneous media, there exists a reflected and a refracted field which, as discussed in Section 3.1, also propagate as plane waves. If the incident field is electrically polarized, $V_z = E_z$ in Fig. 4.1. The reflected and refracted (or transmitted) components are then given specifically as

$$E_z^r = R^e \exp(-jk_1 s^r), \qquad E_z^t = T^e \exp(-jk_2 s^t) \tag{1a}$$

where the incoming field is assumed to have unit strength, and phase is measured from the point of reflection and refraction. Similarly for magnetic polarisation, $V_z = H_z$ and

$$H_z^r = R^m \exp(-jk_1 s^r), \qquad H_z^t = T^m \exp(-jk_2 s^t) \tag{1b}$$

We define a critical angle ν_c given in Section 3.1.3 for slightly lossy media as

$$\sin \nu_c = \sqrt{\left(\frac{\epsilon_2}{\epsilon_1}\right)} \tag{2}$$

such that if the angle of incidence exceeds the critical angle a refracted field does not penetrate into the second medium. If $\epsilon_1 < \epsilon_2$ there is no critical angle and refraction takes place for all angles of incidence. The relationship between the incident, reflected and refracted angle was given in Section 3.1 as

$$\begin{aligned} \nu_i &= \nu_r \\ \sin \nu_t &= \sqrt{\frac{\epsilon_1}{\epsilon_2}} \sin \nu_i \end{aligned} \tag{3}$$

which express Snell's law of reflection and refraction.

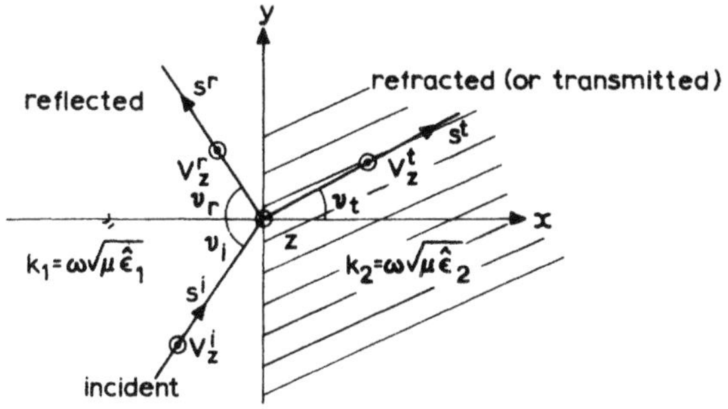

Fig. 4.1 Reflection and refraction at a plane interface

The reflection and transmission coefficients were derived in Section 3.1 and for slightly lossy media become

$$\left.\begin{aligned} R^e &= \frac{\sin(\nu_t - \nu_i)}{\sin(\nu_t + \nu_i)} \qquad & T^e &= \frac{2\cos\nu_i \sin\nu_t}{\sin(\nu_t + \nu_i)} \\ R^m &= \frac{\tan(\nu_i - \nu_t)}{\tan(\nu_i + \nu_t)} \qquad & T^m &= \frac{2\cos\nu_i \sin\nu_t}{\sin(\nu_i + \nu_t)\cos(\nu_i - \nu_t)} \end{aligned}\right] \quad \nu_i < \nu_c \tag{4a}$$

$$\left.\begin{aligned} R^e &= \frac{\cos\nu_i - j(\sin^2\nu_c - \sin^2\nu_i)^{1/2}}{\cos\nu_i + j(\sin^2\nu_c - \sin^2\nu_i)^{1/2}} \\ R^m &= \frac{\sin^2\nu_c\cos\nu_i - j(\sin^2\nu_c - \sin^2\nu_i)^{1/2}}{\sin^2\nu_c\cos\nu_i + j(\sin^2\nu_c - \sin^2\nu_i)^{1/2}} \end{aligned}\right] \quad \nu_i > \nu_c \tag{4b}$$

If the second medium is highly conducting, only reflection occurs, and provided that the conductivity is sufficiently large then

$$R^e = -1 \qquad R^m = 1 \tag{5}$$

We noted earlier that for isotropic media the average flow of energy in a plane wave is in the direction of propagation. Moreover, in homogeneous regions, the propagation is along straight paths. We represent this forward propagation of plane waves (and consequently the flow of energy) graphically by straight lines which we call rays, as shown in Fig. 4.1. The amplitude and direction of the reflected and refracted rays have been given in eqns. 1–5. The geometrical optics method uses this rigorous result to obtain approximate solutions for more complicated problems for which analytically exact solutions do not exist or are mathematically intractable. Thus if a source is near a non-planar interface between two media, as in Fig. 4.2, then the *geometrical optics*

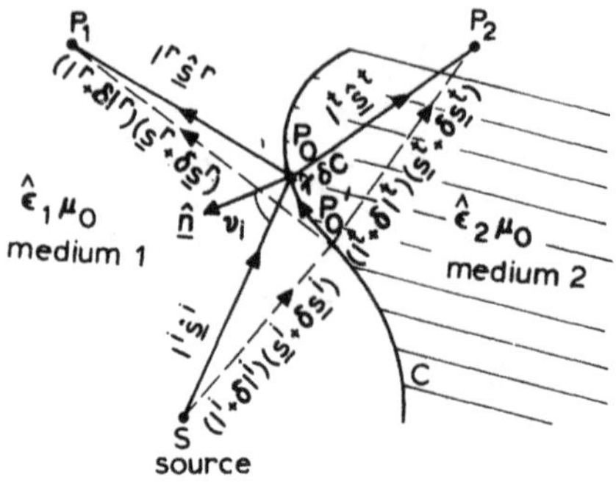

Fig. 4.2 Reflection and refraction at a curved interface

method offers an approximate solution by assuming that each point on the interface behaves locally as if it were part of an infinite planar interface and that the incident field behaves locally as a plane wave. For an incident field, or ray in terms of geometrical optics nomenclature, there will exist at the point P_0 in Fig. 4.2 a reflected ray determined by applying Snell's law for reflection to the outward normal $\hat{n}$ at P_0. At a point P_1 in the medium containing the source the field will be given by the sum of a direct incident ray from the source at S plus a reflected ray from the interface. If refraction takes place a field will exist at a point P_2 in the second medium which is determined by applying Snell's law for refraction to the normal $\hat{n}$ at P_0. The reflection and refraction coefficients for these rays are given directly from eqn. 4*a*. If the incident angle exceeds the critical angle only reflection occurs and the coefficients are then determined from eqn. 4*b*. The amplitude of the reflected and refracted rays will also be modified by the curvature of the interface but we will defer this problem until the next section.

In determining the point of reflection or refraction on the interface between the source and a chosen field point we could set up a search procedure to obtain the point, or points, on the interface which satisfy Snell's laws of reflection and refraction between the incident, reflected and refracted rays. The alternative and commonly used procedure is to use a variational approach based on the *optical path length* defined by nr, where r is the distance along the ray and n is the refractive index. Consider a point P_0' on the interface in Fig. 4.2 which is near the correct point P_0 for reflection between S and P_1. The optical path length vectors SP_0' and $P_0'P_1$ may be written as

$$n_1(l^i + \delta l^i)(s^i + \delta s^i) = n_1 l^i \hat{s}^i - n_1 \hat{\tau}\delta c$$

$$n_1(l^r + \delta l^r)(s^r + \delta s^r) = n_1 l^r \hat{s}^r + n_1 \hat{\tau}\delta c$$

where $(s^i + \delta s^i)$ and $(s^r + \delta s^r)$ are unit vectors. Multiplying the first equation by $\hat{s}^i \cdot$, the second by $\hat{s}^r \cdot$, and adding the resultant equations to the first order only yields

$$\delta(l^i + l^r) = (\hat{s}^r - \hat{s}^i)\cdot\hat{\tau}\delta c$$

The right-hand side of this equation is zero, since from Snell's law of reflection at P_0, we have

$$\hat{s}^i \cdot \hat{\tau} = \hat{s}^r \cdot \hat{\tau} \tag{6}$$

Thus the variational path difference $\delta(l^i + l^r)$ is zero, to first order, which implies that the ray path SP_0P_1 is stationary with respect to infinitesimal variations in path. This result expresses Fermat's principle for reflected rays in homogeneous media.

In a similar manner we find that the refracted path SP_0P_2 is stationary for infinitesimal variations in path *provided* we consider the optical path length which involves the refractive index n. This condition is a direct consequence of applying Snell's law for refraction at P_0 which can be written as

$$n_1\hat{s}^i \cdot \hat{\tau} = n_2\hat{s}^t \cdot \hat{\tau} \tag{7}$$

For reflection, therefore, we need only consider the ray path length r, but for refraction we require the quantity nr which is denoted as the optical path length.

In later chapters we will extend Fermat's principle to encompass various types of rays. So as to differentiate between the various ray phenomena we refer to the above result as *Fermat's principle for reflection and refraction* which may be stated as follows: *reflected or refracted rays from a point S to a point P are those rays for which the optical path length between S and P with one point on the interface between two media is stationary with respect to infinitesimal variations in path.*

So far we have considered an interface between two media. If a third medium exists, as in Fig. 4.3*a*, then we consider the refracted ray into medium 2 from medium 1 as the incident ray to the second interface C_2, and so on for any subsequent interface. There will now be multiple reflection and refraction between the interfaces at C_1 and C_2 which can be determined in the same manner. The total field at a point is then simply the superposition of all the incident, reflected and refracted rays which have been computed to that point. The number of multiple interactions to be considered will depend on the problem and the degree of accuracy desired. For example, if a region consists of stratified layers of homogeneous material with decreasing values of refractive index

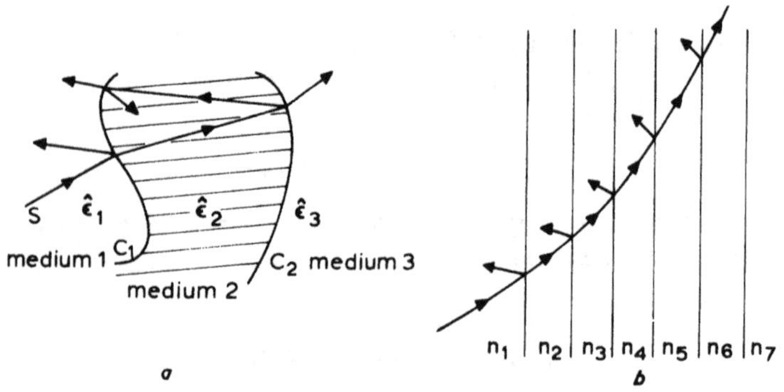

Fig. 4.3 Multiple media regions

which differ only slightly between adjacent media (i.e., $n_m \simeq n_{m+1}$), as illustrated in Fig. 4.3*b*, then for a first order approximation the reflected ray at each interface can be neglected since it will be considerably weaker than the refracted ray. Upon each refraction the transmitted ray will be bent further away from the normal to the interface. In the limit of thin layers with infinitesimal differences between adjacent media the ray path will appear as a smooth curve which bends towards the region of higher refractive index, and the optical path length along this curve C is given by the line integral

$$\int_C n(r)dr$$

which, from Fermat's principle, is stationary for infinitesimal variations in path. We now have a first order geometrical optics solution for propagation in a weakly inhomogeneous medium. For a more detailed exposition of this problem the interested reader is referred to Brekhovskikh (1960).

In discussing the geometrical optics method we have assumed that each medium is only slightly lossy. When one medium is highly conducting then only the reflected ray is of interest. This reflected ray still obeys Fermat's principle and the reflection coefficients are given by eqn. 5.

4.2 Ray tracing

Fermat's principle allows us to determine the network of ray paths throughout a given system. These ray paths trace out the direction of

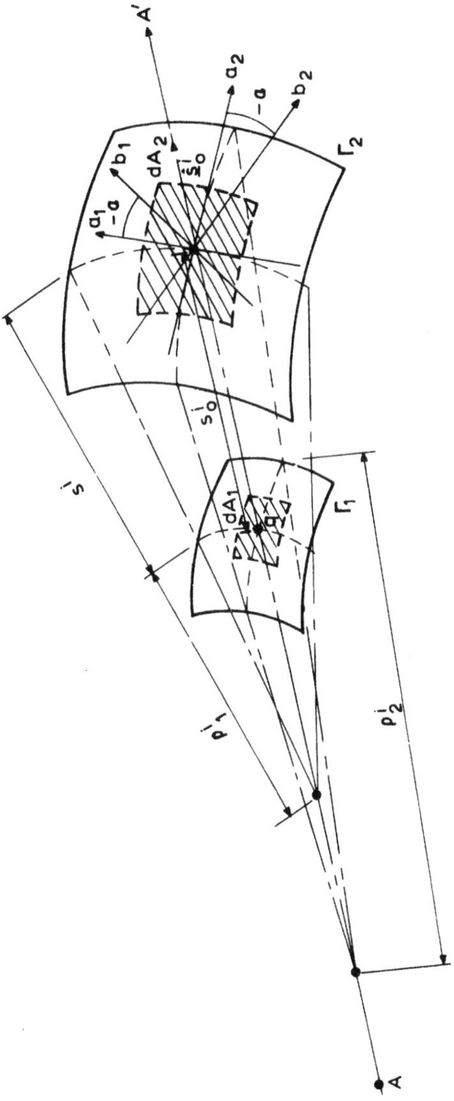

Fig. 4.4 Propagation of the wavefront along the axial ray AA'

power flow of the field, and to complete the task we must determine the amplitude and phase of the field at each point. Consider first the incident field. Let dA_1 be an element of area in an incident wave front Γ_1 about a ray AA' as in Fig. 4.4. Since the outward normal at each point on Γ_1 can also represent a ray path, we call AA' the *axial ray* of the element dA_1 so as to differentiate it from other rays. Rays in the vicinity of the axial ray, i.e., outward normals on dA_1 other than along AA' are referred to as *paraxial rays.* The wavefront will, in general, be curved and dA_1 will have two radii of curvature $\rho_1^i \rho_2^i$ as shown in Fig. 4.4. Paraxial rays taken around the periphery of dA_1 form a tube of rays about the axial ray which will intersect a second parallel wavefront Γ_2 with elemental area dA_2. This ray bundle surrounding the axial ray is called a *pencil* of rays. Since the direction of power flow is entirely along the rays, by the conservation of energy within the pencil we have

$$|E_1^i|^2 \, dA_1 = |E_2^i|^2 \, dA_2$$

where E_1^i, E_2^i are the values of the incident electric field in dA_1, dA_2. Relating the two areas gives

$$\frac{dA_1}{dA_2} = \frac{\rho_1^i \rho_2^i}{(\rho_1^i + s^i)(\rho_2^i + s^i)}$$

Referring the phase and amplitude of the incident field to Γ_1, for the field E_2^i at Γ_2, in terms of the field E_1^i at Γ_1, we have

$$E_2^i = E_1^i \left[\frac{\rho_1^i \rho_2^i}{(\rho_1^i + s^i)(\rho_2^i + s^i)} \right]^{1/2} \exp(-jk_1 s^i) \qquad (8a)$$

This is a familiar equation in geometrical optics. Provided that the field is known over one wavefront we can obtain the field at any other point in a homogeneous medium from this equation. If the radii of curvature are finite but not equal to each other the field is said to be *astigmatic.* When $\rho_1^i = \rho_2^i$ then the field is propagating as a *spherical wave.* If $\rho_1^i = \infty$ the field propagates as a *cylindrical wave,* and if ρ_2^i is also infinite we have a propagating *plane wave.* Thus from eqn. 8*a* we can identify four types of field propagation which we will be concerned with, namely, astigmatic, spherical, cylindrical, and plane wave propagation.

When $s^i = -\rho_1^i$ or $-\rho_2^i$, the geometrical optics method predicts an infinite field, and we refer to such field points as *caustics.* In general there may exist two *caustic surfaces* over which eqn. 8*a* gives an infinite value for the field. For some problems these surfaces will be coincident which, in turn, may degenerate to a single curve (a caustic locus) or a single point (a point caustic). In each case the geometrical optics solution is invalid and we must seek alternative representations for the field in such regions.

As a caustic surface is grazed, $\dfrac{\rho}{\rho+s}$ changes sign from positive to negative in eqn. 8*a*, and introduces a possible phase shift of $\exp\left(\pm\dfrac{j\pi}{2}\right)$ from the square root. The correct branch can be determined from the methods of stationary phase discussed earlier. In fact the appropriate choice is $\exp\left(\dfrac{j\pi}{2}\right)$ (for specific details refer to Section 6.6) so that

$$\begin{aligned}\sqrt{\left(\frac{\rho}{\rho+s}\right)} &= \left|\frac{\rho}{\rho+s}\right|^{1/2}; \qquad \frac{\rho}{\rho+s}>0\\ &= \exp\left(\frac{j\pi}{2}\right)\left|\frac{\rho}{\rho+s}\right|^{1/2}; \qquad \frac{\rho}{\rho+s}<0\end{aligned} \tag{8b}$$

If the field given by eqn. 8 is incident upon a curved boundary as in Fig. 4.2 then, in general, it will split up into a reflected and refracted field. These fields will not only be dependent on the reflection and refraction coefficients of the local environment about a point such as P_0, but will also be dependent on the local curvature of the interface,

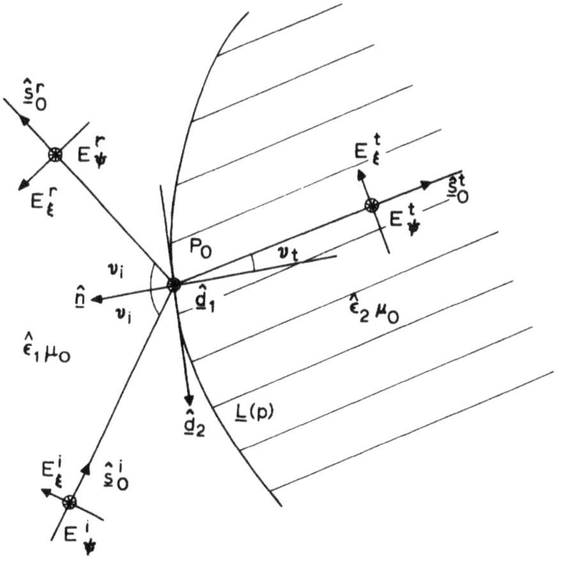

Fig. 4.5 Ray-based co-ordinate system for reflection and refraction

which will modify the values of $\rho_1^i \rho_2^i$ for the curvature of the reflected and refracted wavefronts. Consider first the reflected field. If we refer the phase and amplitude to the reflection point P_0, then we may write the reflected field in the pencil surrounding the reflected axial ray $P_0 s_0^r$ in Fig. 4.5 as

$$E^r(s^r) = \mathbf{R}(P_0) E^i(P_0) \left[\frac{\rho_1^r \rho_2^r}{(\rho_1^r + s^r)(\rho_2^r + s^r)} \right]^{1/2} \exp(-jk_1 s^r) \tag{9}$$

where $\mathbf{R}$ is the *reflection matrix* determined at P_0 which relates the components of the incident field to those of the reflected field, and ρ_1^r, ρ_2^r are the values for the principal radii of curvature of the reflected wavefront. We must now determine the various components of $\mathbf{R}$. This is reduced to its simplest form if we consider a *ray-based co-ordinate system* as illustrated in Fig. 4.5. By taking a spherical co-ordinate system (s_0, ψ, ξ) about each axial ray, where s_0 is in the direction of the ray, we have field components in only two directions, ψ, ξ, since the fields are locally plane waves and no component exists in the direction of propagation. We need only consider the electric field, as the magnetic field is given by

$$H = \sqrt{\left(\frac{\hat{\epsilon}}{\mu}\right)} \hat{s}_0 \times E \tag{10}$$

With the absence of $\hat{s}_0$ components of field the reflection matrix reduces to a 2×2 matrix (as will be the case for transmission and diffraction matrices to be discussed later) and at the point of reflection where $s^r = 0$, eqn. 9 becomes

$$E^r(P_0) = \mathbf{R}(P_0) E^i(P_0) \tag{11}$$

which upon expansion becomes

$$\begin{bmatrix} E_\psi^r(P_0) \\ E_\xi^r(P_0) \end{bmatrix} = \begin{bmatrix} R_{11} & R_{12} \\ R_{21} & R_{22} \end{bmatrix} \begin{bmatrix} E_\psi^i(P_0) \\ E_\xi^i(P_0) \end{bmatrix}$$

The components of $\mathbf{R}$ are now determined from the geometrical optics approximation which assumes that the interface at P_0 behaves locally as if it were part of an infinite planar interface. We can now arbitrarily orientate the electric field components such that the E_ψ component corresponds to electric polarisation for the problem in Section 3.1, and the E_ξ component corresponds to magnetic polarisation. The reflection matrix evaluated at P_0 is therefore

$$\mathbf{R}(P_0) = \begin{bmatrix} R^e & 0 \\ 0 & R^m \end{bmatrix} \tag{12}$$

where R^e, R^m are given by eqn. 4. If the second medium has infinite conductivity (as we will assume for most metals at radio frequencies) then the values for R^e and R^m simplify to -1 and 1, as given in eqn. 5.

By the same procedure we can determine the refraction, or transmission, matrix when both media are slightly lossy. The transmitted electric field in the pencil surrounding the transmitted axial ray $P_0 s_0^t$ in Fig. 4.5 is written as

$$E^t(s^t) = \mathbf{T}(P_0) E^i(P_0) \left[\frac{\rho_1^t \rho_2^t}{(\rho_1^t + s^t)(\rho_2^t + s^t)}\right]^{1/2} \exp(-jk_2 s^t); \qquad \nu_i < \nu_c \tag{13}$$

where $\mathbf{T}$ is the *transmission or refraction matrix*, and $\rho_1^t \rho_2^t$ are the principal radii of curvature for the transmitted wavefront.

By analogy with the reflected field of eqn. 11 we can write down the transmission matrix as

$$\mathbf{T}(P_0) = \begin{bmatrix} T^e & 0 \\ 0 & T^m \end{bmatrix} \tag{14}$$

where the field components are expressed in the ray-based co-ordinates and T^e, T^m are given in eqn. 4.

It now remains to calculate the principal radii of curvature for reflection and refraction. In some procedures, as, for example, found in Chapter 16 of Collin and Zucker (1969), this calculation is quite complicated, with much effort being required to obtain an analytical solution for all but the simplest of problems. However, an elegant and simple method is given by Deschamps (1972) (and a clear exposition of the same approach can be found in Lee (1975)) which involves the approximation of the wavefront and interface surface in the neighbourhood of the reflection (or refraction) point by second-order equations. By phase matching the incident, reflected and refracted field on the interface surface, we obtain Snell's law for reflection and refraction from the linear term and the principal radii of curvature for reflection and refraction from the quadratic term. We shall now discuss this method in some detail.

In Fig. 4.4 local co-ordinates (s_0^i, a_1, a_2) about the axial ray AA' through $\Gamma_{1,2}$ are shown. The axes a_1, a_2 are such that the principal radii of curvature ρ_1^i, ρ_2^i lie in the $a_1 s_0^i, a_2 s_0^i$ planes. In the neighbourhood of the axial ray through Γ_1 the wavefront can be approximated by the second-order equation

$$s^i = -\frac{1}{2}\left(\frac{1}{\rho_1^i} a_1^2 + \frac{1}{\rho_2^i} a_2^2\right)$$

More generally we will have a local co-ordinate system (s_0^i, b_1, b_2) where $b_1 s_0^i$, $b_2 s_0^i$ are not necessarily in the planes containing the principal radii of curvature. In this case we must write the preceeding equation in the more general form

$$s^i = -\tfrac{1}{2}(Q_{11}^i b_1^2 + 2Q_{12}^i b_1 b_2 + Q_{22}^i b_2^2)$$

which in matrix notation can be written as

$$s^i = -\tfrac{1}{2} \boldsymbol{b}^T \mathbf{Q}^i \boldsymbol{b} \tag{15}$$

where

$$\boldsymbol{b} = \begin{bmatrix} b_1 \\ b_2 \end{bmatrix}; \qquad \mathbf{Q}^i = \begin{bmatrix} Q_{11}^i & Q_{12}^i \\ Q_{12}^i & Q_{22}^i \end{bmatrix}$$

The superscript T in eqn. 15 denotes the transpose of the matrix and $\mathbf{Q}^i$ is known as the *curvature matrix* for the incident field.

If we assume temporarily that the co-ordinates (b_1, b_2) at Γ_1 coincide with (a_1, a_2) then we define the curvature matrix as $\mathbf{Q}_0^i$, where

$$\mathbf{Q}_0^i(s_0^i) = \begin{bmatrix} (\rho_1^i + s_0^i)^{-1} & 0 \\ 0 & (\rho_2^i + s_0^i)^{-1} \end{bmatrix} \tag{16}$$

The determinant of this matrix is simply

$$\det \mathbf{Q}_0^i(s_0^i) = \frac{1}{(\rho_1^i + s_0^i)(\rho_2^i + s_0^i)} \tag{17}$$

The inverse $(\mathbf{Q}^i)^{-1}$ of the curvature matrix is defined through

$$\mathbf{Q}^i(\mathbf{Q}^i)^{-1} = \mathbf{I}; \qquad \mathbf{I} = \begin{bmatrix} 1 & 0 \\ 0 & 1 \end{bmatrix} \tag{18}$$

from which we derive

$$(\mathbf{Q}^i)^{-1} = \frac{1}{Q_{11}^i Q_{22}^i - Q_{12}^{i2}} \begin{bmatrix} Q_{22}^i & -Q_{12}^i \\ -Q_{12}^i & Q_{11}^i \end{bmatrix} \tag{19}$$

By straightforward matrix addition we obtain the useful relationship

$$[\mathbf{Q}_0^i(s_0^i)]^{-1} = [\mathbf{Q}_0^i(0)]^{-1} + s_0^i \mathbf{I} \tag{20}$$

For the general $\mathbf{Q}^i$ matrix when (b_1, b_2) do not coincide with (a_1, a_2) we will now show that the relationships of eqns. 17 and 20 remain true. From eqn. 15 we may write

$$s^i = -\tfrac{1}{2}\boldsymbol{a}^T\mathbf{Q}_0^i\boldsymbol{a}, \quad \text{where} \quad \boldsymbol{a} = \begin{bmatrix} a_1 \\ a_2 \end{bmatrix} \tag{21}$$

The co-ordinates (a_1, a_2) and (b_1, b_2) can be related through an angle α as defined in Fig. 4.4, from which we obtain

$$\boldsymbol{a} = \mathbf{J}_\alpha \boldsymbol{b} \tag{22}$$

where

$$\mathbf{J}_\alpha = \begin{bmatrix} \cos\alpha & \sin\alpha \\ -\sin\alpha & \cos\alpha \end{bmatrix}$$

Substitution of eqn. 22 into eqn. 21, and using the reversal rule for a transposed product, see eqn. 2.91, we get

$$s^i = -\tfrac{1}{2}\boldsymbol{b}^T\mathbf{Q}^i\boldsymbol{b} \tag{23}$$

where

$$\mathbf{Q}^i = \mathbf{J}_\alpha^T\mathbf{Q}_0^i\mathbf{J}_\alpha$$

The components of the incident field curvature matrix in terms of $\rho_1^i, \rho_2^i, s_0^i$, and the angle α are given directly from $\mathbf{J}_\alpha^T\mathbf{Q}_0^i\mathbf{J}_\alpha$ as

$$\begin{aligned} Q_{11}^i &= \cos^2\alpha(\rho_1^i + s_0^i)^{-1} + \sin^2\alpha(\rho_2^i + s_0^i)^{-1} \\ Q_{12}^i &= \cos\alpha\sin\alpha\,[(\rho_1^i + s_0^i)^{-1} - (\rho_2^i + s_0^i)^{-1}] \\ Q_{22}^i &= \sin^2\alpha(\rho_1^i + s_0^i)^{-1} + \cos^2\alpha(\rho_2^i + s_0^i)^{-1} \end{aligned} \tag{24}$$

With these components for $\mathbf{Q}^i$ it follows that the expression for the determinant of $\mathbf{Q}_0^i$ as given by eqn. 17 is also true for $\mathbf{Q}^i$. Similarly for the relationship given in eqn. 20. Thus we have for the general curvature matrix

$$\begin{aligned} \det \mathbf{Q}^i(s_0^i) &= \frac{1}{(\rho_1^i + s_0^i)(\rho_2^i + s_0^i)} \\ \{\mathbf{Q}^i(s_0^i)\}^{-1} &= \{\mathbf{Q}^i(0)\}^{-1} + s_0^i\mathbf{I} \end{aligned} \tag{25}$$

The field $E_2(s^i)$, at the point s^i within the pencil about the axial ray, can now be written in terms of the curvature matrix if we express s^i in the local ray-based co-ordinates (s_0^i, b_1, b_2). From eqn. 8 we have, for the field in the pencil

$$E^i(s^i) = E^i(q)\left[\frac{\rho_1^i \rho_2^i}{(\rho_1^i + s^i)(\rho_2^i + s^i)}\right]^{1/2} \exp(-jk_1 s^i) \qquad (26)$$

Since the paraxial rays are almost parallel to the axial ray we may approximate the phase of eqn. 26 at the field point $s^i(s_0^i, b_1, b_2)$ by *adding* $\frac{1}{2}\boldsymbol{b}^T Q^i(s_0^i)\boldsymbol{b}$ to the phase at $(s_0^i, 0, 0)$, and ignore higher order terms. With the amplitude term taken as the value at $(s_0^i, 0, 0)$, eqn. 8 now becomes

$$\begin{aligned} E^i(s^i) &\simeq E^i(q)\left[\frac{\det \mathbf{Q}^i(s_0^i)}{\det \mathbf{Q}^i(0)}\right]^{1/2} \exp(-jk_1 s^i); \\ s^i &= s_0^i + \tfrac{1}{2}\boldsymbol{b}^T \mathbf{Q}^i(s_0^i)\boldsymbol{b} \end{aligned} \qquad (27)$$

This equation expresses the field in the pencil in terms of a local co-ordinate system (s_0^i, b_1, b_2) about the axial ray where we are at liberty to choose any desired direction for b_1, b_2. For reflection and refraction at a curved boundary illustrated in Fig. 4.5, it will be convenient to allow (s_0^i, b_1, b_2) for each axial ray to correspond to the co-ordinates (s_0, ψ, ξ) used earlier in determining the reflection and transmission matrices. We shall still retain the $\boldsymbol{b}$ matrix but its components will now be oriented so that $\hat{\boldsymbol{b}}_1 = \hat{\boldsymbol{\psi}}, \hat{\boldsymbol{b}}_2 = \hat{\boldsymbol{\xi}}$.

With the incident field represented by eqn. 27 we can now write down the reflected and transmitted pencils emanating from P_0 in Fig. 4.5 in the same format. Thus eqns. 9 and 13 can be written alternatively as

$$\begin{aligned} E^r(s^r) &\simeq \mathbf{R}(P_0)E^i(P_0)\left[\frac{\det \mathbf{Q}^r(s_0^r)}{\det \mathbf{Q}^r(0)}\right]^{1/2} \exp(-jk_1 s^r); \\ s^r &= s_0^r + \tfrac{1}{2}\boldsymbol{b}^T \mathbf{Q}^r(s_0^r)\boldsymbol{b} \\ E^t(s^t) &\simeq \mathbf{T}(P_0)E^i(P_0)\left[\frac{\det \mathbf{Q}^t(s_0^t)}{\det \mathbf{Q}^t(0)}\right]^{1/2} \exp(-jk_2 s^t); \\ s^t &= s_0^t + \tfrac{1}{2}\boldsymbol{b}^T \mathbf{Q}^t(s_0^t)\boldsymbol{b} \end{aligned} \qquad (28)$$

The only unknowns in eqn. 28 are the reflection curvature matrix $\mathbf{Q}^r$ and the transmission curvature matrix $\mathbf{Q}^t$. If we express a point P_0' on the interface in the vicinity of P_0 in the local co-ordinates (n, d_1, d_2) shown in Fig. 4.5, then as before for the wavefront, the equation of the interface surface $L(P_0')$ about P_0 may be approximated by a second order equation. Using the matrix notation as in eqn. 15

$$L(P_0') \simeq d[\hat{d}_1 \quad \hat{d}_2] - \tfrac{1}{2}d^T\mathbf{C}d\hat{n} \tag{29}$$

where

$$d = \begin{bmatrix} d_1 \\ d_2 \end{bmatrix}; \qquad \mathbf{C} = \begin{bmatrix} C_{11} & C_{12} \\ C_{12} & C_{22} \end{bmatrix}$$

The matrix $\mathbf{C}$ defines the curvature of the interface in the neighbourhood of P_0 in exactly the same way that the curvature matrix $\mathbf{Q}^i$ defines the curvature of the incident wavefront about the axial ray. With the origin of the incident ray co-ordinates (s_0^i, b_1, b_2) coincident with the interface surface co-ordinates (n, d_1, d_2) origin at P_0 (i.e., the point q in Fig. 4.4 corresponds to the point P_0 in Fig. 4.5) the relationship between these co-ordinates is

$$s_0^i = -d_2 \sin \nu_i - n \cos \nu_i = -d_2 \sin \nu_i + \tfrac{1}{2}d^T\mathbf{C}d \cos \nu_i + 0(d^3)$$

$$b = \mathbf{K}^i d + 0(d^2) \quad \text{where} \quad \mathbf{K}^i = \begin{bmatrix} 1 & 0 \\ 0 & -\cos \nu_i \end{bmatrix} \tag{30}$$

We can now express the phase $k_1 s^i$ of the incident pencil on the interface in the vicinity of P_0 in terms of the (n, d_1, d_2) co-ordinates

$$\begin{aligned} k_1 s^i &= k_1 [s_0^i + \tfrac{1}{2}b^T\mathbf{Q}^i(s_0^i)b] \\ &= k_1 [-d_2 \sin \nu_i + \tfrac{1}{2}d^T\mathbf{C}d \cos \nu_i + \tfrac{1}{2}(\mathbf{K}^i d)^T\mathbf{Q}^i(s_0^i)\mathbf{K}^i d] \\ &= -k_1 d_2 \sin \nu_i + \tfrac{1}{2}k_1 d^T \mathbf{\Gamma}^i d \end{aligned} \tag{31}$$

where

$$\mathbf{\Gamma}^i = \mathbf{C} \cos \nu_i + \mathbf{K}^i\mathbf{Q}^i(s_0^i)\mathbf{K}^i$$

The value of $\mathbf{Q}^i$ is a function of s_0^i but in the vicinity of the surface we may take its value as that given at the point P_0. (This is within the approximations of the method.) Similarly, we may express the phase $k_1 s^r$, $k_2 s^t$ of the reflected and transmitted pencil on the interface in the neighbourhood of P_0, so that

$$\begin{aligned} k_1 s^r &= -k_1 d_2 \sin \nu_r + \tfrac{1}{2}k_1 d^T\mathbf{\Gamma}^r d \\ k_2 s^t &= -k_2 d_2 \sin \nu_t + \tfrac{1}{2}k_2 d^T\mathbf{\Gamma}^t d \end{aligned} \tag{32}$$

where

$$\mathbf{\Gamma}^r = -\mathbf{C} \cos \nu_r + \mathbf{K}^r\mathbf{Q}^r(0)\mathbf{K}^r; \qquad \mathbf{K}^r = \begin{bmatrix} 1 & 0 \\ 0 & \cos \nu_r \end{bmatrix}$$

$$\mathbf{\Gamma}^t = \mathbf{C} \cos \nu_t + \mathbf{K}^t\mathbf{Q}^t(0)\mathbf{K}^t; \qquad \mathbf{K}^t = \begin{bmatrix} 1 & 0 \\ 0 & -\cos \nu_t \end{bmatrix}$$

For the necessary phase match to be achieved between the incident, reflected and transmitted field on the surface at P_0, the linear and quadratic terms in eqn. 32 must agree with those in eqn. 31. The linear terms confirm Snell's law for reflection and refraction while the quadratic terms require that

$$k_1 \mathbf{\Gamma}^i = k_1 \mathbf{\Gamma}^r = k_2 \mathbf{\Gamma}^t$$

Solving for $\mathbf{Q}^r, \mathbf{Q}^t$ at P_0 yields

$$\mathbf{Q}^r = \begin{bmatrix} 2C_{11}\cos\nu_i + Q^i_{11} & 2C_{12} - Q^i_{12} \\ 2C_{12} - Q^i_{12} & 2C_{22}\sec\nu_i + Q^i_{22} \end{bmatrix}$$

$$\mathbf{Q}^t = \begin{bmatrix} k_2^{-1}(k_1 Q^i_{11} + hC_{11}) & \sec\nu_t k_2^{-1}(k_1\cos\nu_i Q^i_{12} - hC_{12}) \\ \sec\nu_t k_2^{-1}(k_1\cos\nu_i Q^i_{12} - hC_{12}) & \sec^2\nu_t k_2^{-1}(k_1\cos^2\nu_i Q^i_{22} + hC_{22}) \end{bmatrix} \quad (33)$$

where

$$h = k_1\cos\nu_i - k_2\cos\nu_t$$

We have now completed the derivation of the geometrical optics field. The solution has been simplified by the choice of co-ordinates given in Fig. 4.5. Note that a positive curvature has been assigned to a diverging pencil and an interface which is locally convex to the chosen outward normal.

Although we have determined the curvature of the reflected and transmitted wavefront in terms of the curvature matrix, it may be desired to obtain the appropriate principal radii of curvature. These are readily derived from the components in the curvature matrix by solving the simultaneous equations in eqn. 24 for ρ^i_1, ρ^i_2 when $s^i_0 = 0$. Thus, for any given curvature matrix $\mathbf{Q}$ (and this may include the interface curvature matrix $\mathbf{C}$ if required) we obtain ρ_1, ρ_2 from the relationship

$$\frac{1}{\rho_{1,2}} = \tfrac{1}{2}[Q_{11} + Q_{22} \pm \sqrt{\{(Q_{11} - Q_{22})^2 + 4Q^2_{12}\}}] \quad (34)$$

4.3 Higher order terms

The geometrical optics field developed above is a solution to the scalar Helmholtz equation

$$\nabla^2 u + k^2 u = 0 \quad (35)$$

Expressing the field about an axial ray AA' as in Fig. 4.4, then at a field point $s(s_0, b_1, b_2)$ within the pencil surrounding the axial ray, we can write u for the geometrical optics solution, from eqns. 27 and 28 as

$$u(s) = \exp(-jks)\, u_0(s_0) \tag{36}$$

where

$$u_0(s_0) = u(q)\left[\frac{\det \mathbf{Q}(s_0)}{\det \mathbf{Q}(0)}\right]^{1/2}$$

$$s = s_0 + \tfrac{1}{2}\boldsymbol{b}^T \mathbf{Q}(s_0)\boldsymbol{b}$$

For the field to be evaluated at s it is required that the field is known in the neighbourhood of the field point at q.

If we now assume that an asymptotic solution to eqn. 35 for a homogeneous medium can be written in the form

$$\begin{aligned} u(s) &\sim k^\tau \exp(-jks) \sum_{m=0}^{\infty} \left(\frac{j}{k}\right)^m u_m(s_0); \\ &-1 < \tau \leqslant 0, \quad k \to \infty \end{aligned} \tag{37}$$

then direct substitution of eqn. 37 into eqn. 35 reveals that the following equations must be satisfied:

$$(\nabla s)^2 = 1 \qquad \text{(eiconal equation)} \tag{38a}$$

$$2\nabla s \cdot \nabla u_m + \nabla^2 s u_m = -\nabla^2 u_{m-1}; \qquad u_{-1} = 0 \quad \text{(transport equations)} \tag{38b}$$

With $\tau = 0$, the leading term in eqn. 37 is the geometrical optics field of eqn. 36, and the remaining higher order terms are obtained recursively through the transport equations of eqn. 38*b*. The introduction of k^τ is to allow for the other types of ray fields that we shall come across in the following chapters. For example, when $\tau = -0{\cdot}5$ the leading term in eqn. 37 can be equated with an edge diffracted field.

If the co-ordinates $\boldsymbol{b}$ in the expression for s given in eqn. 36 are in the direction of the principal radii of curvature for the wavefront at s, it follows that

$$s = s_0 + \tfrac{1}{2}b_1^2 Q_{11}(s_0) + \tfrac{1}{2}b_2^2 Q_{22}(s_0)$$

from which we get

$$\nabla s = \hat{s}_0 + b_1 \hat{b}_1 Q_{11}(s_0) + b_2 \hat{b}_2 Q_{22}(s_0)$$

$$\nabla^2 s = Q_{11}(s_0) + Q_{22}(s_0) = \frac{\rho_1 + \rho_2 + 2s_0}{(\rho_1 + s_0)(\rho_2 + s_0)}$$

We also have

$$\nabla u_m = \frac{\partial u_m}{\partial s_0} \quad \hat{s}_0$$

and the transport equations now become

$$2\frac{\partial u_m}{\partial s_0} + \nabla^2 s u_m = -\nabla^2 u_{m-1}$$

This is a first order linear differential equation to which the solution is

$$u_m(s_0) = u_m(q)\left[\frac{\rho_1\rho_2}{(\rho_1 + s_0)(\rho_2 + s_0)}\right]^{1/2} - \frac{1}{2}\int_0^{s_0}\left[\frac{(\rho_1 + s_0')(\rho_2 + s_0')}{(\rho_1 + s_0)(\rho_2 + s_0)}\right]^{1/2} \nabla^2 u_{m-1}(s_0')\,ds_0' \tag{39}$$

In some circumstances an alternative form to eqn. 39 is required. Rewriting eqn. 39 as

$$u_m(s_0)\left[\frac{(\rho_1 + s_0)(\rho_2 + s_0)}{\rho_1}\right]^{1/2} = u_m(q)\rho_2^{1/2} - \frac{1}{2\sqrt{(\rho_1)}}\int_0^{s_0} \{(\rho_1 + s_0')(\rho_2 + s_0')\}^{1/2}\,\nabla^2 u_{m-1}(s_0')\,ds_0'$$

then, provided that the field point s_0 is not at a caustic, the left-hand side of this equation will remain finite. If, however, the point q approaches a caustic surface so that the radius of curvature $\rho_2 \to 0$, then each term on the right hand side may, in general, diverge. Since their addition must yield a finite result, a useful computational tool is to take the *finite part* of each term as $\rho_2 \to 0$. Thus the above equation becomes

$$u_m(s_0)\left[\frac{(\rho_1 + s_0)s_0}{\rho_1}\right]^{1/2} = \operatorname*{fin\,lim}_{\rho_2 \to 0} u_m(q)\rho_2^{1/2} - \operatorname*{fin\,lim}_{\rho_2 \to 0} \frac{1}{2\sqrt{(\rho_1)}}\int_0^{s_0} \{(\rho_1 + s_0')s_0'\}^{1/2}\,\nabla^2 u_{m-1}(s_0')\,ds_0'$$

Rearranging this equation we have, finally

$$u_m(s_0) = \left[\frac{\rho_1}{(\rho_1 + s_0)s_0}\right]^{1/2}\delta_m - \frac{1}{2}\,⨍_0^{s_0}\left[\frac{(\rho_1 + s_0')s_0'}{(\rho_1 + s_0)s_0}\right]^{1/2} \nabla^2 u_{m-1}(s_0')\,ds_0' \quad \text{for} \quad \rho_2 \to 0 \tag{40}$$

where the slash through the integral denotes the finite part operation, and δ_m is the initial value

$$\delta_m = \operatorname*{fin\,lim}_{\rho_2 \to 0} u_m(q)\rho_2^{1/2}$$

4.4 Summary

The geometrical optics field about an axial ray AA' as in Fig. 4.4 is given by

$$E(s) = E(q)\left[\frac{\rho_1\rho_2}{(\rho_1 + s)(\rho_2 + s)}\right]^{1/2} \exp(-jks); \quad H = \sqrt{\left(\frac{\hat{\epsilon}}{\mu}\right)}\, \hat{s}_0 \times E \tag{41}$$

where
$$\sqrt{\left(\frac{\rho}{\rho + s}\right)} = \left|\frac{\rho}{\rho + s}\right|^{1/2}; \qquad \frac{\rho}{\rho + s} > 0$$
$$= \exp\left(\frac{j\pi}{2}\right)\left|\frac{\rho}{\rho + s}\right|^{1/2}; \qquad \frac{\rho}{\rho + s} < 0$$

The field at the point $s(s_0, b_1, b_2)$ is given in terms of a known field at $q(0, 0, 0)$, and ρ_1, ρ_2 are the principal radii of curvâture of the wavefront at q. This equation may be expressed alternatively as

$$E(s) = E(q)\left[\frac{\det \mathbf{Q}(s_0)}{\det \mathbf{Q}(0)}\right]^{1/2} \exp(-jks); \quad s = s_0 + \tfrac{1}{2}\boldsymbol{b}^T\mathbf{Q}\boldsymbol{b} \tag{42}$$

where $\mathbf{Q}$ is the curvature matrix

$$\mathbf{Q} = \begin{bmatrix} Q_{11} & Q_{12} \\ Q_{12} & Q_{22} \end{bmatrix} \tag{43}$$

If the co-ordinates (a_1, a_2) are in the direction of the principal radii of curvature about the wavefront at s_0, then (b_1, b_2) can be related through an angle α as defined in Fig. 4.4, so that

$$\boldsymbol{a} = \mathbf{J}_\alpha \boldsymbol{b}; \qquad \mathbf{J}_\alpha = \begin{bmatrix} \cos\alpha & \sin\alpha \\ -\sin\alpha & \cos\alpha \end{bmatrix} \tag{44}$$

The components of eqn. 43 become, in terms of ρ_1, ρ_2, α

$$\begin{aligned} Q_{11}(s_0) &= \frac{\cos^2\alpha}{\rho_1 + s_0} + \frac{\sin^2\alpha}{\rho_2 + s_0} \\ Q_{12}(s_0) &= \cos\alpha\sin\alpha\left[\frac{1}{\rho_1 + s_0} - \frac{1}{\rho_2 + s_0}\right] \\ Q_{22}(s_0) &= \frac{\sin^2\alpha}{\rho_1 + s_0} + \frac{\cos^2\alpha}{\rho_2 + s_0} \end{aligned} \tag{45}$$

Alternatively, for a given curvature matrix, we obtain ρ_1, ρ_2 from

$$\frac{1}{\rho_{1,2}} = \tfrac{1}{2}(Q_{11}(0) + Q_{22}(0) \pm \sqrt{[\{Q_{11}(0) - Q_{22}(0)\}^2 + 4Q_{12}^2(0)]}) \quad (46)$$

Two useful properties of the curvature matrix are

$$\det \mathbf{Q}(s_0) = \frac{1}{(\rho_1 + s_0)(\rho_2 + s_0)} \quad (47a)$$

$$[\mathbf{Q}(s_0)]^{-1} = [\mathbf{Q}(0)]^{-1} + s_0 \mathbf{I}; \qquad \mathbf{I} = \begin{bmatrix} 1 & 0 \\ 0 & 1 \end{bmatrix} \quad (47b)$$

When a known geometrical optics field E^i, H^i is incident upon a dielectric interface, then points of reflection and refraction, such as P_0 in Fig. 4.5, are determined from Fermat's principle for reflection and refraction. Reflected or refracted rays from a point S to a point P are those rays for which the optical path length between S and P [i.e., $\sqrt{(\epsilon)}r$ where r is the distance along the ray], with one point on the interface between the two media, is stationary with respect to infinitesimal variations in path. Mathematically, with reference to Fig. 4.5, this means that at a reflection and refraction point

$$\begin{aligned} \hat{s}^i \cdot \hat{d}_2 &= \hat{s}^r \cdot \hat{d}_2 && \text{for reflection} \\ \sqrt{(\epsilon_1)}\hat{s}^i \cdot \hat{d}_2 &= \sqrt{(\epsilon_2)}\hat{s}^t \cdot \hat{d}_2 && \text{for refraction} \end{aligned} \quad (48)$$

From eqn. 42 the incident electric field can be written as

$$\begin{aligned} E^i(s^i) &= E^i(q) \left[\frac{\det \mathbf{Q}^i(s_0^i)}{\det \mathbf{Q}^i(0)}\right]^{1/2} \exp(-jk_1 s^i); \\ s^i &= s_0^i + \tfrac{1}{2} b^{i^T} \mathbf{Q}^i(s_0^i) b^i \end{aligned} \quad (49a)$$

then at the point of reflection and refraction (or transmission) at P_0 the reflected pencil is given by

$$\begin{aligned} E^r(s^r) &= \mathbf{R}(P_0) E^i(P_0) \left[\frac{\det \mathbf{Q}^r(s_0^r)}{\det \mathbf{Q}^r(0)}\right]^{1/2} \exp(-jk_1 s^r); \\ s^r &= s_0^r + \tfrac{1}{2} b^{r^T} \mathbf{Q}^r(s_0^r) b^r \end{aligned} \quad (49b)$$

and the transmission pencil by

$$\begin{aligned} E^t(s^t) &= \mathbf{T}(P_0) E^i(P_0) \left[\frac{\det \mathbf{Q}^t(s_0^t)}{\det \mathbf{Q}^t(0)}\right]^{1/2} \exp(-jk_2 s^t); \\ s^t &= s_0^t + \tfrac{1}{2} b^{t^T} \mathbf{Q}^t(s_0^t) b^t \end{aligned} \quad (49c)$$

provided that the second medium is only slightly lossy. If this medium is highly conducting then there will be only an incident and reflected pencil. Using the local ray-based co-ordinate system (s_0, ψ, ξ) for the field components about each axial ray, so that $\hat{b}_1 = \hat{\psi}$ and $\hat{b}_2 = \hat{\xi}$, the reflection matrix **R** evaluated at P_0 reduces to the 2 x 2 matrix (since there are no components of field in the $\hat{s}_0$-direction)

$$\mathbf{R}(0) = \begin{bmatrix} R^e & 0 \\ 0 & R^m \end{bmatrix} \tag{50}$$

For slightly lossy media

$$\begin{aligned} R^e &= \frac{\sin(\nu_t - \nu_i)}{\sin(\nu_t + \nu_i)}; \qquad R^m = \frac{\tan(\nu_i - \nu_t)}{\tan(\nu_i + \nu_t)}; \qquad \nu_i < \nu_c \\ R^e &= \frac{\cos\nu_i - j\chi}{\cos\nu_i + j\chi}; \qquad R^m = \frac{\sin^2\nu_c \cos\nu_i - j\chi}{\sin^2\nu_c \cos\nu_i + j\chi}; \qquad \nu_i > \nu_c \end{aligned} \tag{51}$$

where $\chi = (\sin^2\nu_c - \sin^2\nu_i)^{1/2}$; ν_i is the angle of incidence to the outward normal at P_0; ν_t, ν_c are the transmission and critical angles defined by

$$\begin{aligned} \sin\nu_t &= \sqrt{\left(\frac{\epsilon_1}{\epsilon_2}\right)} \sin\nu_i; \qquad \nu_i < \nu_c \\ \sin\nu_c &= \sqrt{\left(\frac{\epsilon_2}{\epsilon_1}\right)} \end{aligned} \tag{52}$$

When the second medium is highly conducting, such as a metallic substance, then

$$R^e = -1; \qquad R^m = 1 \tag{53}$$

Similarly, the transmission matrix **T** for slightly lossy media is given by

$$\mathbf{T} = \begin{bmatrix} T^e & 0 \\ 0 & T^m \end{bmatrix} \tag{54a}$$

where

$$T^e = \frac{2\cos\nu_i \sin\nu_t}{\sin(\nu_t + \nu_i)}; \qquad T^m = \frac{2\cos\nu_i \sin\nu_t}{\sin(\nu_i + \nu_t)\cos(\nu_i - \nu_t)}; \qquad \nu_i < \nu_c \tag{54b}$$

with ν_t, ν_c given by eqn. 52. No transmission takes place for $\nu_i > \nu_c$.

To complete the evaluation of eqn. 49 we require the solution for $\mathbf{Q}^r$, $\mathbf{Q}^t$ at the point P_0. If we define a curvature matrix **C** for the interface at P_0 with respect to the co-ordinates $\boldsymbol{d}$ in Fig. 4.5, then at P_0

$$\mathbf{Q}^r = \begin{bmatrix} 2C_{11}\cos\nu_i + Q^i_{11} & 2C_{12} - Q^i_{12} \\ 2C_{12} - Q^i_{12} & 2C_{22}\sec\nu_i + Q^i_{22} \end{bmatrix} \tag{55a}$$

$$\mathbf{Q}^t = \begin{bmatrix} k_2^{-1}(k_1 Q^i_{11} + hC_{11}) & k_2^{-1}\sec\nu_t(k_1\cos\nu_i Q^i_{12} - hC_{12}) \\ k_2^{-1}\sec\nu_t(k_1\cos\nu_i - hC_{12}) & k_2^{-1}\sec^2\nu_t(k_1\cos^2\nu_i Q^i_{22} + hC_{22}) \end{bmatrix} \tag{55b}$$

where $h = k_1 \cos\nu_i - k_2 \cos\nu_t$.

References

BREKHOVSKIKH, L.M. (1960): 'Waves in layered media' (Academic Press)
COLLIN, R.E., and ZUCKER, F.J. (1969): 'Antenna theory' (McGraw-Hill)
DESCHAMPS, G.A. (1972): 'Ray techniques in electromagnetics', *Proc. IEEE*, **60**, pp. 1022–1035
LEE, S.W. (1975): 'Electromagnetic reflection from a conducting surface: geometrical optics solution', *IEEE Trans.*, **AP-23**, pp. 184–191

Chapter 5

Diffraction by straight edges and surfaces

5.1 Plane wave diffraction at a half-plane

We begin our study of edge diffraction with the simplest case, namely, plane wave diffraction at a half-plane. The exact solution to this problem was derived in Section 3.2 and we shall now proceed to discuss the solution in detail.

When a plane wave is normally incident upon a half-plane (i.e., the incident wavefront is parallel to the edge), as shown in Fig. 5.1, and has a field component V_z^i in the z-direction such that

$$V_z^i = \exp\{jk(x\cos\phi_0 + y\sin\phi_0)\} \tag{1}$$

where ϕ_0 is the angle of incidence to the half-plane, then the exact solution for the z-component at any field point (ρ, ϕ) can be written in the compact form

$$V_z(\rho,\phi) = u^i(\rho,\phi) \mp u^r(\rho,\phi) \tag{2}$$

The upper sign is for electric polarisation when the field component $V_z^i = E_z^i$ and hence eqn. 2 expresses the total electric field. For the lower sign we have magnetic polarisation where $V_z^i = H_z^i$ and eqn. 2 now yields the total magnetic field. The superscript i or r signifies that the particular field component is associated with the *incident* or *reflected* geometrical optics field. These components in eqn. 2 are given by

$$u^{i,r}(\rho,\phi) = U(\epsilon^{i,r})u_0^{i,r}(\rho,\phi) + u_d^{i,r}(\rho,\phi) \tag{3}$$

where U is the unit step function (1 for $\epsilon > 0$, 0 otherwise), $u_0^{i,r}$ is the *geometrical optics* field

$$u_0^{i,r}(\rho,\phi) = \exp\{jk\rho\cos(\phi \mp \phi_0)\} \tag{4}$$

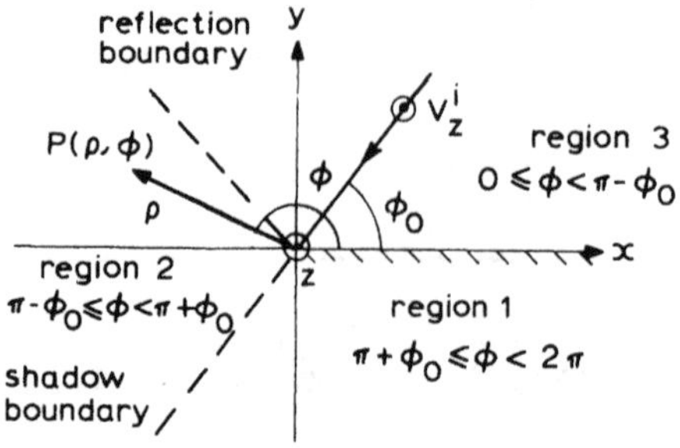

Fig. 5.1 Half-plane

and $u_d^{i,r}$ is the *edge diffracted* field

$$
\begin{aligned}
u_d^{i,r}(\rho,\phi) &= -\epsilon^{i,r}K_-\{|a^{i,r}|\sqrt{(k\rho)}\}\exp(-jk\rho);\\
a^{i,r} &= \sqrt{(2)}\cos\tfrac{1}{2}(\phi \mp \phi_0)\\
\epsilon^{i,r} &= \operatorname{sgn}(a^{i,r});\qquad K_-(s) = \sqrt{\left(\frac{j}{\pi}\right)}\exp(js^2)\int_s^\infty \exp(-jt^2)dt
\end{aligned}
\tag{5}
$$

The half-plane solution as expressed in eqns. 2–5 is seen to be given by the linear sum of four terms; the incident and reflected geometrical optics field u_0^i, u_0^r, and a diffracted field u_d associated with each optical term. These optical terms are multiplied by the unit step function from which we can divide the space surrounding the half-plane into three regions as shown in Fig. 5.1. In region 1 both ϵ^i and ϵ^r are negative so that the only field which may exist in this region is a diffracted field. As the line at $\phi - \phi_0 = \pi$ is crossed we enter region 2 where ϵ^i has changed from negative to positive, and the geometrical optics field u_0^i is now present in addition to the diffracted field. We shall refer to this boundary at $\phi - \phi_0 = \pi$ which separates region 1 from region 2 as the *shadow boundary* since it delineates the regions of light and shadow in the optical sense. Note that geometrical optics alone gives rise to a discontinuity of field across this shadow boundary, as is evident from the behaviour of the step function.

The boundary between region 2 and region 3 is at $\phi + \phi_0 = \pi$ and marks the change in ϵ^r from negative to positive. Thus in region 3 both optical terms u_0^i and u_0^r are present. This second optical term is the reflected wave from the half-plane which obviously does not exist beyond the boundary at $\phi + \phi_0 = \pi$. This latter boundary is the *reflection boundary* and, as for the shadow boundary, has a discontinuity in the geometrical optics field across it.

The geometrical optics field is seen to divide the space surrounding the half-plane into well-defined areas of reflection plus direct illumination, direct illumination alone, and total shadow. For the exact solution of eqn. 2 to be valid, the remaining diffracted field components must compensate for the discontinuities in the geometrical optics field across the reflection and shadow boundaries, to ensure continuity of the total field. Let us first investigate the diffraction term u_d at these boundaries. From eqn. 5 we observe that this term is determined by the behaviour of the modified Fresnel integral $K_-(x)$. At the reflection and shadow boundaries $\cos(\phi \pm \phi_0) = -1$, which causes the modified Fresnel integral argument to be zero. Thus the value of this function from eqn. 2.38 is given as $\frac{1}{2}$ whence

$$u_d^{i,r} = -\tfrac{1}{2}\epsilon^{i,r}\exp(-jk\rho) \tag{6}$$

As expected, the diffracted field on the optical boundaries has a discontinuity (due to the change in sign of $\epsilon^{i,r}$) which exactly compensates for the geometrical optics field. In other words, the superposition of the geometrical optics field $u_0^{i,r}$ and diffraction term $u_d^{i,r}$ gives continuity of field across these boundaries. It is important to note that the amplitude of the diffraction term $u_d^{i,r}$ in eqn. 6 is exactly half the appropriate geometrical optics term. An obvious point perhaps, but one which we will expand on later.

Away from the optical boundaries and the edge the modified Fresnel integral argument is non-zero, and if sufficiently large we may use its asymptotic form, as given in eqn. 2.38. Retaining only the first term of the asymptotic expansion,

$$K_-(s) \sim \frac{1}{2s\sqrt{(j\pi)}}; \qquad s \text{ large} \tag{7}$$

where, for the edge diffraction term $u_d^{i,r}$ given by eqn. 5, we have

$$s^{i,r} = |a^{i,r}|\sqrt{(k\rho)} \tag{8}$$

so that using eqns. 7 and 8 yields

$$\begin{aligned} u_d^{i,r} &= -\epsilon^{i,r}K_-\{|a^{i,r}|\sqrt{(k\rho)}\}\exp(-jk\rho) \\ &\sim -\sec\left(\frac{\phi \mp \phi_0}{2}\right)\frac{\exp(-jk\rho)}{\sqrt{(8j\pi k\rho)}} \end{aligned} \tag{9}$$

The asymptotic form of $u_d^{i,r}$ given in this equation implies that when removed from optical boundaries each *diffracted field component appears to originate from a line source situated along the edge*, having a polar diagram given by $-\sec\frac{1}{2}(\phi \mp \phi_0)$.

So far we have considered only the z-directed component of the diffracted field, but now we must consider how the total electromagnetic diffracted field behaves. From the eqns. 3.65–3.67 for electric polarisation we deduce that on the shadow boundary the electromagnetic diffracted field $\boldsymbol{E}^d, \boldsymbol{H}^d$ (obtained from these equations by replacing $u^{i,r}$ with $u_d^{i,r}$) is of the form

$$\begin{aligned} E_z^d &= -\tfrac{1}{2}\epsilon^i u_0^i + 0\left\{\frac{1}{\sqrt{(k\rho)}}\right\} \\ H_\phi^d &= \sqrt{\left(\frac{\hat{\epsilon}}{\mu}\right)}\,\tfrac{1}{2}\epsilon^i u_0^i + 0\left\{\frac{1}{\sqrt{(k\rho)}}\right\} \qquad (10) \\ H_\rho^d &= 0\left\{\frac{1}{\sqrt{(k\rho)}}\right\} \end{aligned}$$

which is essentially a propagating electromagnetic plane wave. A similar expression exists for the reflection boundary and at the boundaries for magnetic polarisation. Removed from the optical boundaries we may use the asymptotic form of the diffraction term $u_d^{i,r}$ to give, for electric polarisation,

$$\begin{aligned} E_z^d &\sim D^e(\phi, \phi_0)\frac{\exp(-jk\rho)}{\sqrt{(8j\pi k\rho)}} \\ H_\phi^d &\sim -\sqrt{\left(\frac{\hat{\epsilon}}{\mu}\right)}\,E_z^d \qquad (11) \\ H_\rho^d &\sim 0 \end{aligned}$$

and from eqns. 3.70–3.72 for magnetic polarisation

$$\begin{aligned} H_z^d &\sim D^m(\phi, \phi_0)\frac{\exp(-jk\rho)}{\sqrt{(8j\pi k\rho)}} \\ E_\phi^d &\sim \sqrt{\left(\frac{\mu}{\hat{\epsilon}}\right)}\,H_z^d \qquad (12) \\ E_\rho^d &\sim 0 \end{aligned}$$

where $D^e(\phi, \phi_0)$ and $D^m(\phi, \phi_0)$, known as the *edge diffraction coefficients* are

$$D^{e,m} = -\left\{\sec\left(\frac{\phi-\phi_0}{2}\right) \mp \sec\left(\frac{\phi+\phi_0}{2}\right)\right\} \qquad (13)$$

We have now established two distinctive properties in the behaviour of the diffracted field. At the optical boundaries when $a^{i,r} = 0$ its dominant behaviour is that of an electromagnetic plane wave having half the amplitude of the geometrical optics field. Well removed from these boundaries and the edge, the modified Fresnel integral argument $|a^{i,r}|\sqrt{(k\rho)}$ is large, and the diffracted field is seen from eqns. 11 and 12 to behave as an electromagnetic cylindrical wave emanating from a line source situated at the edge, having a polar diagram determined by the edge diffraction coefficients $D^e(\phi, \phi_0)$ and $D^m(\phi, \phi_0)$. This field can be expressed graphically by rays emanating from the edge, since along these *edge diffracted rays*, the field decays by $\rho^{-1/2}$ and has, locally, a plane wave phase front. Between these two extremes there are *transition regions* where the diffracted field has neither plane wave or cylindrical wave characteristics. The extent of these transition regions is dependent on what we mean quantitatively by large values of $|a^{i,r}|\sqrt{(k\rho)}$. In Fig. 2.8 the modified Fresnel integral and the first term of its asymptotic expansion are plotted against the argument x, where for $x > 3$ the leading term in the asymptotic expansion is seen to give a good approximation to the function.

When s is kept at a constant value, say s_c, then from eqn. 8, by squaring both sides we get

$$\rho = F\sec^2\frac{\Phi}{2} \tag{14}$$

where $F = \dfrac{s_c^2}{2k}$ and the angle $\Phi = \phi \mp \phi_0$. Eqn. 14 is the curve of a parabola about each optical boundary with the focus at the edge. This is illustrated in Fig. 5.2*a* for the reflection boundary. By choosing $s_c = 3{\cdot}55$ [i.e. $2\sqrt{(\pi)}$] then the focal distance F simplifies to λ and eqn. 14 becomes

$$\rho = \sec^2\frac{\Phi}{2}; \qquad s_c = 3{\cdot}55 \tag{15}$$

The parabolic curve generated by this equation about each optical boundary is shown in Fig. 5.2*b*. The areas *inside* the parabolic cylinders, which are shaded in the figure, are the transition regions for the diffracted field where we must evaluate the modified Fresnel integrals without approximations. In the area *outside* these cylinders we may use the asymptotic form of the diffracted field given in eqns. 11 and 12. As a summary, then, we have:

$|a^{i,r}|\sqrt{(k\rho)} = 0;$ $\quad u_d^{i,r}$ has plane wave characteristics

$0 < |a^{i,r}|\sqrt{(k\rho)} < 3{\cdot}55;$ $\quad u_d^{i,r}$ is in a transition state

$|a^{i,r}|\sqrt{(k\rho)} \geqslant 3{\cdot}55;$ $\quad u_d^{i,r}$ has cylindrical wave characteristics

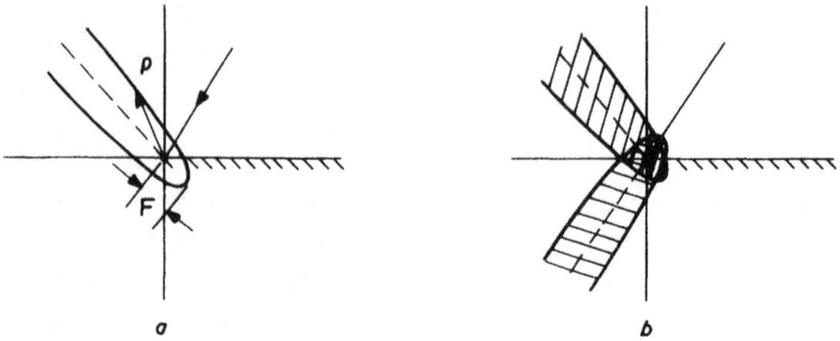

Fig. 5.2 Transition regions about the optical boundaries

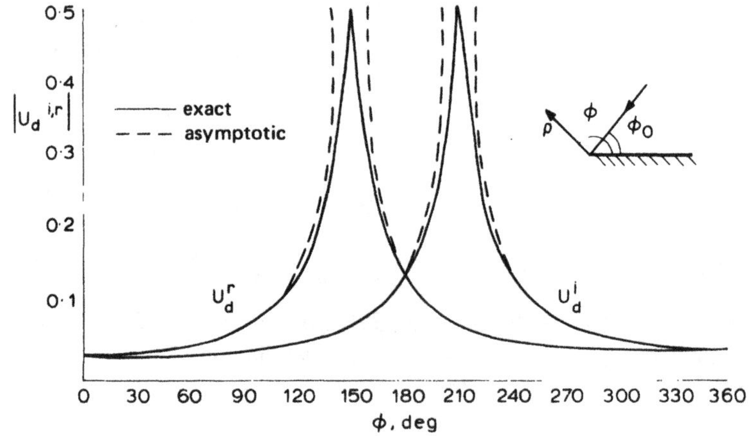

Fig. 5.3 Diffraction terms at $\rho = 5\lambda$ for plane wave incident at a half-plane

The diffracted field may be written for electric polarisation, with an incident field E_z^i, as

$$E_z^d(\rho) = (u_d^i - u_d^r)E_z^i(0); \quad H_\phi^d \simeq -\sqrt{\left(\frac{\hat{\epsilon}}{\mu}\right)}E_z^d; \quad H_\rho^d \simeq 0 \quad (16a)$$

and for magnetic polarisation, with an incident field H_z^i, as

$$H_z^d(\rho) = (u_d^i + u_d^r)H_z^i(0); \quad E_\phi^d \simeq \sqrt{\left(\frac{\mu}{\hat{\epsilon}}\right)}H_z^d; \quad E_\rho^d \simeq 0 \quad (16b)$$

The degree of approximation in eqn. 16 is dependent on the magnitude of $k\rho$. In practice, eqn. 16 has been used successfully for $k\rho > 1{\cdot}0$.

As an example we consider the field at a distance of 5λ from the

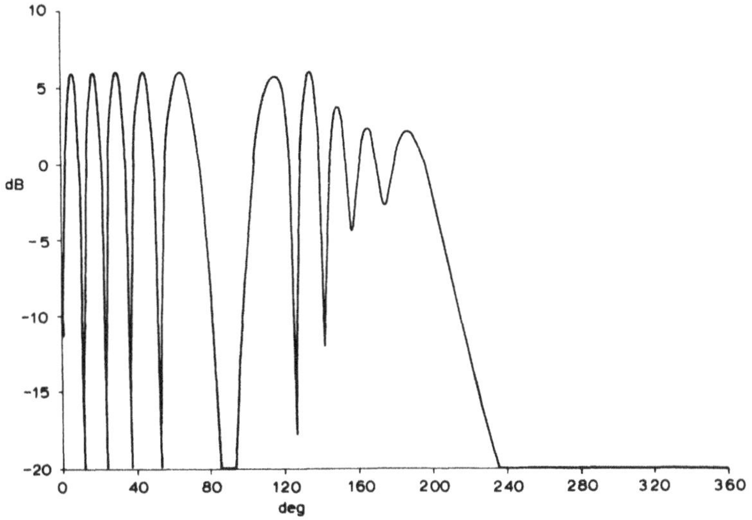

Fig. 5.4 Total field from a semi-infinite half plane illuminated by a plane wave (E-polarisation)
Angle of incidence = 30 deg
Distance of field point from edge = 5λ

edge for both electric and magnetic polarisation when the incident plane wave of unit intensity is at 30° to the half-plane. In Fig. 5.3 the individual diffraction terms u_d^i and u_d^r are shown. When the asymptotic solution of eqn. 9 is used within the transition region where it is invalid, it increases to infinity at the optical boundary. Also note that $u_d^{i,r}$ is an even function about this boundary, i.e.,

$$u_d^{i,r}(\rho, \Phi) = u_d^{i,r}(\rho, -\Phi); \qquad \Phi = \phi \mp \phi_0 \tag{17}$$

The total field for electric polarisation is given in Fig. 5.4 and for magnetic polarisation in Fig. 5.5.

When the angle of incidence $\phi_0 \to 0$, the incident and the reflected field merge and the incoming field is said to be at *grazing incidence.* In this situation $u^i(\rho, \phi) = u^r(\rho; \phi)$ giving $V_z = 0$ for electric polarization (as required by the boundary conditions on the half-plane) and $V_z = 2u^i$ for magnetic polarization. When considering grazing incidence it is common practice to assume unit strength for the geometrical optics field and normalize the equations by a factor of $\frac{1}{2}$. Thus for grazing incidence we have for eqn. 2

$$\begin{aligned} V_z &= 0 \qquad \text{electric polarisation} \\ V_z &= u^i \qquad \text{magnetic polarisation} \end{aligned} \tag{18}$$

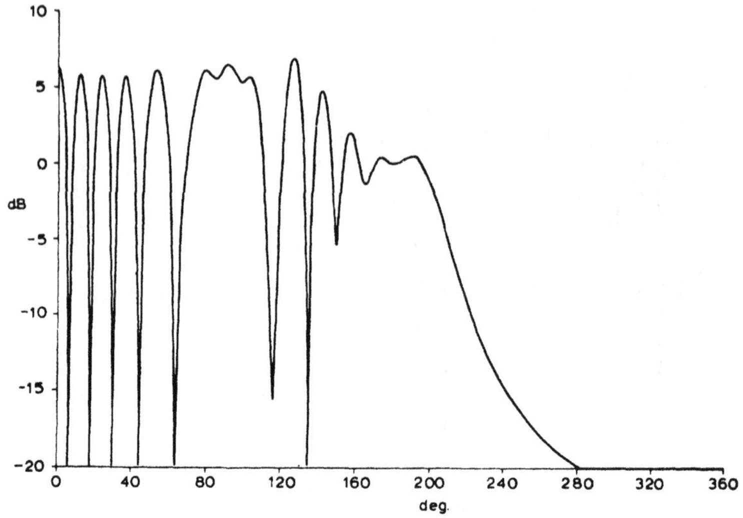

Fig. 5.5 Total field from a semi-finite half-plane illuminated by a plane wave (H-polarisation)
Angle of incidence = 30 deg
Distance of field point from edge = 5λ

5.2 Plane wave diffraction at a wedge

When a plane wave is normally incident upon a wedge, as in Fig. 5.6, and has a field component V_z^i in the z-direction such that

$$V_z^i = \exp\{jk(x\cos\phi_0 + y\sin\phi_0)\}$$

where ϕ_0 is the angle of incidence measured from the wedge face along the positive x-axis, then the exact solution for the z-component at any field point (ρ, ϕ) can be written in the compact form used previously for the half-plane, viz.,

$$V_z = u^i \mp u^r \tag{19}$$

As before, the upper (lower) sign is for electric (magnetic) polarisation when $V_z^i = E_z^i(H_z^i)$, and eqn. 19 expresses the total electric (magnetic) field where the superscript $i(r)$ relates to the incident (reflected) geometrical optics field. The components in eqn. 19 are given by

$$u^{i,r} = \sum_n U(\epsilon^{i,r})\, u_0^{i,r} + u_d^{i,r} \tag{20}$$

where, for the *geometrical optics* field

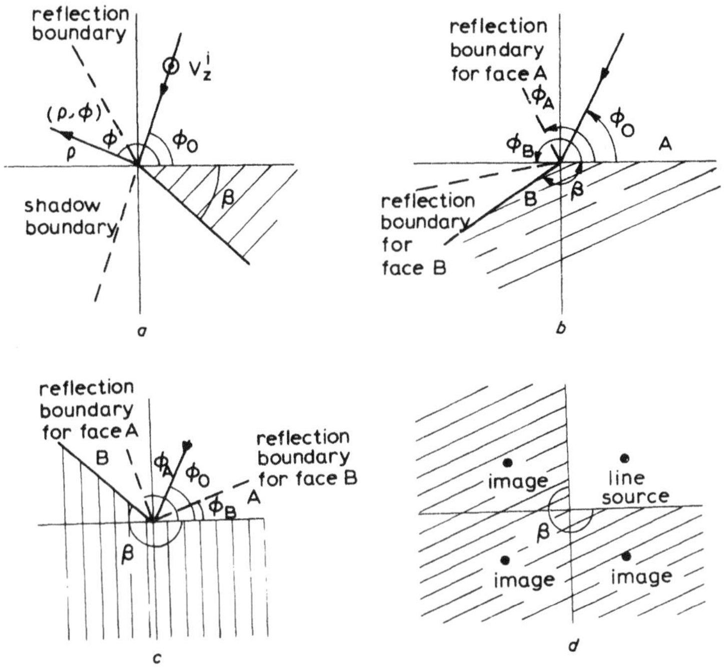

Fig. 5.6 Wedge configurations

$$
\begin{aligned}
u_0^{i,r} &= \exp\{jk\rho\cos(\Phi^{i,r} + 2n\pi N)\} \\
\epsilon^{i,r} &= \begin{bmatrix} +1 & \text{for illuminated region} \\ -1 & \text{for shadow region} \end{bmatrix} \\
\Phi^{i,r} &= \phi \mp \phi_0; \qquad N = \frac{2\pi - \beta}{\pi}
\end{aligned}
\tag{21}
$$

The value of n is any integer which satisfies

$$|\Phi + 2n\pi N| < \pi$$

It will be noted that this condition makes the inclusion of the unit step function $U(\epsilon^{i,r})$ superfluous, but we have retained it at this stage simply to stress the similarity with the half-plane solution.

The geometrical optics field in the half-plane solution was seen to give rise to shadow and reflection boundaries across which the geometrical optics field was discontinuous. An identical situation arises for the field incident upon a wedge as in Fig. 5.6*a*. If, however, the incident field illuminates *both* wedge faces, as in Fig. 5.6*b*, then there will be no shadow boundary in the space surrounding the wedge. Instead there will exist *two* reflection boundaries resulting from a reflected optical field from each wedge face. This can be seen mathematically from eqn. 21 where for the wedge angle equal to, say, 135° and the incident angle ϕ_0 equal to 120°, a reflection boundary occurs at $\phi = 60°$ with $n = 0$, and at $\phi = 150°$ with $n = -1$. The geometrical optics field is discontinuous across these two boundaries, and it is necessary for the edge diffracted field to have a compensating discontinuity in its leading term to ensure that the total field is smoothly varying at these boundaries.

The exact solution for the *edge diffracted* field $u_d^{i,r}$ in eqn. 20 is given from Section 3.3 by the integral expression

$$u_d^{i,r} = f(\Phi^{i,r}) + f(-\Phi^{i,r})$$

where (22*a*)

$$f(\Phi) = -\frac{1}{4\pi j N}\int_{-j\infty}^{j\infty} \cot\left(\frac{\xi + \pi + \Phi}{2N}\right) \exp(-jk\rho\cos\xi)$$

Alternatively $u_d^{i,r}$ may be deduced directly from the eigenfunction solution by subtracting out the geometrical optics field, viz,

$$u_d^{i,r} = \frac{1}{N}\sum_{m=0}^{\infty} \epsilon_\nu j^\nu J_\nu(k\rho)\cos\nu\Phi^{i,r} - \sum_n U(\epsilon^{i,r})u_0^{i,r}; \quad \nu = \frac{m}{N} \qquad (22b)$$

Provided that the distance ρ from the edge is not much greater than a wavelength, it should not be necessary to compute more than about 20 terms in this series. For larger values of ρ (i.e., $k\rho$ greater than about 10) the diffraction term $f(\Phi^{i,r})$ can be approximated by

$$f(\Phi^{i,r}) \simeq -\epsilon^{i,r}K_{-}\{|a^{i,r}|\sqrt{(k\rho)}\}\Lambda^{i,r}\exp(-jk\rho) \qquad (23)$$

where

$$\Lambda^{i,r} = \frac{a^{i,r}}{\sqrt{(2)}N}\cot\left(\frac{\pi + \Phi^{i,r}}{2N}\right)$$

$$a^{i,r} = \sqrt{(2)}\cos\tfrac{1}{2}(\Phi^{i,r} + 2n\pi N), \quad \epsilon^{i,r} = \operatorname{sgn}(a^{i,r})$$

and the value of n is chosen to satisfy the following two conditions on the optical boundaries:

$$|\Phi^{i,r} + 2n\pi N| = \pi \qquad (24a)$$

$$\Lambda^{i,r} = 1 \qquad (24b)$$

Consider first the reflection boundaries. As mentioned above we may have, in general, two reflection boundaries as illustrated in Figs. 5.6*b* and 5.6*c*. If we measure the incident angle ϕ_0 from face A, then on the reflection boundary A due to this wedge face, we note that

$$\phi_A + \phi_0 = \pi$$

On the reflection boundary B due to the reflected wave from face B we have

$$\phi_B + \phi_0 = 3\pi - 2\beta = -\pi + 2N\pi$$

Since it is required on each boundary that eqn. 24 should be satisfied, then at reflection boundary A it is seen that this is achieved by $f(-\Phi^r)$ with $n = 0$. Similarly on reflection boundary B we have $f(\Phi^r)$ with $n = -1$.

With two reflection boundaries there exists no shadow boundary in the space surrounding the wedge. It is important, however, to choose the correct values for n in $f(\pm\Phi^i)$ so as to satisfy the boundary conditions when a reflection boundary is near a wedge face. In other words, for electric polarisation the field must vanish on the wedge faces. On face A from where the angles are being measured, this condition is met if

$$\begin{aligned} u_d^i &= f(\Phi^i)_{n=0} + f(-\Phi^i)_{n=-1} \\ u_d^r &= f(\Phi^r)_{n=-1} + f(-\Phi^r)_{n=0} \end{aligned} \tag{25a}$$

and on face B the boundary conditions require that

$$\begin{aligned} u_d^i &= f(\Phi^i)_{n=-1} + f(-\Phi^i)_{n=0} \\ u_d^r &= f(\Phi^r)_{n=-1} + f(-\Phi^r)_{n=0} \end{aligned} \tag{25b}$$

Where a shadow boundary exists in visible space we have $f(\mp\Phi^i)_{n=0}$ depending on whether $\Phi^i = \pm\pi$ on the shadow boundary, with the remaining term given as $f(\pm\Phi^i)_{n=-1}$ (as in eqn. 25).

As for the half plane, when the modified Fresnel integral argument $|a^{i,r}|\sqrt{(k\rho)} > 3$ we can use its asymptotic value to give

$$f(\Phi^{i,r}) \sim -\frac{1}{N}\cot\left(\frac{\pi + \Phi^{i,r}}{2N}\right)\frac{\exp(-jk\rho)}{\sqrt{(8j\pi k\rho)}} \tag{26}$$

On an optical boundary $a^{i,r} = 0$ and

$$\Lambda^{i,r} \xrightarrow[a^{i,r}\to 0]{} 1 \tag{27a}$$

which reduces eqn. 23 to

$$f(\Phi^{i,r}) = -\tfrac{1}{2}\epsilon^{i,r}\exp(-jk\rho) \tag{27b}$$

When both diffraction terms $f(\Phi)$ and $f(-\Phi)$ are outside the transition regions we may use eqn. 26 for each term to give

$$u_d^{i,r} \sim \frac{\dfrac{2}{N}\sin\dfrac{\pi}{N}}{\cos\dfrac{\pi}{N}-\cos\dfrac{\Phi^{i,r}}{N}}\,\frac{\exp(-jk\rho)}{\sqrt{(8j\pi k\rho)}} \tag{28}$$

From this expression we note that the diffracted field vanishes when $\sin\dfrac{\pi}{N}=0$. For *interior* wedge angles, such as in Figs. 5.6*c* and *d* where $0<N<1$, this occurs when $N=\dfrac{1}{m}$, $m=1,2,3,\ldots$. In such cases the *exact* solution is given entirely in terms of the geometrical optics field. This can be seen by the example of the interior right angle wedge illustrated in Fig. 5.6*d* where $N=\frac{1}{2}$. If a line source is situated at P then the field is determined from the finite number of images in the wedge faces as shown in the figure.

As $N\to 0$ the solution is increasingly dominated by the geometrical optics field. For this reason we will be mainly concerned with wedge angles less than 270° (where $\frac{1}{2}<N\leqslant 2$) in which the diffracted field forms an important part of the total solution.

If we have an optical boundary at $\Phi^{i,r}=\pi$ at which the leading term is

$$u_d^{i,r} = -\tfrac{1}{2}\epsilon^{i,r}\exp(-jk\rho)$$

then by satisfying this condition and that of eqn. 28 for field points removed from the boundary, it follows that the diffracted field $u_d^{i,r}$ may be represented by

$$u_d^{i,r} = \frac{2}{N}\,\frac{\sin\dfrac{\pi}{N}\left|\cos\dfrac{\Phi^{i,r}}{2}\right|}{\cos\dfrac{\pi}{N}-\cos\dfrac{\Phi^{i,r}}{N}}\,K_-\left\{\sqrt{(2k\rho)}\left|\cos\dfrac{\Phi^{i,r}}{2}\right|\right\}\exp(-jk\rho) \tag{29}$$

This is the first term in the asymptotic series for wedge diffraction as given by Pauli (1938). The appeal of this equation is that we have only one term in place of two terms for constructing $u_d^{i,r}$, and this term has an obvious similarity to the corresponding expression for half-plane diffraction, so that when $N=2$ (i.e., the half-plane) we are seen to retrieve the exact solution for the half-plane given by eqn. 5. Extensive use has been made of eqn. 28 in wedge diffraction problems and it is perfectly adequate provided that we have only a single reflection boundary as in Fig. 5.6*a*. When two reflection boundaries exist, as in Fig. 5.6*b* however, it cannot be made to satisfactorily distinguish between the two boundaries. For such cases we must use our previous solution given by eqn. 25.

We have discussed so far only the z-directed component of the diffracted field for wedge diffraction. The behaviour of the total electromagnetic diffracted field is analogous to the half-plane solution and the comments in Section 5.1 apply here. Thus provided that $k\rho > 1{\cdot}0$, eqn. 16 is a good approximation for wedge diffraction. The problem of grazing incidence also applies and eqn. 18 is used for grazing incidence to a wedge face.

5.3 Oblique incidence

The incident plane wave assumed in Sections 5.1 and 5.2 was of the form

$$u_0(\rho, \phi - \phi_0, k) = \exp\{jk\rho\cos(\phi - \phi_0)\}$$

where ϕ_0 is the angle of incidence to the wedge face along the positive x-axis. Such a plane wave is at normal incidence to the z-axis (i.e., the edge). The subsequent edge diffraction at field points removed from the optical boundaries and the edge was seen to be given by rays emanating from an apparent line source along the edge, as illustrated in Fig. 5.7*a*. From these results we may deduce the quasi-2-dimensional problem of oblique incidence. A plane wave at oblique incidence of angle θ_0 to the edge as shown in Fig. 5.7*b* may be written as

$$\begin{aligned} U_0 &= \exp\left[jk\{\rho\cos(\phi - \phi_0)\sin\theta_0 - z\cos\theta_0\}\right] \\ &= u_0(\rho, \phi - \phi_0, k\sin\theta_0)\exp(-jkz\cos\theta_0) \end{aligned} \tag{30}$$

Thus the incident field for oblique incidence is given by replacing k with $k\sin\theta_0$ in u_0 for normal incidence and multiplying the latter by $\exp(-jkz\cos\theta_0)$. Eqn. 30 is also true for the reflected optical field with $(\phi - \phi_0)$ replaced by $(\phi + \phi_0 + 2n\pi N)$. By performing this simple operation on the diffraction term $u_d^{i,r}$, as given by eqn. 25 for the wedge, we obtain for oblique incidence

$$\begin{aligned} u_d^{i,r} &= f(\Phi^{i,r}) + f(-\Phi^{i,r}) \\ f(\Phi^{i,r}) &= -\epsilon^{i,r} K_-\{|a^{i,r}|\sqrt{(k\rho\sin\theta_0)}\}\Lambda^{i,r} \\ &\quad \exp\{-jk(\rho\sin\theta_0 + z\cos\theta_0)\} \end{aligned}$$

By considering the diffracted field at a point $\rho = s^d\sin\theta_0, z = s^d\cos\theta_0$,

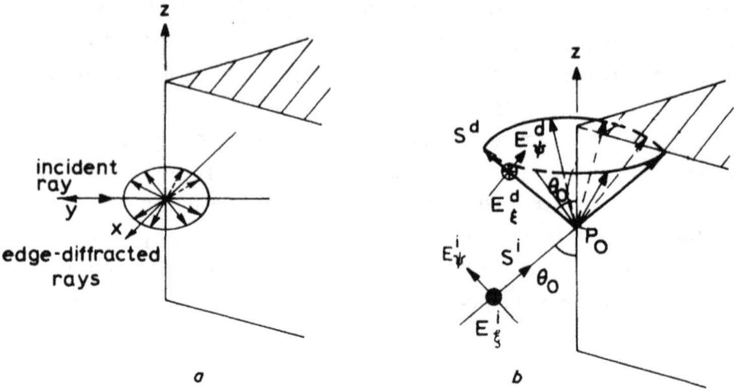

Fig. 5.7 Normal and oblique incidence
a Normal incidence
b Oblique incidence

where $\hat{s}^d$ will be seen to be the direction of the diffracted rays, the diffraction term $f(\Phi^{i,r})$ simplifies to

$$f(\Phi^{i,r}) = -\epsilon^{i,r} K_{-}\{|a^{i,r}| \sqrt{(ks^d)} \sin\theta_0\} \Lambda^{i,r} \exp(-jks^d) \quad (31a)$$

and hence

$$u_d^{i,r} = f(\Phi^{i,r}) + f(-\Phi^{i,r}) \sim \frac{\frac{2}{N}\sin\frac{\pi}{N}\csc\theta_0}{\cos\frac{\pi}{N} - \cos\frac{\Phi^{i,r}}{N}} \frac{\exp(-jks^d)}{\sqrt{(8j\pi ks^d)}} \quad (31b)$$

As before the diffracted field component for regions removed from optical boundaries appears to originate from a line source situated at the edge. In this case, with the field travelling along the $\hat{s}^d$-vector from the source, the variation of ϕ between $0-2\pi$ generates a *cone* of edge diffracted rays of semi-angle θ_0, whose apex is on the edge as illustrated in Fig. 5.7*b*. When the field is at normal incidence, then $\theta_0 = \frac{\pi}{2}$, and the cone degenerates into a disc so that the diffracted rays remain within the plane of incidence.

Previously we defined electric polarisation as the case when the incident electric field is parallel to the edge, i.e., when $\boldsymbol{E}^i = \hat{z}E_z^i$. For oblique incidence, however, the pertinent component of incident field is now $\hat{\psi}E_\psi^i$. To be consistent with the geometrical optics formulation in Chapter 4, we use the ray-based co-ordinate system illustrated in Fig. 4.5 and shown in Fig. 5.7*b* as it applies to the straight edge, so that along each diffracted ray we get from eqn. 16*a*

$$E_\psi^d(s^d) = (u_d^i - u_d^r) E_\psi^i(P_0); \qquad H_\xi^d \simeq \sqrt{\left(\frac{\hat{\epsilon}}{\mu}\right)} E_\psi^d;$$

$$H_s^d \simeq 0 \quad \text{for} \quad ks^d > 1{\cdot}0$$

Similarly, for magnetic polarisation we get from eqn. 16*b* the field along the diffracted ray as

$$H_\psi^d(s^d) = (u_d^i + u_d^r) H_\psi^i(P_0); \quad E_\xi^d \simeq -\sqrt{\left(\frac{\mu}{\hat{\epsilon}}\right)} H_\psi^d; \qquad E_s^d \simeq 0$$

For a generally incident plane wave we can now combine these two equations into the single equation

$$\begin{bmatrix} E_\psi^d(s^d) \\ E_\xi^d(s^d) \end{bmatrix} = \begin{bmatrix} (u_d^i - u_d^r) & 0 \\ 0 & (u_d^i + u_d^r) \end{bmatrix} \begin{bmatrix} E_\psi^i(P_0) \\ E_\xi^i(P_0) \end{bmatrix}$$

$$= \begin{bmatrix} D^e & 0 \\ 0 & D^m \end{bmatrix} \begin{bmatrix} E_\psi^i(P_0) \\ E_\xi^i(P_0) \end{bmatrix} \frac{\exp(-jks^d)}{\sqrt{(s^d)}}$$

which may be written in the compact form

$$\boldsymbol{E}^d(s^d) = \mathbf{D}\boldsymbol{E}^i(P_0)\frac{\exp(-jks^d)}{\sqrt{(s^d)}}; \quad \text{also} \quad \boldsymbol{H}^d = \sqrt{\left(\frac{\hat{\epsilon}}{\mu}\right)} \hat{\boldsymbol{s}}^d \times \boldsymbol{E}^d \qquad (32)$$

The quantity $\mathbf{D}$ is known as the *edge diffraction matrix* by analogy with the geometrical optics formulation in Section 4.2. The components in $\mathbf{D}$ are determined from eqn. 31. Removed from optical boundaries we have

$$D^{e,m} \sim \frac{\frac{2}{N}\sin\frac{\pi}{N}\csc\theta_0}{\sqrt{(8j\pi k)}} \left\{ \left(\cos\frac{\pi}{N} - \cos\frac{\phi-\phi_0}{N} \right)^{-1} \mp \left(\cos\frac{\pi}{N} - \cos\frac{\phi+\phi_0}{N} \right)^{-1} \right\} \qquad (33)$$

Eqn. 32 expresses the edge diffraction in terms of a ray based co-ordinate system about the diffracted rays for $ks^d > 1{\cdot}0$. The total field is given by the addition of the geometrical optics field. For more general configurations, to be considered shortly, we will only be concerned with the diffracted field since the geometrical optics field can be determined from the methods discussed earlier in Chapter 4.

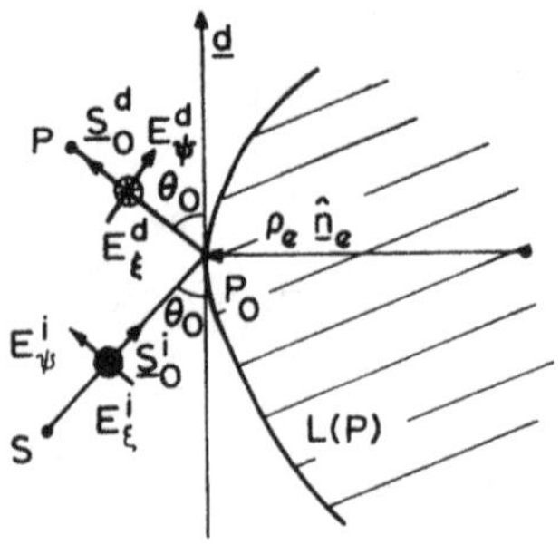

Fig. 5.8 Source near a curved edge

5.4 GTD formulation for edge diffraction

We have seen that the exact solution for plane wave diffraction at a straight edge gives rise to a cone of diffracted rays (provided that we are not in the vicinity of the edge, i.e., the condition that $ks^d > 1{\cdot}0$) emanating from each point along the edge, as illustrated in Fig. 5.7*b*. By analogy with geometrical optics, the GTD method uses this rigorous result to obtain approximate solutions for more complicated problems involving edge geometries. Thus if an arbitrary field is incident upon a body having on some part of it an edge in a perfectly conducting surface, then *the GTD method offers an approximate solution to the edge diffracted component of the solution by assuming that each point on the edge behaves locally as if it were part of an infinite straight edge and that the incident field behaves locally as a plane wave.* Thus if a source is near an edge, as in Fig. 5.8, then a cone of rays will be generated at each point along the edge. The semi-angle of each cone is determined by the angle of the incident ray to the edge-tangent at the point of diffraction. At a field point *P* the total field may consist of an incident and reflected geometrical optics field plus an edge diffracted field. The geometrical optics field may be determined by the ray tracing procedures discussed earlier in Section 4.2. We represent the diffracted field by an edge ray emanating from a point along the edge, determined from the condition that the incident and diffracted rays must make equal angles to the tangent at the edge. This condition can be written mathematically as

$$\hat{s}^i \cdot \hat{d} = \hat{s}^d \cdot \hat{d} \tag{34}$$

at the edge diffraction point. We may proceed further with the analogy with geometrical optics by stating a Fermat's principle applicable to edge diffracted rays. Thus, *Fermat's principle for edge diffraction states that edge diffracted rays from a point S to a point P are those rays for which the optical path length between S and P with one point on the diffracting edge is stationary with respect to infinitesimal variations in path.*

Fermat's principle gives us the edge diffracted ray paths and it is necessary now to determine the amplitude and phase of the field along these rays. We allow the incident field, in general, to be an astigmatic wave, as represented by eqn. 4.26, i.e.,

$$E^i(s^i) = E^i(q)\left[\frac{\rho_1^i\rho_2^i}{(\rho_1^i+s^i)(\rho_2^i+s^i)}\right]^{1/2}\exp(-jks^i)$$

where, as before, the distance s^i along the ray is measured from the field point q. Now by using a similar phase matching procedure at the edge, as we did on the surface of a dielectric interface for the geometrical optics field in Section 4.2, the amplitude along the diffracted ray can be readily found. In fact we do not require to derive new expressions as we may use the results obtained in Section 4.2. Thus from eqn. 4.27 we write the incident phase term s^i as

$$s^i = s_0^i + \tfrac{1}{2}\boldsymbol{b}^T\mathbf{Q}^i(s_0^i)\boldsymbol{b}$$

where the wavefront co-ordinate system is oriented so that $\hat{\boldsymbol{b}}_1 = \hat{\boldsymbol{\psi}}$ and $\hat{\boldsymbol{b}}_2 = \hat{\boldsymbol{\xi}}$. Since we are dealing with a line curvature for the edge and not a surface curvature as in Section 4.2, the equation for the edge $L(P_0')$ about the edge point P_0 (see Fig. 5.8) approximated by a second order equation is, by analogy with eqn. 4.29,

$$L(P_0') \simeq d\hat{\boldsymbol{d}} - \tfrac{1}{2}\frac{d^2}{\rho_e}\hat{\boldsymbol{n}}_e \tag{35}$$

where $\hat{\boldsymbol{n}}_e$ is the outward normal from the edge along the direction of edge curvature ρ_e at P_0. With the origin of the incident ray co-ordinates (s_0^i, b_1, b_2) coincident with the edge co-ordinates (n_e, d) origin at P_0, the relationship between them is

$$\begin{aligned} s_0^i &= d\cos\theta_0 + n_e\hat{\boldsymbol{s}}^i\cdot\hat{\boldsymbol{n}}_e = d\cos\theta_0 - \tfrac{1}{2}\frac{d^2}{\rho_e}\hat{\boldsymbol{s}}^i\cdot\hat{\boldsymbol{n}}_e + 0(d^3) \\ b_1 &= -d\sin\theta_0 + 0(d^2) \end{aligned} \tag{36}$$

Expressing the phase of the incident pencil on the edge in the vicinity of P_0 in terms of (n_e, d) through eqn. 36 gives

$$ks^i = kd\cos\theta_0 + \tfrac{1}{2}kd^2\left(\sin^2\theta_0 Q_{11}^i - \frac{\hat{\boldsymbol{s}}^i\cdot\hat{\boldsymbol{n}}_e}{\rho_e}\right) \tag{37}$$

Removed from the optical boundaries and the edge, the s^d dependency of the diffracted field for plane wave diffraction at a straight edge is seen from eqn. 32 to be $(s^d)^{-1/2}$. Using the GTD assumption that each point on the edge behaves locally as part of an infinite straight edge under plane wave illumination, then the edge becomes a caustic of the edge diffracted rays. This caustic is associated with the principal radius of curvature in the $\hat{\boldsymbol{\xi}} - \hat{s}^d$ plane. The other principal radius of curvature which is in the $\hat{\boldsymbol{\psi}} - \hat{s}^d$ plane is obtained by phase matching ks^i, given by eqn. 37, with a similar expression for ks^d at the diffraction point P_0. Thus in the neighbourhood of P_0 we have

$$ks^d = kd\cos\theta_0 + \tfrac{1}{2}kd^2\left(\frac{\sin^2\theta_0}{\rho_3} - \frac{\hat{s}^d \cdot \hat{n}_e}{\rho_e}\right) \tag{38}$$

where ρ_3 is the principal radius of curvature in the $\hat{\boldsymbol{\psi}} - s^d$ plane of the diffracted field from P_0 when measured from the edge. For eqns. 37 and 38 to be equal requires

$$\frac{1}{\rho_3} = Q_{11}^i - \frac{1}{\rho_e \sin^2\theta_0}(\hat{s}^i \cdot \hat{n}_e - \hat{s}^d \cdot \hat{n}_e) \tag{39}$$

and the field along diffracted rays removed from optical boundaries and the edge becomes from the straight edge formulation of eqn. 32

$$\begin{aligned} E^d(s^d) &\sim \mathbf{D}E^i(P_0)\left[\frac{\rho_3}{s^d(\rho_3 + s^d)}\right]^{1/2} \exp(-jks^d) \\ H^d(s^d) &\sim \sqrt{\frac{\hat{\epsilon}}{\mu}}\, s^d \times E^d(s^d) \end{aligned} \tag{40}$$

where the components of the $\mathbf{D}$ matrix are given by eqn. 33. Note that the diffraction coefficients D^e and D^m in eqn. 33 are invalid in transition regions (and in fact go to infinity on the optical boundaries) and thus give a non-uniform solution to the edge diffracted field. The result given in eqn. 40 is essentially the Keller GTD solution for edge diffraction. Despite the non-uniformity of this solution it was successfully applied to many engineering problems. Eventually its limitations began to be felt and means of providing a uniform solution were investigated. Within the context of GTD two main uniform theories have been developed which demand our attention. The two theories became known as the Uniform Geometrical Theory of Diffraction (UTD) and the Uniform Asymptotic Theory (UAT). We will compare these two theories later in Section 5.7. Here (as elsewhere in this book) we will concentrate mainly on the UTD approach, as it is somewhat simpler to use in practice and has been extended to embrace a wider variety of diffraction phenomena than UAT. For example, UTD has not only been applied to edge diffraction, as we shall show shortly, but also to phenomena such as convex surface diffraction and diffraction by a discontinuity in curvature, as will be discussed in the next chapter.

UTD is a rather straight-forward attempt at providing useful uniform diffraction formulations for engineering applications. It was developed in the early 1970's in at least two institutions simultaneously, namely the Ohio State University in the USA (Kouyoumjian and Pathak, 1974; Pathak and Kouyoumjian, 1974) and Queen Mary College in the U.K. (James and Poulton, 1973; James, 1974). UTD has little claim to mathematical rigour, but relies on the reasonableness of the following assumptions:

(i) Outside transition regions, the non-uniform Keller-type GTD solutions remain valid.
(ii) The total field must be continuous across the optical boundaries.
(iii) The dominant behaviour of the diffracted field in transition regions is correctly described by appropriate integral functions suggested by the canonical problem. In the case of edge diffraction this is the Fresnel integral. (Other transition functions will be described later for different diffraction phenomenon). To satisfy (i) and (ii) the argument(s) of the integral function is (are) suitably modified in a completely heuristic manner.

The application of this approach to edge diffraction will now be given.

We noted earlier that on optical boundaries the diffraction term $f(\Phi)$ was equal to half the associated geometrical optics term u_0 to ensure continuity of the total field through the transition regions. When removed from these regions, the asymptotic form for plane wave incidence upon a straight edge is now modified by the curvature ρ_3, so that

$$f(\Phi^{i,r}) = -\tfrac{1}{2}\epsilon^{i,r}\, u_0^{i,r} \qquad \text{on optical boundaries} \qquad (41a)$$

$$f(\Phi^{i,r}) \sim -\frac{\csc\theta_0}{N}\cot\left(\frac{\pi+\Phi^{i,r}}{2N}\right)\left(\frac{\rho_3}{\rho_3+s^d}\right)^{1/2}\frac{\exp(-jks^d)}{\sqrt{(8j\pi ks^d)}} \qquad (41b)$$

If our geometrical optics field is given by

$$u_0^{i,r}(s) = \left[\frac{\rho_1^{i,r}\rho_2^{i,r}}{(\rho_1^{i,r}+s)(\rho_2^{i,r}+s)}\right]^{1/2}\exp(-jks) \qquad (42)$$

then a uniform solution to diffraction at a curved edge is obtained if we write the diffraction term as

$$f(\Phi^{i,r}) = -\epsilon^{i,r}u_0^{i,r}(s^d)K_-(v^{i,r})\Lambda^{i,r} \tag{43a}$$

where $v^{i,r}$ is given by

$$v^{i,r} = \sqrt{(k\sigma^{i,r})}\,|a^{i,r}|\sin\theta_0$$

This is seen to be closely similar to the expression for $f(\Phi^{i,r})$ for oblique plane wave incidence given by eqn. 31*a*. The difference is that the geometrical optics term is now in the more general form as given by eqn. 42, and in the modified Fresnel integral argument s^d has been replaced by $\sigma^{i,r}$ where

$$\sigma^{i,r} = \frac{\rho_1^{i,r}\rho_2^{i,r}(\rho_3 + s^d)s^d}{(\rho_1^{i,r} + s^d)(\rho_2^{i,r} + s^d)\rho_3} \tag{43b}$$

and $\rho_1^{i,r}$, $\rho_2^{i,r}$ are the radii of curvature computed at the edge. That this expression for $f(\Phi)$ satisfies eqn. 41*a* when $v = 0$, and eqn. 41*b* when v is large can be readily verified for $\sigma > 0$.

If a caustic surface is grazed then σ changes sign and the expression for the diffraction term $f(\Phi)$ given in eqn. 43 is not valid. We shall consider this aspect, along with other problems in curved edge diffraction, later, see Section 6.6. For the remainder of this chapter we shall only consider straight edge diffraction so that the edge curvature $\rho_e \to \infty$ and

$$\frac{1}{\rho_3} = Q_{11}^i \qquad \text{(for a straight edge)} \tag{44}$$

Also

$$\begin{aligned} \rho_1^i &= \rho_1^r = \rho_1 \\ \rho_2^i &= \rho_2^r = \rho_2 \end{aligned} \tag{45}$$

The diffracted electric field for an astigmatic pencil incident upon a wedge where the edge and surfaces are straight may be written about the diffracted rays from the edge point P_0 as

$$\begin{aligned} E^d(s^d) &= \mathbf{D}E^i(P_0)\left[\frac{\rho_3}{s^d(\rho_3 + s^d)}\right]^{1/2} \exp(-jks^d); \\ \mathbf{D} &= \begin{bmatrix} D^e & 0 \\ 0 & D^m \end{bmatrix} \end{aligned} \tag{46}$$

where the components D^e and D^m of the edge diffraction matrix are given as

$$D^{e,m} = \{h(\Phi^i) + h(-\Phi^i)\} \mp \{h(\Phi^r) + h(-\Phi^r)\} \qquad (47a)$$

where

$$h(\Phi^{i,r}) = -\epsilon^{i,r}\{\sigma^{i,r}\}^{1/2}K_-(v^{i,r})\Lambda^{i,r}; \qquad \sigma^{i,r} > 0 \qquad (47b)$$

and $\sigma^i = \sigma^r$ for straight edge and surface diffraction. For large values of $v^{i,r}$ the asymptotic value of the modified Fresnel integral reduces $h(\Phi^{i,r})$ to

$$h(\Phi^{i,r}) \sim \frac{-\csc\theta_0 \cot\left(\dfrac{\pi + \Phi^{i,r}}{2N}\right)}{N\sqrt{(8j\pi k)}} \qquad (47c)$$

In the important special case of the half-plane where $N = 2$ the diffraction coefficients simplify to

$$D^{e,m} = -\{\epsilon^i(\sigma^i)^{1/2}K_-(v^i) \mp \epsilon^r(\sigma^r)^{1/2}K_-(v^r)\}; \qquad n = 0 \qquad (47d)$$

The case when $\sigma^{i,r} < 0$ will be considered in Section 6.6.

5.5 Higher order edge diffraction terms

The GTD solution obtained so far for wedge diffraction as given by eqns. 46 and 47 takes no account of any higher order derivatives of the incident field. If the incident field is non-uniform at the edge then the effect is to create higher order terms in the asymptotic solution of the diffracted field. Attempts to evaluate these higher order terms have, in the main, been restricted to the half-plane. With the similarity between edge diffraction for the half-plane and the wedge we then heuristically extend the half-plane results to the wedge. To simplify the analysis it will be assumed initially that the incident field is at normal incidence to the half-plane edge (i.e., $\theta_0 = \frac{\pi}{2}$). Once the answer is obtained for this case the simple transformation given in Section 5.3 will yield the result for oblique incidence.

Using the ray-based co-ordinate system, the electric field components E^d_ψ, E^d_ξ about a ray normally diffracted from a half-plane at the point P_0 are, from the previous section,

$$E^d_{\substack{\psi\\ \xi}}(s^d) = \left\{E^i_{\substack{\psi\\ \xi}}(P_0)u^i_d + E^r_{\substack{\psi\\ \xi}}(P_0)u^r_d\right\} \qquad (48)$$

where

$$E^r_{\psi}(P_0) = \mp E^i_{\psi}(P_0) \qquad \text{for electric or magnetic polarisation}$$
$$\xi \qquad\qquad \xi$$

$$u_d^{i,r} = -\epsilon^{i,r} u_0 K_-(v^{i,r})$$

$$u_0 = \left[\frac{\rho_1\rho_2}{(\rho_1+s^d)(\rho_2+s^d)}\right]^{1/2} \exp(-jks^d) \tag{49a}$$

$$v^{i,r} = \sqrt{(k\sigma)}\,|a^{i,r}|; \qquad \sigma = \frac{\rho_1\rho_2(\rho_3+s^d)s^d}{(\rho_1+s^d)(\rho_2+s^d)\rho_3}$$

$$a^{i,r} = \sqrt{2}\cos\tfrac{1}{2}(\phi\mp\phi_0); \qquad \epsilon^{i,r} = \operatorname{sgn}(a^{i,r})$$

Using the asymptotic expansion for the modified Fresnel integral as given in eqn. 2.38 then

$$u_d^{i,r} \sim -\sec\frac{\Phi^{i,r}}{2}\left[1+\frac{s^d}{\rho_3}\right]^{-1/2} \frac{\exp(-jks^d)}{\sqrt{(8j\pi ks^d)}} \sum_{m=0}^{\infty} j^m(\tfrac{1}{2})_m \left(2k\sigma\cos^2\frac{\Phi^{i,r}}{2}\right)^{-m}; \qquad v^{i,r}>0, \quad k\to\infty \tag{49b}$$

At grazing incidence, when the incident angle to the half plane $\phi_0 = 0$, the diffracted field component E^d_{ψ} (i.e., the component associated with electric polarisation at the half plane) as given by eqn. 48 will be zero, since the incident field component E^i_{ψ} must be zero to satisfy the boundary conditions. Similarly, when the incident angle $\phi_0 = \pi$ the diffracted field component E^d_{ξ} will be zero. For these two conditions diffraction associated with the first derivative of the incident field across the edge will now become the leading term in the asymptotic expansion of the diffracted field for these field components. Suitable diffraction coefficients for these two cases were obtained by Keller (1962) by solving a special diffraction problem. A more general solution for all angles of incidence was obtained by Lewis and Boersma (1969) using their UAT approach (see Section 5.7), where not only were all the higher order diffraction terms included, but also higher order terms in both the incident and reflected fields. Their approach was motivated by the earlier work of Wolfe (1966) and a subsequent paper by Ahluwalia *et al.* (1968) gives a refinement to their method. We shall only be concerned with diffraction due to the first derivative of the incident field at the edge. Since this term is dependent on the slope of the incoming field it is referred to as the *slope-diffraction* term. Our derivation is based on the work of Lewis and Boersma (1969).

The edge diffracted field at a half-plane as given by eqn. 48 does not predict the field if $E^{i}_{\substack{\psi\\ \xi}}(P_0) = 0$. In an attempt to account for higher order diffraction we define a modified diffraction term $v_d^{i,r}$ constructed by the addition to $u_d^{i,r}$ of a term in the form of eqn. 4.37.

$$v_d^{i,r} \sim E^{i,r}_{\substack{\psi\\ \xi}} \left\{ u_d^{i,r} + k^{-1/2} \exp(-jks^d) \sum_{m=0}^{\infty} \left(\frac{j}{k}\right)^m \zeta_m^{i,r} \right\}; \quad k \to \infty \tag{50}$$

The evaluation of $\zeta_m^{i,r}$ is made considerably easier if we solve for the region outside the optical boundaries. Thus using eqn. 49*b*, $v_d^{i,r}$ can be written as

$$v_d^{i,r} \sim k^{-1/2} \exp(-jks^d) \sum_{m=0}^{\infty} \left(\frac{j}{k}\right)^m w_m^{i,r} \tag{51}$$

where

$$w_m^{i,r} = E^{i,r}_{\substack{\psi\\ \xi}} \left\{ \zeta_m^{i,r} - \sec\frac{\Phi^{i,r}}{2} \left\{\left(1 + \frac{s^d}{\rho_3}\right) s^d\right\}^{-1/2} \frac{1}{\sqrt{(8j\pi)}} \left(\frac{1}{2}\right)_m \left(2\sigma\cos^2\frac{\Phi^{i,r}}{2}\right)^{-m} \right\} \qquad v^{i,r} > 0, \qquad k \to \infty$$

For eqn. 51 to be a solution to the scalar Helmholtz equation we must satisfy the eikonal and transport equations. If we solve these latter equations in terms of the field at the edge then we must use the form as given by eqn. 4.40 since the edge is a caustic of the diffracted rays. Along a diffracted ray, therefore, we have from eqn. 4.40

$$w_m^{i,r}(s^d) = \left[\left(1 + \frac{s^d}{\rho_3}\right) s^d\right]^{-1/2} \delta_m^{i,r} - \frac{1}{2} \int_0^{s_d} \left[\frac{(\rho_3 + s)s}{(\rho_3 + s^d)s^d}\right]^{1/2} \nabla^2 w_{m-1}^{i,r}(s)\,ds \tag{52}$$

where the slash denotes the finite part of the integral as described earlier.

In solving this equation we make use of the edge condition given in eqn. 3.86 for the half-plane. Noting the conditions on the *z*-components of field in eqn. 3.86, and that $u_d^{i,r}$ in eqn. 50 obeys the edge condition, then we require

$$\lim_{s^d \to 0} |\zeta_m^{i,r}| < \infty$$

for the edge condition on $v_d^{i,r}$ to be satisfied.

We begin by solving for $\delta_0^{i,r}$ and applying the edge condition on $\zeta_0^{i,r}$ as $s^d \to 0$. For this case the integral in eqn. 52 is zero (since $w_{-1}^{i,r} = 0$) so that

$$w_0^{i,r}(s^d) = \frac{\delta_0^{i,r}}{\sqrt{(s^d)}} + 0\{(s^d)^{1/2}\}$$

Expanding $E_{\psi_\xi}^{i,r}$ in a Taylor series about the edge diffraction point at P_0 in terms of s^d gives

$$E_{\psi_\xi}^{i,r} = E_{\psi_\xi}^{i,r}(P_0) + \hat{s}^d \cdot \nabla E_{\psi_\xi}^{i,r}(P_0) s^d + 0\{(s^d)^2\} \tag{53}$$

We have from the last two equations and eqn. 51

$$E_{\psi_\xi}^{i,r}(P_0)\zeta_0^{i,r} = \frac{\delta_0^{i,r}}{\sqrt{(s^d)}} + E_{\psi_\xi}^{i,r}(P_0)\frac{\sec\dfrac{\Phi^{i,r}}{2}}{\sqrt{(8j\pi s^d)}} + 0\{(s^d)^{1/2}\}$$

It is clear from this equation that to satisfy the edge condition requires

$$\delta_0^{i,r} = -E_{\psi_\xi}^{i,r}(P_0)\frac{1}{\sqrt{(8j\pi)}}\sec\frac{\Phi^{i,r}}{2}.$$

so that from eqn. 52

$$w_0^{i,r}(s^d) = -E_{\psi_\xi}^{i,r}(P_0)\left\{\left(1+\frac{s^d}{\rho_3}\right)s^d\right\}^{-1/2}\frac{\sec\dfrac{\Phi^{i,r}}{2}}{\sqrt{(8j\pi)}} \tag{54}$$

and the term $\zeta_0^{i,r}$ in eqns. 50 and eqn. 51 is zero.

To solve for the next higher order term $\delta_1^{i,r}$ in eqn. 52 we require first the evaluation of $\nabla^2 w_0^{i,r}(s)$ where

$$\nabla^2 w_0 = \frac{1}{s}\frac{\partial}{\partial s}\left(s\frac{\partial w_0}{\partial s}\right) + \frac{1}{s^2}\frac{\partial^2 w_0}{\partial\phi^2}$$

since w_0 is independent of the z-direction. Performing this differentiation on eqn. 54 we get for small s

$$\nabla^2 w_0^{i,r} = -E_{\psi_\xi}^{i,r}(P_0)\frac{1}{2\sqrt{(8j\pi)}}\sec^3\frac{\Phi^{i,r}}{2}\left\{s^{-5/2} - \frac{1}{2\rho_3}s^{-3/2} + 0(s^{-1/2})\right\}$$

Substituting into eqn. 52 yields

$$w_1^{i,r}(s^d) = \left[\left(1+\frac{s^d}{\rho_3}\right)s^d\right]^{-1/2}\left[\delta_1^{i,r} + E_{\psi_\xi}^{i,r}(P_0)\frac{\sec^3\dfrac{\Phi^{i,r}}{2}}{4\sqrt{(8j\pi)}}\int_0^{s^d}\{s^{-2}+0(1)\}\,ds\right]$$

The finite part integral in this equation is readily evaluated so that for small s^d we get

$$w_1^{i,r}(s^d) = \frac{\delta_1^{i,r}}{\sqrt{(s^d)}} - E_{\underset{\xi}{\psi}}^{i,r}(P_0)\frac{\sec^3\dfrac{\Phi^{i,r}}{2}}{4\sqrt{(8j\pi)}} \left\{(s^d)^{-3/2} - \frac{(s^d)^{-1/2}}{2\rho_3}\right\} + 0\{(s^d)^{1/2}\} \tag{55}$$

The value of $w_1^{i,r}$ from eqn. 51 is

$$w_1^{i,r} = E_{\underset{\xi}{\psi}}^{i,r}\left[\zeta_1^{i,r} - \frac{\sec^3\dfrac{\Phi^{i,r}}{2}}{4\sqrt{(8j\pi)}}\left\{\left(1+\frac{s^d}{\rho_3}\right)s^d\right\}^{-1/2}_{\sigma^{-1}}\right]$$

and upon expanding for small s^d using eqn. 53 gives

$$w_1^{i,r} = E_{\underset{\xi}{\psi}}^{i,r}(P_0)\zeta_1^{i,r} - \frac{\sec^3\dfrac{\Phi^{i,r}}{2}}{4\sqrt{(8j\pi)}} \left[E_{\underset{\xi}{\psi}}^{i,r}(P_0)\left\{(s^d)^{-3/2} - \left(\frac{3}{2\rho_3} - \frac{\rho_1+\rho_2}{\rho_1\rho_2}\right)(s^d)^{-1/2}\right\} + \hat{s}^d\cdot\nabla E_{\underset{\xi}{\psi}}^{i,r}(P_0)(s^d)^{-1/2}\right] + 0\{(s^d)^{1/2}\}$$

Equating this equation with that given in eqn. 55 we find that the terms in $(s^d)^{-3/2}$ cancel, and for $\zeta_1^{i,r}$ to remain bounded as the edge is approached requires the term in $(s^d)^{-1/2}$ to be zero, giving

$$\delta_1^{i,r} = -\frac{\sec^3\dfrac{\Phi^{i,r}}{2}}{4\sqrt{(8j\pi)}}\left\{E_{\underset{\xi}{\psi}}^{i,r}(P_0)\left(\frac{\rho_1+\rho_2}{\rho_1\rho_2} - \frac{1}{\rho_3}\right) + \hat{s}^d\cdot\nabla E_{\underset{\xi}{\psi}}^{i,r}(P_0)\right\} \tag{56}$$

We now have the required terms necessary to evaluate the next higher term $w_1^{i,r}$ from eqn. 52. The integral in this equation for $w_1^{i,r}$ becomes

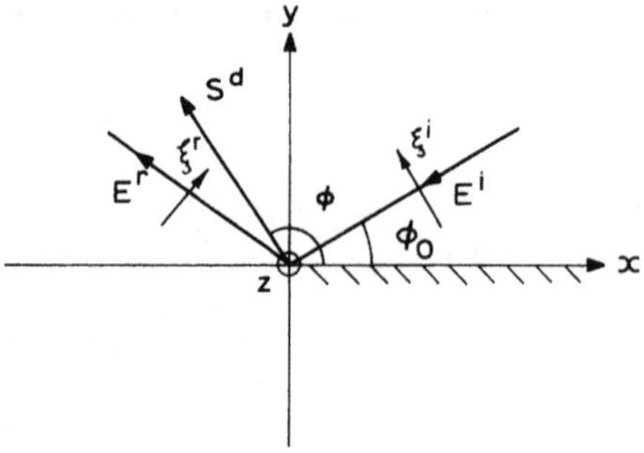

Fig. 5.9 Half-plane

$$\frac{1}{2}\int_0^{s^d}\left[\frac{(\rho_3+s)s}{(\rho_3+s^d)s^d}\right]^{1/2}\nabla^2 w_0^{i,r}(s)\,ds$$

$$= -E^{i,r}_{\substack{\psi\\ \xi}}(P_0)\,\frac{\sec^3\dfrac{\Phi^{i,r}}{2}}{4\sqrt{(8j\pi)}}\left\{\left(1+\frac{s^d}{\rho_3}\right)s^d\right\}^{-1/2} \qquad (57)$$

$$\int_0^{s^d}\frac{\rho_3^2+2s\rho_3+s^2\left(1+3\cos^2\dfrac{\Phi^{i,r}}{2}\right)}{s^2(\rho_3+s)^2}\,ds$$

which, in general, cannot be explicitly evaluated. Since $w_1^{i,r}$ is a higher order term it will be of particular interest mainly when the incident field $E^{i}_{\substack{\psi\\ \xi}}(P_0)$ is very small or zero. Under these conditions $w_1^{i,r}$ becomes the leading term in the asymptotic evaluation of the edge diffracted field. Also, when $E^{i,r}_{\substack{\psi\\ \xi}}(P_0)=0$ the contribution from the integral in eqn. 52 is zero, and we need only the first term. Thus from eqns. 52 and 56 we have for $E^{i,r}_{\substack{\psi\\ \xi}}(P_0)=0$

$$w_1^{i,r}(s^d) = -\left\{\left(1+\frac{s^d}{\rho_3}\right)s^d\right\}^{-1/2}\frac{\sec^3\dfrac{\Phi^{i,r}}{2}}{4\sqrt{(8j\pi)}}\,\hat{s}^d\cdot\nabla E^{i,r}_{\substack{\psi\\ \xi}}(P_0)$$

This equation simplifies by noting from Fig. 5.9

$$\hat{s}^d = \hat{x}\cos\phi + \hat{y}\sin\phi$$

$$\hat{\xi}^{i,r} = \mp\hat{x}\sin\phi_0 + \hat{y}\cos\phi_0$$

and

$$\nabla E^{i,r} = \hat{x}\frac{\partial E^{i,r}}{\partial x} + \hat{y}\frac{\partial E^{i,r}}{\partial y} = (-\hat{x}\sin\phi_0 \pm \hat{y}\cos\phi_0)\frac{\partial E^{i,r}}{\partial \xi^{i,r}}$$

so that

$$\hat{s}^d \cdot \nabla E^{i,r}_{\substack{\psi\\ \xi}}(P_0) = \pm\sin(\phi \mp \phi_0)\frac{\partial}{\partial \xi^{i,r}} E^{i,r}_{\substack{\psi\\ \xi}}(P_0)$$

The equation for $w_1^{i,r}$ now becomes

$$w_1^{i,r}(s^d) = \mp\left\{\left(1+\frac{s^d}{\rho_3}\right)s^d\right\}^{-1/2} \frac{1}{2\sqrt{(8j\pi)}}\sin\tfrac{1}{2}\Phi^{i,r}\sec^2\tfrac{1}{2}\Phi^{i,r}\frac{\partial}{\partial\phi}E^{i,r}_{\substack{\psi\\ \xi}}(P_0) \qquad (58a)$$

where

$$\frac{\partial}{\partial\xi^r}E^r_{\substack{\psi\\ \xi}}(P_0) = \mp\frac{\partial}{\partial\xi^i}E^i_{\substack{\psi\\ \xi}}(P_0) \qquad (58b)$$

and the upper (lower) sign is for electric (magnetic) polarisation.

To recapitulate, the leading diffraction term removed from the shadow boundaries is, from eqns. 51 and 54

$$v_d^{i,r} \sim E^{i,r}_{\substack{\psi\\ \xi}}(P_0)\,u_d^{i,r} + 0(k^{-3/2}) \qquad (59a)$$

where $u_d^{i,r}$ is the leading term in the asymptotic expansion for the half-plane given in eqn. 49*b*, i.e.,

$$-\sec\tfrac{1}{2}\Phi^{i,r}\left\{\left(1+\frac{s^d}{\rho_3}\right)s^d\right\}^{-1/2}\frac{\exp(-jks^d)}{\sqrt{(8j\pi k)}}$$

If the incident field $E^i_{\substack{\psi\\ \xi}}(P_0) = 0$, then the leading term for the diffracted field is now given by eqns. 51 and 58 as

$$v_d^{i,r} \sim \pm\frac{1}{jk}\frac{\partial}{\partial\xi^{i,r}}E^{i,r}_{\substack{\psi\\ \xi}}(P_0)\left\{\left(1+\frac{s^d}{\rho_3}\right)s^d\right\}^{-1/2} \sin\tfrac{1}{2}\Phi^{i,r}\sec^2\tfrac{1}{2}\Phi^{i,r}\frac{\exp(-jks^d)}{2\sqrt{(8j\pi k)}} \qquad (59b)$$

This term is referred to as the slope-diffraction term. There are close similarities between this term and that given by eqn. 59*a* for the first order term. In fact it is easily seen that eqn. 59*b* may be written alternatively, by relating it to $u_d^{i,r}$ in eqn. 59*a*, as

$$v_d^{i,r} \sim \frac{1}{jk}\frac{\partial}{\partial\xi^{i,r}} \underset{\xi}{E_\psi^{i,r}}(P_0)\frac{\partial}{\partial\phi_0}u_d^{i,r} + 0(k^{-5/2}) \tag{60}$$

As an example consider the case of magnetic polarisation when the incident angle $\phi_0 = \pi$. With the diffracted electric field given by

$$\underset{\xi}{E_\psi^d}(s^d) \sim v_d^i + v_d^r \tag{61}$$

then using the leading term as given by eqn. 59*a* predicts no diffracted field. The slope-diffraction term given by eqn. 60 now becomes the leading term in the asymptotic expansion of the diffracted field. Thus with

$$\frac{\partial}{\partial\xi^r}\underset{\xi}{E_\psi^r}(P_0) = \frac{\partial}{\partial\xi^i}\underset{\xi}{E_\psi^i}(P_0)$$

we get

$$\underset{\xi}{E_\psi^d} \sim k^{-3/2}\exp(-jks^d)\frac{\partial}{\partial\xi^i}\underset{\xi}{E_\psi^i}(P_0)\left\{\left(1+\frac{s^d}{\rho_3}\right)s^d\right\}^{-1/2}$$

$$\sqrt{\left(\frac{j}{8\pi}\right)}\cos\tfrac{1}{2}\phi\csc^2\tfrac{1}{2}\phi + 0(k^{-5/2})$$

Similarly, when we have electric polarisation at grazing incidence (i.e., $\phi_0 = 0$), the first order term is zero and the leading term is given by

$$\underset{\xi}{E_\psi^d} \sim -k^{-3/2}\exp(-jks^d)\frac{\partial}{\partial\xi^i}\underset{\xi}{E_\psi^i}(P_0)\left\{\left(1+\frac{s^d}{\rho_3}\right)s^d\right\}^{-1/2}$$

$$\sqrt{\left(\frac{j}{8\pi}\right)}\sin\tfrac{1}{2}\phi\sec^2\tfrac{1}{2}\phi + 0(k^{-5/2})$$

Both these results agree with that obtained by Keller (1962) [except that he has a sign wrong as pointed out by Lewis and Boersma (1969)] who solved a special diffraction problem for these two particular cases.

The above results have been restricted to field points outside the transition regions. To extend the results to these regions we simply use the uniform solution for $u_d^{i,r}$, as given by eqn. 49*a*, so that for the slope-diffraction term differentiation of this function with respect to the incident angle ϕ_0

$$\frac{\partial}{\partial\phi_0}u_d^{i,r} = \pm\sqrt{\left(\frac{k\sigma}{2}\right)}u_0\sin\tfrac{1}{2}\Phi^{i,r}\left\{\sqrt{\left(\frac{j}{\pi}\right)} - 2jv^{i,r}K_-(v^{i,r})\right\} \tag{62}$$

which is asymptotic to

$$\pm \frac{\exp(-jks^d)}{2\sqrt{8j\pi k)}} \left\{ \left(1 + \frac{s^d}{\rho_3}\right) s^d \right\}^{-1/2} \sin \tfrac{1}{2}\Phi^{i,r} \sec^2 \tfrac{1}{2}\Phi^{i,r} + 0(k^{-5/2})$$

by taking the first two terms in the asymptotic expansion of the modified Fresnel integral. Use of eqn. 62 ensures a uniform solution for slope-diffraction sometimes referred to as the Modified Slope-Diffraction (MSD) solution. On an optical boundary when $v^{i,r} = 0$ the dominant term for slope diffraction from eqns. 60 and 62 is seen to be proportional to

$$\sqrt{\left(\frac{\sigma}{2j\pi k}\right)} u_0 \frac{\partial}{\partial \xi^{i,r}} E^{i,r}$$

So far we have been concerned with higher order diffraction for normal incidence at a half-plane. The extension to the wedge and oblique incidence can be derived from the above formulations. For oblique incidence we perform the operation described in Section 5.3 on eqn. 60 so that

$$v_d^{i,r} \sim \frac{1}{jk \sin\theta_0} \frac{\partial}{\partial \xi^{i,r}} E_{\underset{\xi}{\psi}}^{i,r}(P_0) \frac{\partial}{\partial \phi_0} u_d^{i,r}$$

when $u_d^{i,r}$ is the diffraction function for oblique incidence. The general slope-diffracted field E^{sd} can now be derived from the matrix equation eqn. 46 for the first order term* and eqn. 58*b*. With $\hat{n} = \hat{\Phi}_0$,

$$E^{sd} = \frac{1}{jk \sin\theta_0} \frac{\partial}{\partial \phi_0} \mathbf{D} \frac{\partial}{\partial n} E^i \left[\frac{\rho_3}{s^d(\rho_3 + s^d)}\right]^{1/2} \exp(-jks^d) \quad (63a)$$

where

$$\frac{\partial}{\partial \phi_0} D^{e,m} = -[h'(\Phi^i) - h'(-\Phi^i)] \mp h'[(\Phi^r) - h'(-\Phi^r)] \quad (63b)$$

$$h'(\Phi) = \frac{\partial}{\partial \Phi} h(\Phi)$$

$$= \sqrt{(\sigma)} \Bigg[\sqrt{\left(\frac{k\sigma}{2}\right)} \sin \tfrac{1}{2}(\Phi + 2n\pi N) \sin\theta_0 \left\{ 2jvK_-(v) - \sqrt{\left(\frac{j}{\pi}\right)} \right\} \Lambda$$

$$+ \frac{\epsilon}{2N} K_-(v) \left\{ \sin \tfrac{1}{2}(\Phi + 2n\pi N) \cot\left(\frac{\Phi + \pi}{2N}\right)\right. \quad (63c)$$

$$\left. + \frac{a}{\sqrt{(2)}N} \csc^2 \left(\frac{\pi + \Phi}{2N}\right) \right\} \Bigg]$$

* It has been pointed out by T.S. Bird (private communication) that one cannot obtain an MSD solution for the wedge by taking the derivative of the Pauli expansion given by eqn. 29.

When v is large

$$2jvK_{-}(v) - \sqrt{\left(\frac{j}{\pi}\right)} \sim -\frac{1}{2v^2\sqrt{(j\pi)}}$$

$$K_{-}(v) \sim \frac{1}{2v\sqrt{(j\pi)}}$$

and the first two terms in eqn. 63c cancel to give

$$\frac{\partial}{\partial\Phi} h(\Phi) \sim \frac{1}{2N^2} \frac{1}{\sqrt{(8j\pi k)}} \csc^2\left(\frac{\pi + \Phi}{2N}\right) \csc\theta_0 \qquad (63d)$$

5.6 Physical optics approximation

In solving electromagnetic scattering from perfectly conducting bodies, extensive use is made of the physical optics approximation to the induced currents. We will consider the relationship between this method and GTD as it applies to straight edge diffraction. It will be seen that there is a close relationship between these methods for edge diffraction and this gives us a better understanding of the analysis of problems involving such diffraction. We will conclude by very briefly considering Ufimtsev's PTD method where correction currents are added to the physical optics currents in order to give an improved assessment of diffraction.

Consider first a *magnetically polarised* incoming plane wave $\boldsymbol{H}^i$ at normal incidence to a half plane as shown in Fig. 5.10 where

$$\boldsymbol{H}^i = \hat{z}h(\rho, \phi)\exp\{jk\rho\cos(\phi - \phi_0)\} \qquad (64)$$

and $h(\rho, \phi)$ is the amplitude dependence of the incident field. The physical optics approximation for a perfectly conducting obstacle gives the electric and magnetic source currents $\boldsymbol{J}_s, \boldsymbol{M}_s$ on the surface of the obstacle as

$$\boldsymbol{J}_s = 2\hat{n} \times \boldsymbol{H}^i; \qquad \boldsymbol{M}_s = 0 \qquad (65)$$

where $\hat{n}$ is the outward normal from the obstacle. The scattered magnetic field $\boldsymbol{H}^s$ for a 2-dimensional source distribution given by these currents is, from eqn. 2.16

$$\boldsymbol{H}^s(\rho, \phi) = \sqrt{\left(\frac{jk}{8\pi}\right)} \int \boldsymbol{J}_s \times \hat{\boldsymbol{P}} \frac{\exp(-jkP)}{\sqrt{(P)}} ds; \qquad kP \gg 1 \quad (66)$$

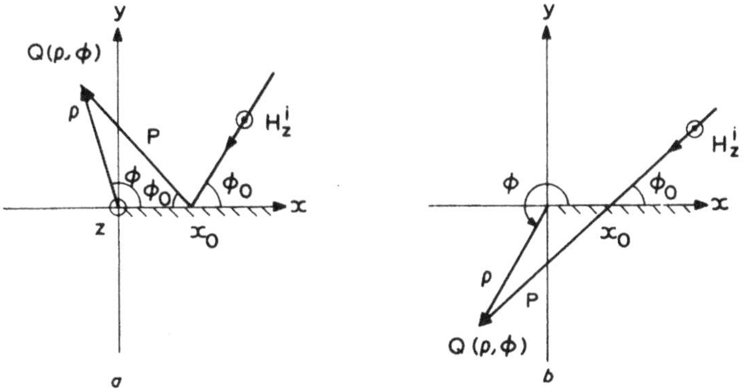

Fig. 5.10 Physical optics notation for half-plane
a $\phi < \pi$
b $\phi > \pi$

where P is the distance from the source point to the field point at Q and $\hat{P}$ is the unit vector in this direction.

Applying eqn. 65 to the half-plane problem in Fig. 5.10 with $0 < \phi_0 < \pi$ which, due to the symmetry of the problem does not restrict generality, we have

$$\begin{aligned} \boldsymbol{J}_s &= 2\hat{x}h(x)\exp(jkx\cos\phi_0) \qquad 0<\phi_0<\pi \\ \boldsymbol{P} &= P\hat{\boldsymbol{P}} = \boldsymbol{\rho} - \boldsymbol{x} = \rho\{\hat{x}\cos\phi + \hat{y}\sin\phi\} - \boldsymbol{x} \end{aligned} \tag{67}$$

where

$$P = \{\rho^2 + x^2 - 2\rho x\cos\phi\}^{1/2}$$

Substitution of these quantities into eqn. 66 yields

$$\begin{aligned} \boldsymbol{H}^s(\rho,\phi) = \hat{z}\sqrt{\left(\frac{jk}{8\pi}\right)}\, 2\rho\sin\phi \int_0^\infty h(x)P^{-3/2} \\ \exp\{jk(x\cos\phi_0 - P)\}dx \end{aligned} \tag{68}$$

This integral may now be evaluated asymptotically by the methods discussed in Section 2.3. We can rewrite eqn. 68 as

$$H_z^s(\rho,\phi) = \int_0^\infty f(x)\exp\{jkg(x)\}dx \tag{69}$$

where

$$f(x) = \sqrt{\left(\frac{jk}{8\pi}\right)}\, 2\rho\sin\phi h(x)P^{-3/2}$$

$$g(x) = x\cos\phi_0 - P$$

The asymptotic evaluation of this integral requires higher order derivatives of the phase function $g(x)$, viz,

$$\begin{aligned} g'(x) &= \cos\phi_0 - P^{-1}(x-\rho\cos\phi) \\ g''(x) &= -P^{-1} + P^{-3}(x-\rho\cos\phi)^2 \end{aligned} \tag{70}$$

At stationary phase points x_0, $g'(x_0)=0$ and from eqn. 70, we derive x_0 as

$$x_0 = P\cos\phi_0 + \rho\cos\phi \tag{71}$$

from which we get

$$\begin{aligned} g(x_0) &= \rho\cos\phi\cos\phi_0 - P\sin^2\phi_0 \\ g''(x_0) &= -P^{-1}\sin^2\phi_0 \end{aligned} \tag{72}$$

Note that for $\phi<\pi$, we construct from eqn. 71 a reflected ray path from the incident field to the field point at Q, as seen from Fig. 5.10*a*. For $\phi>\pi$ the construction is as for a direct optical ray through the half-plane to the field point at Q as seen in Fig. 5.10*b*. The contribution I_0 from an isolated stationary phase point is given from eqn. 2.75 as

$$I_0 = \sqrt{\left(\frac{2\pi}{k\,|g''(x_0)|}\right)}\, f(x_0)\exp\left(j\left[kg(x_0)+\frac{\pi}{4}\,\mathrm{sgn}\,\{g''(x_0)\}\right]\right)$$

so that for eqn. 69 this becomes

$$I_0 = \frac{\rho\sin\phi}{P\sin\phi_0}\, h(x_0)\exp\{jk(\rho\cos\phi\cos\phi_0 - P\sin^2\phi_0)\}$$

From Fig. 5.10 we can see that

$$P\sin\phi_0 = \mp\rho\sin\phi; \qquad \phi \lessgtr \pi$$

so that I_0 simplifies to

$$I_0 = \mp h(x_0)\exp\{jk\rho\cos(\phi\mp\phi_0)\}; \qquad \phi\lessgtr\pi \tag{73}$$

which is just the incident and reflected geometrical optics field at x_0. To include the contribution from the end point of the integral at $x=0$ we use the formulation given by eqn. 2.86 so that

$$H_z^s(\rho,\phi) \sim U(-\epsilon_1)I_0 + \epsilon_1 f(0)\exp\{jkg(0)\}$$
$$\sqrt{\left(\frac{2}{k\,|g''(0)|}\right)}\exp(\mp jv^2)F_\pm(v); \qquad g''(0)\gtrless 0$$

where

$$\epsilon_1 = \operatorname{sgn}(-x_0)$$

$$v = \sqrt{\left(\frac{k}{2\,|g''(0)|}\right)}\,|g'(0)|$$

The evaluation of this equation for the half-plane gives

$$H_z^s(\rho, \phi) \sim U(-\epsilon_1) I_0 + \epsilon_1 \sqrt{\left(\frac{j}{\pi}\right)}\, h(0) \exp(-jk\rho) \exp(jv^2) F_-(v);$$
$$v = \sqrt{\left(\frac{k\rho}{2}\right)} \left|\frac{\cos\phi_0 + \cos\phi}{\sin\phi}\right| \tag{74}$$

The first term of this equation gives the reflected geometrical optics field in the upper half-space. For the lower half-space it gives the incident geometrical optics field. When the incident field H_z^i is added to eqn. 74 to give the total field, then in the lower half-space it cancels with the first term in eqn. 74 to give a zero geometrical optics field in the shadow region of the half-plane.

Of particular interest is the diffracted field component H_z^d in eqn. 74

$$H_z^d(\rho, \phi) = \epsilon_1 \sqrt{\left(\frac{j}{\pi}\right)}\, h(0) \exp(-jk\rho) \exp(jv^2) F_-(v)$$
$$= \epsilon_1 h(0) K_-(v) \exp(-jk\rho) \tag{75}$$

The diffracted field given by the exact solution to the half-plane was seen earlier to be given by the sum of two terms which were both in the same form as eqn. 75. The difference with the physical optics approximation is that we have only one term, and with a different Fresnel integral argument v. On the optical boundaries $v = 0$ in eqn. 75 and the diffracted field is half the corresponding geometrical optics field. Thus the physical optics approximation is in agreement with the leading term of the exact solution. It is only removed from the optical boundaries when v is large that the physical optics result is in error in its leading diffraction term. For large v, eqn. 75 becomes

$$H_z^d(\rho, \phi) \sim -h(0) \frac{2\sin\phi}{\cos\phi_0 + \cos\phi} \frac{\exp(-jk\rho)}{\sqrt{(8j\pi k\rho)}}$$

and the corresponding expression for the exact solution to the half-plane is obtained from eqns. 12 and 13 as

$$H_z^d(\rho, \phi) \sim -h(0)\{\sec\tfrac{1}{2}(\phi - \phi_0) + \sec\tfrac{1}{2}(\phi + \phi_0)\} \frac{\exp(-jk\rho)}{\sqrt{(8j\pi k\rho)}}$$

Both expressions are in the form of a line source at the edge

$$H_z^d(\rho,\phi) \sim h(0)\,D^m(\phi,\phi_0)\frac{\exp(-jk\rho)}{\sqrt{(8j\pi k\rho)}} \tag{76}$$

where for the exact solution the diffraction coefficient D^m may be written as

$$D^m(\phi,\phi_0) = \frac{-4\cos\frac{1}{2}\phi\cos\frac{1}{2}\phi_0}{\cos\phi_0+\cos\phi} \qquad \text{(exact solution)} \tag{77a}$$

and for the physical optics approximation

$$D^m(\phi,\phi_0) = \frac{-2\sin\phi}{\cos\phi_0+\cos\phi} \qquad \text{(physical optics solution)} \tag{77b}$$

Comparing these last two equations we can write down an edge correction multiplication term

$$C^m = \cos\tfrac{1}{2}\phi_0\csc\tfrac{1}{2}\phi \tag{78}$$

which will correct the physical optics result so that it yields the exact solution outside the optical boundaries.

We may proceed in the same way for an *electrically polarised* incident plane wave to the half-plane. The z-component of diffracted field at field points removed from the optical boundaries is given by

$$E_z^d \sim h(0)\,D^e(\phi,\phi_0)\frac{\exp(-jk\rho)}{\sqrt{(8j\pi k\rho)}} \tag{79}$$

where for the exact half-plane solution

$$D^e(\phi,\phi_0) = \frac{4\sin\frac{1}{2}\phi\sin\frac{1}{2}\phi_0}{\cos\phi_0+\cos\phi} \qquad \text{(exact solution)} \tag{80a}$$

and for the physical optics approximation

$$D^e(\phi,\phi_0) = \frac{2\sin\phi_0}{\cos\phi_0+\cos\phi} \qquad \text{(physics optics solution)} \tag{80b}$$

From these equations we obtain the edge correction multiplication factor for electric polarisation C^e as

$$C^e = \sin\tfrac{1}{2}\phi\sec\tfrac{1}{2}\phi_0 \tag{81}$$

Fig. 5.11 gives examples of half-plane diffraction for both electric and magnetic polarisation to illustrate the difference between the physical optics approximation and the exact solution. It is seen, for example,

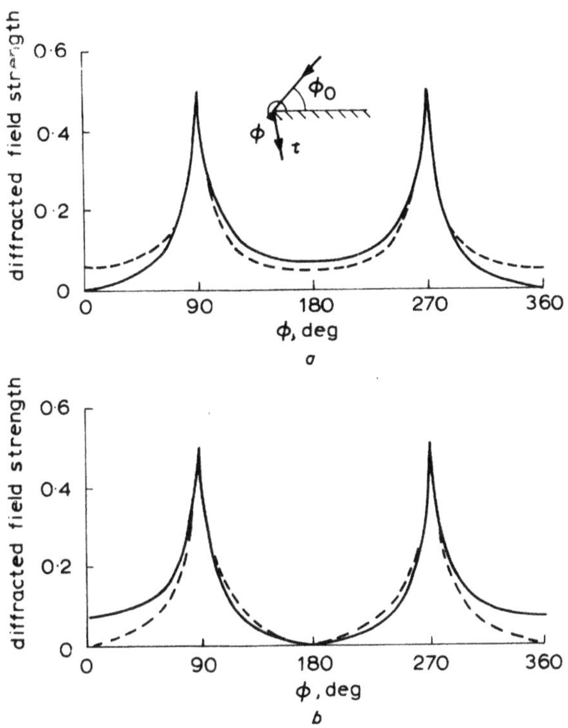

Fig. 5.11 Physical optics and exact solution for diffraction at a half-plane ($\tau = 10\,\lambda, \phi_0 = 90$ deg.)
a Electric polarisation
b Magnetic polarisation
——— exact
– – – asymptotic physical optics

that the former method gives a result that does not satisfy the boundary condition on the half-plane for electric polarisation.

If the half-plane is now replaced by a wedge and one wedge face only is illuminated, then the physical optics method will yield the same results as above. These will at least be correct in the leading term at an optical boundary. Removed from these boundaries we may, as in eqns. 78 and 81, derive edge-correction multiplication factors which will correct the physical optics result. If both wedge faces are illuminated, then the current termination at the edge for each wedge face will provide an end point contribution, to give the correct behaviour at the appropriate reflection boundary.

For the more general case of an astigmatic field incident on a curved screen we have the half-plane solution for the diffracted field outside the optical boundaries multiplied by the factor [e.g. see eqn. 40]

$$\left[\frac{\rho_3}{\rho_3 + s^d}\right]^{1/2}$$

where ρ_3 is determined from eqn. 39. To extend the results to the transition regions, we earlier imposed the condition that each diffracted term must be equal to half the associated geometrical optics term on the optical boundary. This condition is implicit for the diffracted field evaluated from the physical optics integral using the methods of stationary phase. Hence the comparability between the physical optics approximation and the GTD formulation for the half-plane is unchanged for the more general edge diffraction problem.

We return now to the half-plane under plane wave illumination and consider higher order diffraction effects in the physical optics approximation. For a non-uniform plane wave we may evaluate the slope-diffraction field at field points removed from optical boundaries and the edge by solving the second term in the asymptotic expansion about the end point given by eqn. 2.80. Thus for the magnetic polarisation integral of eqn. 69 the non-uniform slope-diffraction term is given by

$$\frac{1}{(jk)^2}\,\frac{f'(0)g'(0)-f(0)g''(0)}{\{g'(0)\}^3}\exp\{jkg(0)\} \tag{82}$$

For the particular case when the incident field at the edge $h(0) = 0$, this now becomes the leading term in the diffracted field to yield

$$H_z^d \sim \frac{1}{jk}\,\frac{\partial h(0)}{\partial x}\,\frac{2\sin\phi}{(\cos\phi + \cos\phi_0)^2}\,\frac{\exp(-jk\rho)}{\sqrt{(8j\pi k\rho)}} \tag{83}$$

This equation simplifies by noting that (where $\hat{n} = \hat{\boldsymbol{\varphi}}_0$)

$$\frac{\partial h(0)}{\partial x} = -\frac{\partial h(0)}{\partial n}\sin\phi_0$$

and from eqn. 77*b*, for the physical optics approximation,

$$\frac{\partial D^m}{\partial \phi_0} = -\frac{2\sin\phi\sin\phi_0}{(\cos\phi + \cos\phi_0)^2}$$

so we can write eqn. 83 as

$$H_z^d \sim \frac{1}{jk}\,\frac{\partial h(0)}{\partial n}\,\frac{\partial}{\partial\phi_0}D^m(\phi,\phi_0)\,\frac{\exp(-jk\rho)}{\sqrt{(8j\pi k\rho)}} \tag{84}$$

A similar equation can be derived for electric polarisation slope-diffraction. Comparing eqns. 84 and 76 with the corresponding eqns. 60 and 59*a* of the previous section (where $\rho_3 = \infty$ for plane wave diffraction at a straight edge) we see that they are in exactly the same form.

The relationship between the currents predicted from the physical optics approximation and the exact currents on the half-plane for plane wave illumination will now be considered. Assume an incident field given by

$$E^i = (\hat{\phi}E^i_\phi + \hat{z}E^i_z)\exp\{jk\rho\cos(\phi-\phi_0)\};$$
$$H^i = \sqrt{\left(\frac{\hat{\epsilon}}{\mu}\right)}\,\hat{s}^i \times E^i \tag{85}$$

The total magnetic field, which is independent of the z-co-ordinate, can be written in the scalar potential formulation of eqn. 2.26, viz,

$$H = -\hat{z} \times \nabla u_a - j\omega\hat{\epsilon}\hat{z}u_f \tag{86}$$

where the potentials u_a and u_f are the solutions for electric and magnetic polarisation which were, for the half-plane, solved exactly in Section 3.2. From eqn. 2.28 we have

$$u_a = -\frac{1}{j\omega\mu}E_z; \qquad u_f = -\frac{1}{j\omega\hat{\epsilon}}H_z$$

where E_z and H_z are given from Section 3.2 as

$$E_z = E^i_z(u^i - u^r)$$
$$H_z = H^i_z(u^i + u^r)$$

where, from eqn. 3.65c,

$$u^{i,r} = K_-\{-\sqrt{(2k\rho)}\cos\tfrac{1}{2}(\phi \mp \phi_0)\}\exp(-jk\rho)$$

Eqn. 86 can now be written as

$$H = \frac{E^i_z}{j\omega\mu}\hat{z} \times \nabla(u^i - u^r) - \frac{k}{\omega\mu}E^i_\phi(u^i + u^r)\hat{z} \tag{87}$$

Previously we have only been concerned with the electromagnetic field at field points removed from the edge when $k\rho$ is large. To derive the exact currents J_s flowing on the screen however, we are interested in the magnetic field over the entire half-plane, and particularly for small values of ρ. These currents are derived from the discontinuity in the magnetic field across the half-plane, so that from eqn. 2.18

$$J_s = \hat{y} \times (H(\rho, 0) - H(\rho, 2\pi)) \tag{88}$$

Evaluating the magnetic field on the half-plane we have $H_\rho = H_x$ and $H_\phi = 0$ and eqn. 87 reduces to

$$H = -\frac{E_z^i}{j\omega\mu}\frac{1}{x}\frac{\partial}{\partial\phi}(u^i - u^r)\hat{x} - \frac{k}{\omega\mu}E_\phi^i(u^i + u^r)\hat{z} \tag{89}$$

Solving eqn. 88 with eqn. 89 yields the exact currents J_x, J_z for the half-plane which may be written as

$$\begin{aligned} J_x &= J_x^{PO} + J_x^c \\ J_z &= J_z^{PO} + J_z^c \end{aligned} \tag{90}$$

where J_x^{PO}, J_z^{PO} are the currents due to the physical optics approximation and are given by

$$\begin{aligned} J_x^{PO} &= -\frac{2k}{\omega\mu}E_\phi^i \exp(jkx\cos\phi_0) \\ J_z^{PO} &= \frac{2k}{\omega\mu}E_z^i \sin\phi_0 \exp(jkx\cos\phi_0) \end{aligned} \tag{91}$$

The remaining terms J_x^c, J_z^c are the correction currents made to physical optics to yield the exact currents on the half-plane. These currents are given by

$$\begin{aligned} J_x^c &= -2J_x^{PO}\sqrt{\left(\frac{j}{\pi}\right)}F_-(v); \qquad v = \sqrt{(2kx)}\,|\cos\tfrac{1}{2}\phi_0| \\ J_z^c &= -2J_z^{PO}\left\{F_-(v) - \frac{\exp(-jv^2)}{2jv}\right\} \end{aligned} \tag{92}$$

The nature of the exact currents is shown by an example in Fig. 5.12. In the same way we could obtain correction currents for the wedge, but the expressions are not as tractable as for the half-plane. The approximation to the wedge currents given by Schretter and Bolle (1959) would, however, considerably simplify the expressions for the correction currents J_x^c, J_z^c.

As GTD extends the ray concepts of geometrical optics to include diffraction, the physical theory of diffraction (PTD) extends physical optics by an additional surface current to account for diffraction. For half-plane diffraction these additional currents are given by eqn. 92. In those cases where an asymptotic evaluation of the PTD integral is permissible, the GTD approach yields the same result much more simply and quickly. This is one reason why GTD has had considerably

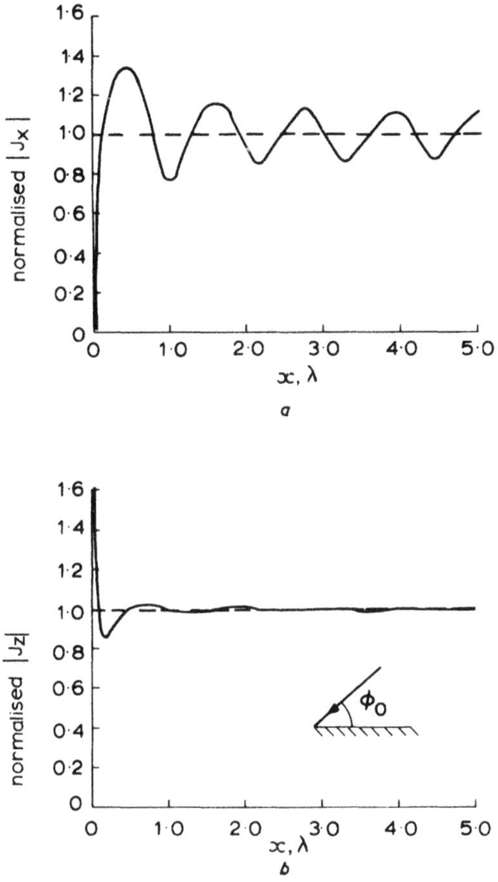

Fig. 5.12 Currents on the half-plane for plane wave incidence at $\phi_0 = 30$ deg.
——— exact
– – – – physical optics

more use than the PTD approach. A situation where this method finds use in place of GTD is at or near caustics, where infinities occur in the ray optical formulations. This question will be pursued further in Section 6.7.

5.7 Comparison of uniform theories

We noted earlier in Section 5.4, that the original GTD solution proposed by Keller gave edge diffraction coefficients that were non-uniform;

that is, they were invalid in the transition regions. In an attempt to provide a uniform solution that could be applied through the transition regions, the UTD approach discussed in Section 5.4 made a number of assumptions about the field behaviour. These were: (i) The Fresnel integral correctly describes the behaviour of the edge diffracted field in the transition regions; (ii) outside the transition regions the non-uniform (e.g. Keller GTD) solution remains valid; (iii) the total field must be continuous across the optical boundaries.

In the non-uniform theory the total electric field $\boldsymbol{E}^t$ can be written as

$$\boldsymbol{E}^t = \boldsymbol{E}^g + \boldsymbol{E}^d \tag{93}$$

where $\boldsymbol{E}^g$ is the geometrical optics field and $\boldsymbol{E}^d$ is the diffracted ray field as, for example, given by eqn. 40. The geometrical optics field $\boldsymbol{E}^g$ can be further split-up into incident and reflected ray fields $\boldsymbol{E}^i$ and $\boldsymbol{E}^r$. Furthermore, as noted earlier, the diffracted field consists of field components which can be associated with these incident and reflected ray fields. For our purposes here, it is convenient to rewrite eqn. 93 as

$$\boldsymbol{E}^t = \sum_{i,r} (U(\epsilon^{i,r})\boldsymbol{E}^{i,r} + \boldsymbol{E}^{d_{i,r}}) \tag{94}$$

where $\epsilon = 1$ for the illuminated region and -1 for the shadow region.

In order to satisfy the assumptions (i)–(iii) above, the approach taken in the UTD solution was to alter the argument ν in the modified Fresnel integral function $K_-(\nu)$ to ensure continuity of the field across the optical boundaries and to retrieve the Keller GTD expression outside the transition regions. The modification to eqn. 94 to achieve this is essentially a multiplication term applied to the non-uniform diffracted field components $\boldsymbol{E}^{d_{i,r}}$. If we define $\tilde{K}_-(\nu)$ as the asymptotic value of $K_-(\nu)$ for large values of ν, then the UTD solution of Section 5.4 is given in essence by

$$\boldsymbol{E}^t_{\mathrm{UTD}} = \sum_{i,r} \left(U(\epsilon^{i,r})\boldsymbol{E}^{i,r} + \frac{K_-(\nu^{i,r})}{\tilde{K}_-(\nu^{i,r})} \boldsymbol{E}^{d_{i,r}} \right). \tag{95}$$

This modification obviously retrieves the Keller non-uniform solution for large $\nu^{i,r}$. With $\nu^{i,r}$ defined by eqn. 43, ensures that the total field $\boldsymbol{E}^t_{\mathrm{UTD}}$ is continuous through the optical boundaries.

An alternative uniform solution to provide continuity of the field through the transitions was developed by Ahluwalia *et al.* (1968), Lewis and Boersma (1969) for a thin screen. It has subsequently been applied to a curved wedge by Ahluwalia (1970) and Lee and Deschamps (1976). This solution, referred to as the Uniform Asymptotic Theory (UAT), begins with the same assumptions (i)–(iii) above. However, a different hypothesis, or anzatz as it is sometimes called, is used to formulate the total field. In place of modifying the diffracted component (as in the UTD solution) the optical term in eqn. 94 is replaced by a new term so that the total field reads as

$$\boldsymbol{E}^t_{\mathrm{UAT}} = \sum_{i,r} (e^{-j\xi^2_{i,r}}[K_-(\xi_{i,r}) - \tilde{K}_-(\xi_{i,r})]\boldsymbol{E}^{i,r} + \boldsymbol{E}^{d_{i,r}}). \qquad (96)$$

The modified Fresnel integral argument $\xi^{i,r}$ is not the same as $\nu^{i,r}$ in UTD but is given by

$$\xi^{i,r} = -\epsilon^{i,r}k^{1/2}|\sqrt{s^i_0 + s^d - s^{i,r}}|$$

and is known as the detour parameter since it measures the excess ray path from the source or image to the observation point via the edge. In the illuminated region $\xi^{i,r} < 0$ and using $K_-(-x) = \exp(jx^2) - K_-(x)$ we obtain the geometrical optics field $\boldsymbol{E}^{i,r}$. Removed from the transition regions $[K_-(\xi^{i,r}) - \tilde{K}_-(\xi^{i,r})] \to 0$ and the non-uniform solution of eqn. (94) is obtained. On the shadow boundary the infinity in the Keller diffracted field $\boldsymbol{E}^{d_{i,r}}$ is exactly cancelled by the infinity in the term $\tilde{K}(\xi^{i,r})\boldsymbol{E}^{i,r}$ so that the total field is continuous through the shadow boundary. Another feature of UAT is the provision of a systematic method for determining the higher order terms in the asymptotic expansion of the field. This aspect of UAT was made use of in Section 5.5 to obtain the slope-diffraction term for use with UTD. In general however the computations required to obtain these higher-order terms become rapidly unmanageable, except in those situations where it becomes the leading term in the asymptotic expansion. In any case, taking more terms in an asymptotic expansion can sometimes give poorer results as a consequence of the divergent nature of asymptotic series.

For general wedge diffraction neither solution can claim superiority over the other. Both techniques are based on unproved hypotheses and both solutions are invalid in the region of caustics including the vicinity of the edge. In the transition region the solutions agree in the dominant k^0-term but give differing solutions for the $k^{-1/2}$ term; see Lee and Deschamp (1976). Outside the transition regions they both recover the Keller GTD non-uniform solution.

For the special case of diffraction by a half-plane a number of studies have been undertaken to compare UAT with UTD. When the incident field is an electromagnetic plane wave, the UAT solution was shown by Deschamps *et al.* (1984) to recover the exact solution including the field in the vicinity of the edge. On the other hand, the UTD solution as developed in Section 5.4 is only valid in the vicinity of the half-plane edge for the electric (magnetic) field when the plane-wave is at normal incidence with the electric (magnetic) field parallel to the edge. In either case the local plane wave approximation to the diffracted rays is no longer valid in the vicinity of the edge, and it is necessary to apply the two-dimensional field equations given by eqn. 2.29 otherwise significant errors in the field calculation near the edge may result, as demonstrated in Figure 11 of Deschamps *et al.* (1984). A similar comment was made by Aas (1979) when comparing the exact solution with the UTD solution for the field in the vicinity of the edge of a 90° wedge. Provided eqn. 2.29 is utilized in calculating the field near the edge, Aas demonstrated that UTD is capable of providing useful results at distances very close to the edge. (The comparison with UAT for this case was not given.) Note that the error in UTD increases for oblique incidence.

Returning to the half-plane, the two-dimensional problem of diffraction by an arbitrary cylindrical wave has been used to compare the uniform theories; Boersma and Lee (1977), Rahmat-Samii and Mittra (1978), Boersma and Rahmat-Samii (1980). In the last of these papers it is shown that the UAT solution is identical to a rigorous uniform asymptotic expansion for the total field, whereas the UTD solution only provides the complete leading term. The inclusion in the UTD formulation of the slope-diffraction term of Section 5.5 does provide the dominant part of the second order term. Specific calculations of the field in the vicinity of the edge and the shadow boundary undertaken by Rahmat-Samii and Mittra (1978), compared the exact solution with UAT and UTD. Provided that UTD included the slope-diffraction term the differences between the three solutions were shown to be numerically very small.

In the practical implementation of the two techniques, UTD is somewhat easier to use and has found much wider application in practice. A difficulty with the UAT solution is the subtraction of two large numbers that tend to infinity at the optical boundaries. For computational purposes it is necessary to take special limiting procedures at these boundaries. Another difficulty with UAT is the erroneous prediction of an infinite field at the image point of the source. Further, UAT to date has been restricted to edge diffraction whereas the UTD approach is applicable to diffraction by curved surfaces as we shall see in the next chapter. For non-edge diffraction the only change in the UTD hypotheses, as pointed out in Section 5.4, is to replace the Fresnel integral with the appropriate integral function to describe the field behaviour in transition regions.

Finally, for more background reading on UAT, the reader is referred to the review article by Lee in Chapter 3 of Uslenghi (1978), and for some recent applications see Menendez and Lee (1982), Sanyal and Bhattacharyya (1983).

5.8 Multiple edge diffraction

As we saw earlier, the edge diffracted field removed for the transition regions had the characteristics of a ray field. If this diffracted ray field is itself incident upon a second edge, then we can apply the GTD formulation as developed in Section 5.4. The first diffracted field becomes the incident field for the diffraction at the second edge. This process can be extended to include any number of subsequent diffractions. The ability to account for multiple, or higher order, edge diffractions in this way has enabled GTD to be successfully used for bodies somewhat less than a wavelength in extent where the theory will not normally be expected to work.

If the second edge lies in the transition region of the diffracted field from the first edge, then we have a difficulty. In the transition regions the field diffracted by an edge is described in terms of the Fresnel integral. In such regions the diffracted field does not have the characteristics of a ray field and therefore any subsequent diffraction, reflection or refraction of this field cannot be treated by a straightforward application of GTD as formulated so far. Diffraction of the transition region field arises in a number of practical situations and therefore a solution to this problem is of considerable interest.

One of the first successful solutions to this problem was obtained by Jones (1973) who used a Wiener–Hopf technique to solve for plane wave diffraction incident upon two staggered parallel half-planes as in Fig. 5.13. The solution of Jones was subsequently verified by Tischenko and Khestanov (1974) who also gave a solution for cylindrical wave incidence. By expanding the Fresnel integral into an infinite series of elementary waves whose individual solutions are known, Boersma (1975) was (via a UAT approach) able to obtain the result of Jones for the case where the transition region of the two edges coincided ($\xi_2 = 0$) in Fig. 5.13). Rahmat-Samii and Mittra (1977) applied STD to the problem illustrated in Fig. 5.13 and gave results for specific cases. In a similar approach to Boersma (1975), Tiberio and Kouyoumjian (1982) expressed the singly diffracted field from a wedge, as calculated by the UTD formulation, as a superposition of slowly varying inhomogeneous waves, and summed the individual

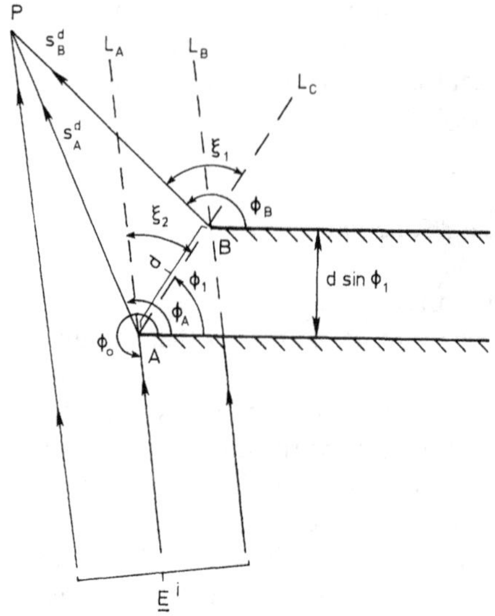

Fig. 5.13 Two staggered parallel plates illuminated by a normally incident plane wave.

diffraction of these waves by UTD to account for double diffraction between wedges. They refer to this technique as the *extended spectral ray method.* The method is applied to a number of examples in Tiberio and Kouyoumjian (1984) and the results compared with those obtained by other methods.

None of the above solutions are exact – they all ignore multiple interaction between the plates – and in some instances only specific cases are given. The most complete solution is that of Jones (1973) and we take his results as the solution to the canonical problem of double knife-edge diffraction. Based on this canonical problem we obtain UTD-type solutions for diffraction by an astigmatic ray-optics field incident on two nearby planar wedges. The extension to curved wedges will be discussed in the next chapter.

The canonical problem is illustrated in Fig. 5.13. A two-dimensional plane wave $\boldsymbol{E}^i$ is incident from the bottom half-space (i.e. $\pi < \phi_0 < 2\pi$) upon two parallel staggered half planes A and B. The line L_C normal to the edges that connects with the two planes is at an angle ϕ_1 to the two parallel surfaces separated by the distance $d \sin \phi_1$. The lower plate A creates a shadow boundary along the line L_A, and similarly the upper plate B, when directly illuminated by the incident plane wave, creates the shadow boundary L_B. We are interested here in the behaviour of the field where ϕ_0 approaches $\pi + \phi_1$ so that the three boundaries L_A, L_B and L_C gradually coalesce.

We begin by considering the diffracted field at the point P from the edge of screen A in the absence of the upper half-plane. From Section 5.4 this edge diffracted field E_A^d for plane wave illumination can be written (where the subscript A refers to the diffracting edge) as

$$E_A^d(s_A^d) = \mathbf{D}_A E^i(A) \frac{\exp(-jks_A^d)}{\sqrt{s_A^d}}. \tag{97a}$$

The elements D_A^e and D_A^m of the edge diffracting matrix are given by

$$D_A^{e,m} = -\sqrt{s_A^d}\,\{\epsilon_A^i K_-(\nu_A^i) \mp \epsilon_A^r K_-(\nu_A^r)\} \tag{97b}$$

where

$$\nu_A^{i,r} = \sqrt{2ks_A^d}\,|\cos\tfrac{1}{2}(\phi_A \mp \phi_0)|, \qquad \epsilon_A^{i,r} = \text{sgn}\,[\cos\tfrac{1}{2}(\phi_A \mp \phi_0)]$$

Similarly, diffraction from the edge of screen B to the field point P in the absence of the lower half-plane is given by

$$E_B^d(s_B^d) = \mathbf{D}_B E^i(B) \frac{\exp(-jks_B^d)}{\sqrt{s_B^d}} \tag{98}$$

where

$$D_B^{e,m} = -\sqrt{s_B^d}\,\{\epsilon_B^i K_-(\nu_B^i) \mp \epsilon_B^r K_-(\nu_B^r)\}$$

$$\nu_B^{i,r} = \sqrt{2ks_B^d}\,|\cos\tfrac{1}{2}(\phi_B \mp \phi_0)|, \qquad \epsilon_B^{i,r} = \text{sgn}\,[\cos\tfrac{1}{2}(\phi_B \mp \phi_0)]$$

Removed from the optical boundaries the leading term of these diffracted fields is of order $k^{-1/2}$.

When the two screens are present we make an initial approximation to the total field by the superposition of eqns. 97 and 98 together with the geometrical optics field. For the region above screen B, taking into account the various shadowing effects that can occur when the two screens are present, we write our first approximation to the total field E^t at P as

$$E^t(P) = U(\epsilon_1)E_A^d + U(\epsilon_2)E_B^d + U(\epsilon_3)E^i, \tag{99}$$

where the unit amplitude quantity $\epsilon_1 = \text{sgn}\,\xi_1$ and is used to account for the blockage effect of screen B on the diffracted field from screen A. The unit amplitude term $\epsilon_2 = +1$ if edge B is directly illuminated, -1 otherwise and accounts for the blockage effects of screen A. Finally, $\epsilon_3 = \epsilon_A^i(\epsilon_B^i)$ when $\epsilon_2 = -1(+1)$.

Equation 99 does not account for any interaction between the two screens. Provided the edge at B is well-removed from the transition regions of the diffracted field from the lower screen edge, i.e., $|\xi_2| \gg 0$, higher-order diffraction from A to B can be included by a straightforward application of the GTD formulation given in Section 5.4. For this higher-order diffraction we take the incident field on screen B as coming from the line source diffracted field situated at edge A. This subsequent diffracted field from this mechanism will give terms of order $k^{-1/2}$ along the boundary L_C and of order k^{-1} when well-removed from this boundary.

To provide a valid and uniform GTD solution when the edge B is illuminated by the transition region of the diffracted field from A, we use the approach taken in James and Poulton (1979) and make direct use of the asymptotic solution to this problem as given by eqns. 9 and 44 in Jones (1973). Although at first sight this solution by Jones appears formidable, it simplifies significantly if we restrict the result to the dominant diffraction effects by including only those terms up to $k^{-1/2}$. This yields a solution sufficient for many engineering applications, and involves the generalized Fresnel integral function $L_-(x, y)$ described in Section 2.2.1. The result for the diffracted field is given by the sum of eqns. 97 and 98 where the transition functions $K_-(x)$ are replaced by the transition function $L_-(x, y)$. Specifically the total field is written as

$$E^t(P) = \mathbf{D}_A E^i(A)\frac{\exp(-jks_A^d)}{\sqrt{s_A^d}} + \mathbf{D}_B E^i(B)\frac{\exp(-jks_B^d)}{\sqrt{s_B^d}} + U(\epsilon_3)E^i(P), \tag{100}$$

with the terms $D^{e,m}$ now given as

$$\begin{aligned} D_A^{e,m} &= -\sqrt{s_A^d}\,\{\epsilon_A^i L_-(x_1, \nu_A^i) \mp \epsilon_A^r L_-(x_1, \nu_A^r)\} \\ D_B^{e,m} &= -\sqrt{s_B^d}\,\{\epsilon_B^i L_-(x_2, \nu_B^i) \mp \epsilon_B^r L_-(x_2, \nu_B^r)\} \end{aligned} \tag{101a}$$

where

$$\begin{aligned} x_1 &= \epsilon_1 \left[\frac{2kds_B^d}{d + s_B^d}\right]^{1/2} |\sin \tfrac{1}{2}\xi_1| \\ x_2 &= \epsilon_2 \sqrt{2kd}\,|\sin \tfrac{1}{2}\xi_2| \end{aligned} \tag{101b}$$

The angles ξ_1 and ξ_2 are defined in Fig. 5.13. In eqn. 100 we have a uniform solution complete to the $k^{-1/2}$ term for the total field in the region above screen B.

In discussing this solution we need to know the properties of the function $L_-(x, y)$. From Section 2.2.1 we note the following: For x large

$$L_-(x, y) \sim U(x)K_-(y) + 0(k^{-1}) \tag{102a}$$

and for $x = 0$

$$L_-(0, y) = \tfrac{1}{2}K_-(y) + \frac{1}{2\pi} \lim_{x, y \to 0} \tan^{-1} \frac{x}{y}. \tag{102b}$$

The second term in eqn. 102*b* is only required when y goes to zero with x.

If the boundaries L_A, L_B and L_C are well separated and the absolute value of the angle ξ_1 is also large, then substituting eqn. 102*a* for $L_-(x, y)$ into eqn. 100 retrieves the solution given by eqn. 99. When the field point P lies along the line L_C then $x_1 = 0$ and substituting eqn. 102*b* into the expression for $\mathbf{D}_A$ yields the half-plane diffracted field multiplied by the factor of $\frac{1}{2}$ necessary to ensure continuity of the $k^{-1/2}$ term. Although this continuity does not occur in eqn. 99 it could have been included by accounting for the higher-order diffraction from A to B as discussed earlier.

Of more immediate interest is the situation where the lines L_A, L_B and L_C coalesce. In this case with $x_2 = 0$ the factor of $\frac{1}{2}$ applies to the half-plane diffracted field from edge B. When the field point P lies along the common boundary line of L_A, L_B and L_C then $x_1 = x_2 = \nu_A^i = \nu_B^i = 0$. Since x_1 approaches zero with ν_A^i it is necessary to take the limit case of eqn. 102*b*. This yields

$$L_-(x_1, \nu_A^i) \xrightarrow[x_1, \nu_A^i \to 0]{} \frac{1}{4} + \frac{\epsilon_1}{2\pi} \lim_{\xi_1 \to 0} \tan^{-1}\left\{\left[\frac{d s_B^d}{(d + s_B^d) s_A^d}\right]^{1/2} \left|\frac{\xi_1}{\phi_A - \phi_1}\right|\right\}$$

$$= \frac{1}{4} + \frac{\epsilon_1}{2\pi} \tan^{-1}\left[\frac{d}{s_B^d}\right]^{1/2}.$$

as a consequence of $d + s_B^d = s_A^d$ and

$$\left|\frac{\xi_1}{\phi_A - \phi_1}\right| \xrightarrow[\xi_1 \to 0]{} \frac{s_A^d}{s_B^d}.$$

Substituting into eqn. 100 we get

$$E^t(P) = -\epsilon_A^i \left(\frac{1}{4} + \frac{\epsilon_1}{2\pi} \tan^{-1} \sqrt{\frac{d}{s_B^d}}\right) E^i(A) \exp(-jk s_A^d)$$
$$- \epsilon_B^i \tfrac{1}{4} E^i(B) \exp(-jk s_B^d) + U(\epsilon_A^i) E^i(P) + 0(k^{-1/2})$$

which, for the current situation and using the relationship $\tan^{-1} 1/\chi = \pi/2 - \tan^{-1} \chi$, finally reduces to

$$E^t(P) = E^i(P)\left[\frac{1}{4} + \frac{1}{2\pi} \tan^{-1} \sqrt{\frac{s_B^d}{d}}\right] + 0(k^{-1}). \qquad (103)$$

In this solution the dominant field component is not simply $\frac{1}{4}E^i(P)$ as would be predicted by a straightforward application of GTD formulated for a single edge, but has an additional term that only goes to zero with s_B^d/d. In fact we see from eqn. 103 that in the far field as $s_B^d \to \infty$ the dominant field component is $\frac{1}{2}\boldsymbol{E}^i$ as is the case for a single edge.

From the canonical solution for the problem depicted in Fig. 5.13 we now use the approach used in Section 5.4 to obtain a uniform GTD solution for the more general case illustrated in Fig. 5.14. That is, by demanding continuity of the field across the shadow boundaries we get a UTD-type solution for diffraction by an astigmatic ray optics field incident on double knife-edge screens. Assume an incident field as given by eqn. 4.26 (in Fig. 5.14 we assume for the moment normal incidence where $\theta_0 = \pi/2$)

$$\boldsymbol{E}^i(s^i) = \boldsymbol{E}^i(q)\left[\frac{\rho_1\rho_2}{(\rho_1 + s^i)(\rho_2 + s^i)}\right]^{1/2} \exp(-jks^i),$$

where the distance s^i along the ray is measured from a given field point q, and ρ_1, ρ_2 are principal radii of curvature of the incident field. (Note that as we are dealing here with planar surfaces we do not require the superscript i to distinguish between the incident and reflected geometrical optics field principal radii of curvature; see Section 5.4.) The total field is similar to eqn. 100, and using the notation of Section 5.4, is given by

$$\boldsymbol{E}^t(P) = \mathbf{D}_A \boldsymbol{E}^i(A)\left[\frac{\rho_{3A}}{s_A^d(\rho_{3A} + s_A^d)}\right]^{1/2} \exp(-jks_A^d)$$

$$+ \mathbf{D}_B \boldsymbol{E}^i(B)\left[\frac{\rho_{3B}}{s_B^d(\rho_{3B} + s_B^d)}\right]^{1/2} \exp(-jks_B^d) + \boldsymbol{E}_{\text{g.o.}}(P) \tag{104}$$

where, as before, the subscripts A and B refer to the appropriate edge, and $\boldsymbol{E}_{\text{g.o.}}$ is the geometrical optics field. In this equation the terms $D^{e,m}$ in the edge diffraction matrices are given by

$$\begin{aligned} D_A^{e,m} &= -\sqrt{\sigma_A}\,\{\epsilon_A^i L_-(x_1, \nu_A^i) \mp \epsilon_A^r L_-(x_1, \nu_A^r)\} \\ D_B^{e,m} &= -\sqrt{\sigma_B}\,\{\epsilon_B^i L_-(x_2, \nu_B^i) \mp \epsilon_B^r L_-(x_2, \nu_B^r)\} \end{aligned} \tag{105}$$

The quantities σ, $\nu^{i,r}$, ρ_3 for each edge are as defined in Section 5.4, where for a straight edge $\sigma^i = \sigma^r = \sigma$. As in the previous example for plane wave illumination, we see that in this more general case *we initially formulate the problem with the edge diffracted field from each screen considered in isolation. To account for the effect of transition field diffraction from edge A to edge B in Fig. 5.14, we replace the*

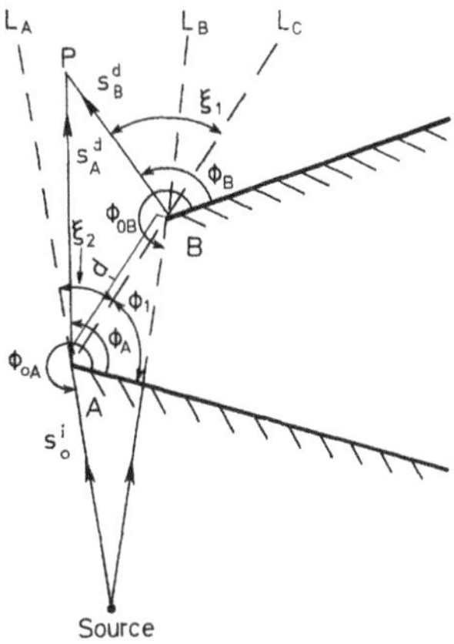

Fig. 5.14 Double knife-edge diffraction.

modified Fresnel integral function $K_-(\chi)$ with the generalised Fresnel integral function $L_-(x, y)$. Specifically the substitutions are

$$K_-(\nu_A^{i,r}) \to L_-(x_1, \nu_A^{i,r})$$
$$K_-(\nu_B^{i,r}) \to L_-(x_2, \nu_B^{i,r}) \tag{106}$$

where $\nu^{i,r}$ is as defined earlier for single edge diffraction and x_1 and x_2 are given by

$$x_1 = \epsilon_1 \left[\frac{2kds_B^d}{d + s_B^d}\right]^{1/2} |\sin \tfrac{1}{2}\xi_1|$$
$$x_2 = \epsilon_2 \left[\frac{2kds_0^i}{d + s_0^i}\right]^{1/2} |\sin \tfrac{1}{2}\xi_2| \tag{107}$$

As before ϵ_1 is used to account for the blockage effect of screen B on the diffracted field from screen A, i.e. $\epsilon_1 = +1$ when no blockage takes place, otherwise $\epsilon_1 = -1$. The term $\epsilon_2 = +1$ if edge B is directly illuminated by the source and is -1 otherwise. The only change in the quantities given in eqn. 101*b* is in the value of x_2. Here we have made use of the expression given by Tishchenko and Khestanov (1974) for diffraction by two half-planes illuminated by a line source. The distance s_0^i is measured along the ray from the source to the edge at A. If the incident field is astigmatic then the value of s_0^i is taken (in a heuristic sense since it has no bases from a rigorous solution) as the curvature of the incident wavefront in the plane of the diffracting edges. Furthermore, although the current formulation could in principle be adopted for oblique incidence when $\theta_0 \neq \pi/2$ (see Section 5.3), most cases of interest are covered by the present formulation.

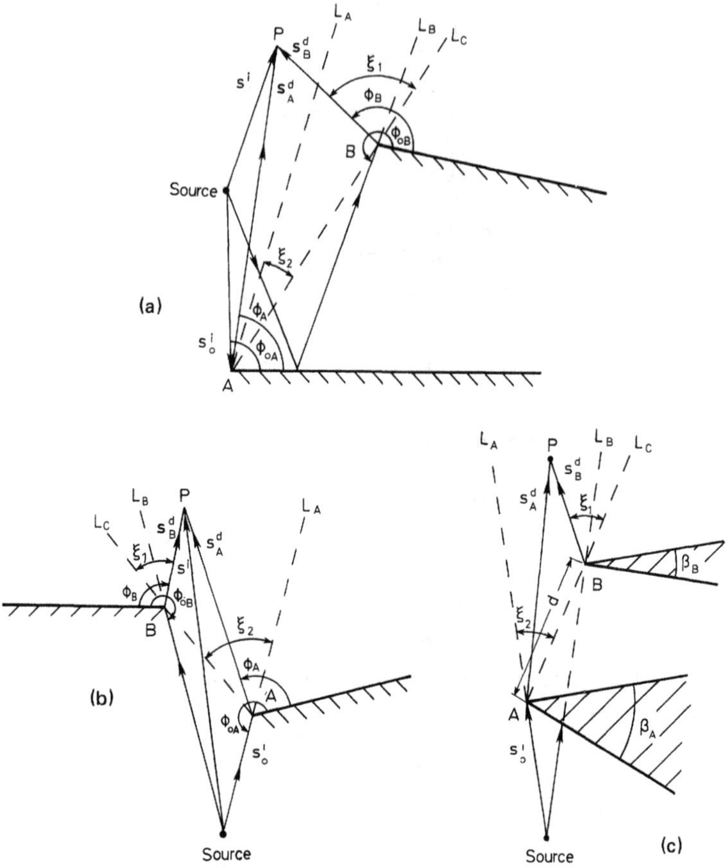

Fig. 5.15 Double knife-edge configurations.

When the shadow boundaries coalesce and the field point is also on this common shadow boundary, we can proceed as above for plane-wave incidence and obtain the total field for cylindrical or spherical wave incidence as

$$E^t(P) = E^i(P)\left[\frac{1}{4} + \frac{1}{2\pi}\tan^{-1}\left[\frac{s_0^i s_B^d}{d(s_0^i + s_A^d)}\right]^{1/2}\right] + 0(k^{-1/2}). \quad (108)$$

The formulation developed in eqns. 104–107 can be applied directly to the situations illustrated in Fig. 5.15. In the case of Fig. 5.15*a*, L_A is now the reflection boundary from screen A and L_B is the shadow boundary created by the reflected geometrical optics field from screen

A. Thus in the initial formulation for the diffracted field from edge B in eqn. 104, the incident field is taken as this reflected field, or image of the source, from the lower screen. In this case to complete the solution to the $k^{-1/2}$ term it will be necessary to superimpose the diffracted field resulting from the direct illumination of edge B by the source. For the configuration in Fig. 5.15b eqns. 104–107 can be applied directly. Here the equivalent expression to eqn. 108 gives

$$E^t(P) = E^i(P)\left[\frac{1}{4} - \frac{1}{2\pi}\tan^{-1}\left[\frac{s_0^i s_B^d}{d(s_0^i + s_A^d)}\right]^{1/2}\right] + 0(k^{-1/2}), \quad (109)$$

which is in agreement with the result given by Lee (1978). In the case of double edge diffraction involving wedges – an example is shown in Fig. 5.15c – the substitutions of eqns. 106–107 can be applied to the wedge diffraction expressions without further modification. An application of eqns. 104–107 to a practical problem is given by James (1980) where they are used in analysing the radiation pattern of a Cassegrain antenna.

Solutions for diffraction by more than two knife-edges are not yet as well-formulated as the double knife-edge case. Lee (1978), for example, has given results for the leading k^0-term for the field along the common shadow boundary for certain special cases. A solution for any number of knife edges has been developed by both Kaloshin (1982) and Vogler (1982) and involves a multiple integral which in general requires to be evaluated numerically. More in the context of ray methods is the solution of Whitteker (1984) where the Fresnel–Kirchhoff approximation is used to calculate the diffraction attenuation due to multiple knife-edges.

5.9 Diffraction by an impedance wedge

We have so far considered edge diffraction from perfectly conducting wedges. However in many applications diffraction from edges in surfaces other than perfectly conducting is of considerable interest. In the situation where the wedge faces have an impedance boundary condition, a rigorous solution for plane wave diffraction at normal incidence was given in the succinct paper of Maliuzhinets (1958). The solution is of necessity more complicated than for a perfectly conducting wedge and the functions describing the field are not in general readily evaluated. However in a number of cases, specifically the half-plane, the 90° wedge, the flat plane, and the 270° wedge, the solution reduces to a form readily amenable to evaluation. The work of Maliuzhinets has been taken up by several authors and applied to a number of problems. The impedance half-plane result was adapted by Bowman (1967) to analyse the backscatter properties of infinite strips with arbitrary face impedances under broadside illumination. His results were supported by measured data. The diffracted field solution used by Bowman was non-uniform giving, as in the Keller GTD formulation, infinite values for the field along optical boundaries. A uniform solution for an impedance half-plane was given by Bucci and Franceschetti (1976) who also considered the case of oblique incidence. Note that with impedance boundary conditions the extension to oblique incidence presents a major difficulty since the field components are coupled. This is in contrast to the perfectly conducting case where, as shown in Section 5.3, the problem of oblique incidence is a simple extension of the solution for normal incidence.

Scattering by an impedance strip was also considered by Tiberio *et al.* (1982) but for the more difficult case of edge-on incidence. Here higher-order diffraction between the edges of the strip involves the diffraction of transition region fields discussed in the previous section. The solution used was the extended spectral ray method mentioned in that Section. The same approach, in association with the flat-plane solution of Maliuzhinets, was used by Tiberio and Pelosi (1983) to analyse scattering from the edges of impedance discontinuities on a flat plane. Surface wave diffraction was also included in this paper.

For the general wedge diffraction problem, uniform asymptotic solutions have been obtained by Mitsmakher (1976), James (1977) and Tiberio *et al.* (1985). We will consider these uniform solutions after studying the Maliuzhinets solution in some detail.

Consider a plane wave to be at normal incidence to the edge at an angle ϕ_0 to a face of the impedance wedge of angle β shown in Fig. 5.16. The wedge is independent of the $\hat{z}$-direction (which lies along the edge) and has given constant surface impedances Z_A, Z_B along the wedge faces as shown. From Section 3.1.5 we can define complex Brewster angles $\nu^e(\nu^m)$, for the case when the electric (magnetic) field of the incident wave is parallel to the edge, as

$$\nu^e_{A,B} = \cos^{-1}(Z_0/Z_{A,B}); \qquad \nu^m_{A,B} = \cos^{-1}(Z_{A,B}/Z_0), \quad (110)$$

where Z_0 is the impedance of the medium surrounding the wedge. The field component V^i_z of the incident plane wave in the $\hat{z}$-direction is given by

$$V^i_z = \exp(jk\rho\cos(\phi - \phi_0))$$

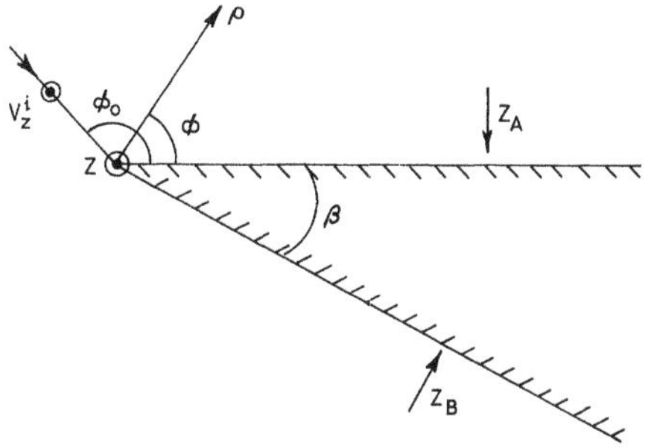

Fig. 5.16 Wedge with impedance faces illuminated by a normally incident plane wave.

where for electric polarization $V_z^i = E_z^i$ and for magnetic polarization $V_z^i = H_z^i$. The exact solution for the total field in the region surrounding the wedge can be obtained from Maliuzhinets (1958) as

$$V_z(\rho, \phi) = \frac{\sin(\phi_0/N)}{2\pi j N \Psi^{e,m}(\phi_0)} \int_C \frac{\Psi^{e,m}(\alpha+\phi)\exp(jk\rho\cos\alpha)d\alpha}{\cos[(\alpha+\phi)/N] - \cos(\phi_0/N)} \tag{111}$$

where $N = (2\pi - \beta)/\pi$, C is the contour in the complex α-plane as shown in Fig. 5.17, and

$$\Psi^{e,m}(z) = \psi_N(z + \nu_B^{e,m})\psi_N(z - \nu_B^{e,m})\psi_N(z - N\pi + \nu_A^{e,m}) \times \psi_N(z - N\pi - \nu_A^{e,m}).$$

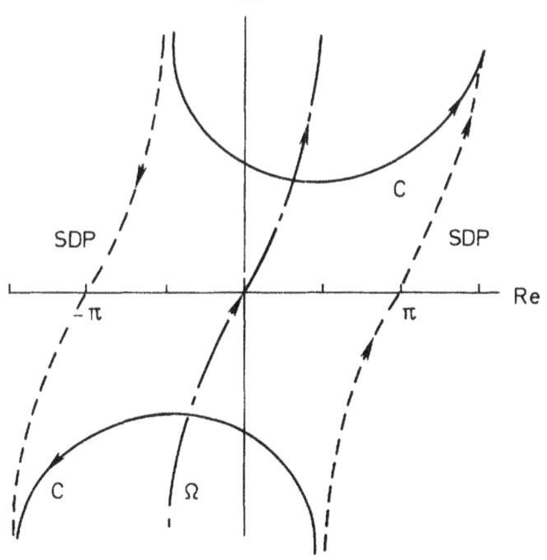

Fig. 5.17 Integration contours in the complex plane.

The function $\psi_N(z)$ ($\Psi_\Phi(z)$ in Maliuzhinets) has the following properties:

$$\psi_N(z) = \exp\left[-\frac{1}{j4\pi N}\int_0^z\int_{-j\infty}^{j\infty}\tan\frac{\alpha}{2N}\frac{d\alpha\, d\mu}{\cos(\alpha-\mu)}\right] \tag{112a}$$

$$\psi_N(z) = \psi_N(-z) \tag{112b}$$

$$\psi_N(z+\pi/2)\psi_N(z-\pi/2) = \psi_N^2(\pi/2)\cos(z/(2N)) \tag{112c}$$

$$\psi_N(z+N\pi)/\psi_N(z-N\pi) = \mathrm{ctg}\,\tfrac{1}{2}(z+\pi/2) \tag{112d}$$

The function $\psi_N(z)$ is, in general, difficult to calculate. However, in some important special cases, namely when $N = 2, \frac{3}{2}, 1, \frac{1}{2}$, the function simplifies considerably. These are as follows:

(i) *Half-plane* ($N = 2$)

$$\psi_2(z) = \exp\left[-\frac{1}{8\pi}\int_0^z \frac{\pi\sin\alpha - 2\sqrt{2}\pi\sin(\alpha/2) + 2\alpha}{\cos\alpha}\, d\alpha\right]$$

Simple approximate but accurate solutions have been given for this function by Volakis and Senior (1985). They begin by noting the recurrence relations

$$\psi_2(z) = 0.932422\cos\left(\frac{z}{4}-\frac{\pi}{8}\right)\Big/\psi_2(z-\pi)$$

$$\psi_2(-z) = \psi_2(z) \tag{113}$$

$$\psi_2(z^*) = \psi_2^*(z)$$

With $z = x + jy$ and using eqn. 113, $\psi_2(z)$ can be determined throughout the entire complex plane from a knowledge of the function in the region $0 \leqslant x \leqslant \pi/2$, $y \geqslant 0$. Within this region the function is given by approximate expressions where the amplitude error is kept within 0.8% and the phase error within 2.4%. The approximations are

$$\psi_2(z) \approx 1 - 0.0139z^2, \qquad y \leqslant 8.$$

$$\psi_2(z) \approx 1.05302\{\cos\tfrac{1}{4}(z - j0.69315)\}^{1/2}, \quad y > 8 \quad 0 \leqslant x \leqslant \pi/2 \tag{114}$$

(ii) *90° wedge* ($N = 3/2$)

$$\psi_{3/2}(z) = \left[\frac{4}{3}\cos\left(\frac{z-\pi}{6}\right)\cos\left(\frac{z+\pi}{6}\right)\right]\Big/\cos\frac{z}{6} \tag{115}$$

(iii) *Flat-plane* ($N = 1$)

$$\psi_1(z) = \exp\left[\frac{1}{4\pi}\int_0^z \frac{2\alpha - \pi \sin\alpha}{\cos\alpha}\, d\alpha\right] \tag{116}$$

(iv) *270° wedge* ($N = \frac{1}{2}$)

$$\psi_{1/2}(z) = \cos\frac{z}{2} \tag{117}$$

The solution for the total field given by eqn. 111 is not particularly convenient as it stands. To cast it into a more useful form, the contour C is deformed by Maliuzhinets into the paths of steepest descent (SDP as shown in Fig. 5.17) to yield

$$V_z(\rho, \phi) = \frac{\sin\dfrac{\phi_0}{N}}{2\pi j N \Psi^{e,m}(\phi_0)} \int_{\text{SDP}} \frac{\Psi^{e,m}(\alpha+\phi)\exp(jk\rho\cos\alpha)\,d\alpha}{\cos\dfrac{\alpha+\phi}{N} - \cos\dfrac{\phi_0}{N}}$$

$$+ u_0^i + u_A^r + u_B^r + S_A + S_B \tag{118}$$

The first term is the diffracted field u^d while the remaining five terms are pole residue terms giving the geometrical optics incident field u_0^i, the geometrical optics reflected fields from the wedge faces u_A^r, u_B^r, and surface waves S_A, S_B, travelling away from the edge along each wedge face. The surface wave terms will not be considered further here. Furthermore, although the incident angle ϕ_0 can in general be complex, we will restrict our discussion to real ϕ_0.

The geometrical optics components can be written as

$$u_0^i = \exp(jk\rho\cos(\phi - \phi_0)) \quad \text{for} \quad |\phi - \phi_0| < \pi$$

$$u_A^r = R_A^{e,m}\exp(jk\rho\cos(\phi + \phi_0)) \quad \text{for} \quad |\phi + \phi_0| < \pi$$

$$u_B^r = R_B^{e,m}\exp(jk\rho\cos(\phi + \phi_0 - 2\pi N)) \quad \text{for} \quad |\phi + \phi_0 - 2\pi N| < \pi \tag{119a}$$

where (as in section 3.1.5)

$$R_A^{e,m} = \frac{\sin\phi_0 - \cos\nu_A^{e,m}}{\sin\phi_0 + \cos\nu_A^{e,m}}$$

$$R_B^{e,m} = \frac{\sin(N\pi - \phi_0) - \cos\nu_B^{e,m}}{\sin(N\pi - \phi_0) + \cos\nu_B^{e,m}} \tag{119b}$$

Turning now to the diffracted field, an asymptotic expansion of the integral in eqn. 118 yields

$$u^d \sim \frac{\frac{2}{N}\sin\frac{\phi_0}{N}}{\Psi^{e,m}(\phi_0)}\left\{\frac{\Psi^{e,m}(\phi+\pi)}{\cos\frac{\phi+\pi}{N}-\cos\frac{\phi_0}{N}}-\frac{\Psi^{e,m}(\phi-\pi)}{\cos\frac{\phi-\pi}{N}-\cos\frac{\phi_0}{N}}\right\}\frac{\exp(-jk\rho)}{\sqrt{8j\pi k\rho}} \tag{120}$$

which is the non-uniform result for the diffracted field given by Maliuzhinets. For a perfectly conducting wedge eqn. 120 can be shown to be identical to the non-uniform solution for plane wave diffraction at a wedge given in Section 5.2.

To obtain a uniform solution to the diffracted field we return to the integral in eqn. 118. For the SDP through $\alpha = \pi(-\pi)$ we make the substitution $\alpha = \pi + \xi(\xi - \pi)$, and using the relation $\cot(x+y) + \cot(x-y) = 2\sin 2x/(\cos 2y - \cos 2x)$ the diffracted field term u^d from eqn. 118 becomes

$$u^d = \frac{1}{4\pi j N\Psi^{e,m}(\phi_0)}\int_\Omega \left\{\Psi^{e,m}(\phi+\xi+\pi)\left[\cot\frac{\pi+\xi+\Phi^r}{2N} - \cot\frac{\pi+\xi+\Phi^i}{2N}\right] + \Psi^{e,m}(\phi+\xi-\pi)\right.$$
$$\left.\times\left[\cot\frac{\pi-\xi-\Phi^r}{2N} - \cot\frac{\pi-\xi-\Phi^i}{2N}\right]\right\}$$
$$\times \exp(-jk\rho\cos\xi)d\xi \tag{121}$$

where $\Phi^{i,r} = \phi \mp \phi_0$, and Ω is the integration contour shown in Fig. 5.17. When $k\rho \gg 1$ the major contribution to this integral will occur in the vicinity of the saddle point where $\xi = 0$. Making this approximation for the amplitude terms in eqn. 121 we can note an immediate similarity of the four terms in this equation with the function $f(\Phi^{i,r})$ that occurs in diffraction by a perfectly conducting wedge. This function is defined in eqn. 3.107 and its subsequent asymptotic evaluation given by eqn. 3.110. Making use of this function we can express the diffracted field for the impedance wedge as

$$u^d \approx [\Psi^{e,m}(\phi_0)]^{-1}\{\Psi^{e,m}(\phi+\pi)[f(\Phi^i)_{n=-1} - f(\Phi^r)_{n=-1}]$$
$$+ \Psi^{e,m}(\phi-\pi)[f(-\Phi^i)_{n=0} - f(-\Phi^r)_{n=0}]\} \tag{122}$$

which is the uniform expression for the diffracted field given by both Mitsmalcher (1976) and James (1977). Removed from the optical boundaries the asymptotic value of $f(x)$ is

$$f(x) \sim -\frac{1}{N}\cot\left(\frac{\pi+x}{2N}\right)\frac{\exp(-jk\rho)}{\sqrt{8j\pi k\rho}}$$

and substitution into eqn. 122 retrieves the non-uniform solution given by eqn. 120.

The solution was for a plane wave normally incident upon the wedge. We can extend this result to an astigmatic field incident on an impedance wedge by the approach taken in Section 5.4 for a perfectly conducting wedge. This gives the edge-diffracted field as in eqn. 46 but with the diffraction matrix components now given by

$$\begin{aligned} D^{e,m} = {} & [\Psi^{e,m}(\phi_0)]^{-1}\{\Psi^{e,m}(\phi+\pi)[h(\Phi^i)-h(\Phi^r)] \\ & + \Psi^{e,m}(\phi-\pi)[h(-\Phi^i)-h(-\Phi^r)]\} \end{aligned} \tag{123}$$

where the function $h(\Phi^{i,r})$ is given by eqn. 47*b*. Eqn. 123 only applies for normal incidence ($\theta_0 = \pi/2$) as the extension to oblique incidence for impedance boundary conditions is not, as noted above, the simple matter it is for perfectly conducting surfaces.

The uniform solution of eqn. 123 reduces to the perfectly conducting wedge uniform solution for electric polarization. For magnetic polarization, however, deficiencies occur in the solution when the surface impedance is small or zero. The main difficulty arises if any two optical boundaries are close together. Compared to the results of Section 5.4 for a perfectly conducting wedge illuminated by a magnetically polarized wave, eqn. 123 (and also eqn. 122) gives increasingly large errors (although still remaining a 'uniform' solution) in the vicinity of optical boundaries as the angular separation between adjacent boundaries becomes less than about 60°. Figure 5.18 shows an example of a 90° wedge illuminated by a magnetic line source. Note that this discrepancy only occurs in the transition regions and that removed from optical boundaries the two solutions are in complete agreement.

Recently Tiberio *et al.* (1985) have undertaken an alternative asymptotic expansion of the Maliuzhinets diffraction integral which overcomes this difficulty with the D^m term in eqn. 123. Although the result is more complicated than eqn. 123 it may be necessary to use their solution in the above circumstances. Only plane-wave incidence is considered in Tiberio *et al.* (1985). However, for a general astigmatic incident wave, their solution can be made uniform in the same way as for eqn. 123 by a change of argument in the Fresnel integral transition function. Specifically we let D^m be given by eqns. 9 to 15 in Tiberio

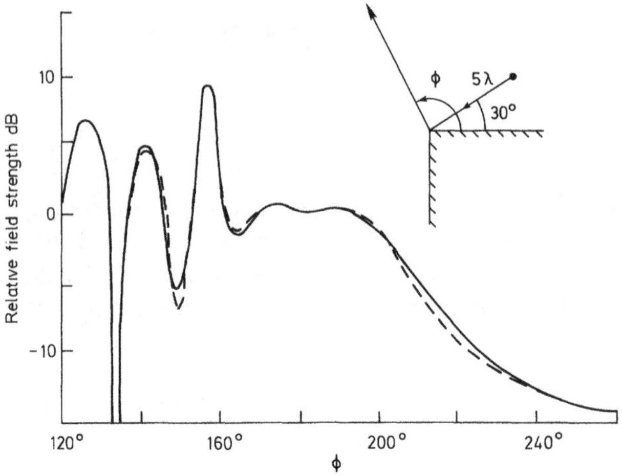

Fig. 5.18 Total field for a magnetically polarized line source near a perfectly conducting wedge with alternative expressions used to calculate the diffracted field component.
——— usual GTD (as in section 5.4)
– – – using eqn. 123 for the diffracted field.

et al. (1985) and substituting $K = k\sigma$ in eqn. 15 where, for straight wedge faces, $\sigma = \sigma^i = \sigma^r$ and given by eqn. 43*b* in Section 5.4.

The geometrical optics field reflected from the wedge faces are determined in the usual way (see Chapter 4) using the reflection coefficients of eqn. 119*b*. To complete the solution the surface wave, if excited, should be included in the field. However, if the field point is well-removed from the wedge faces, the surface wave contribution will be insignificant since its intensity decays exponentially with distance from the surface.

As a simple example of the uniform solution for the diffracted coefficients for an impedance wedge given by eqn. 123, Fig. 5.19 shows the results for a 90° wedge illuminated by a line source near the edge. The Brewster angle for face B is fixed at $\pi/2$ while the other face assumes various impedance values related through the Brewster angle ν_A. With $\nu_A = \pi/2 - \vartheta_A$ we note an initial rapid change in the diffracted field as the imaginary value of ϑ_A increases from zero. For large values of ϑ_A, however, the change is very small with $\vartheta_A = j100$ (not shown) little different from the result for $\vartheta_A = j10$. With $\vartheta_A = \phi_0$ (the angle of incidence) the reflected field from face A is zero and the behaviour of the diffracted field is quite different since the only discontinuity it has to compensate for is at the shadow boundary.

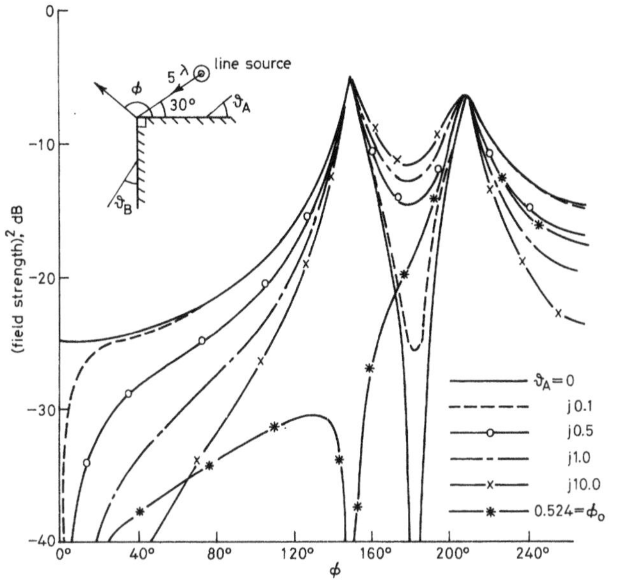

Fig. 5.19 Diffracted far field for line source of unit strength in the far field illuminating 90° impedance wedge. Brewster angle ν_B for face B is fixed at $\pi/2$ while $\nu_A = \pi/2 - \vartheta_A$ varies

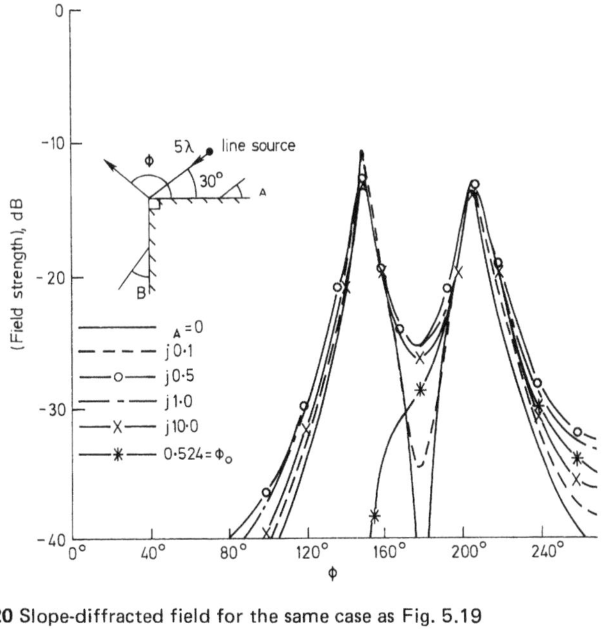

Fig. 5.20 Slope-diffracted field for the same case as Fig. 5.19

As for the perfectly conducting wedge, we can derive a slope-diffraction coefficient for the impedance wedge. The slope-diffraction term is proportional to $(\partial/\partial\phi_0)D^{e,m}$ (see Section 5.5) and this differentiation is readily performed on eqn. 123 although the result is rather lengthy and will not be given explicitly here. An example of the slope-diffraction term is shown in Fig. 5.20 for the same configuration in Fig. 5.19. The solution is also uniform through the optical boundaries.

5.10 Diffraction by a dielectric wedge

The uniform GTD methods developed so far to treat edge diffraction – initially for perfectly conducting bodies and then for the more complex problem of bodies that satisfy impedance boundary conditions – are of great practical importance in solving a large variety of electromagnetic engineering problems. These methods were based on the known solutions to canonical problems in wedge diffraction. The problem of diffraction by a dielectric, or penetrable wedge, is also of considerable practical interest. As yet no exact solution has been found for diffraction by a dielectric wedge and indeed is unlikely ever to be found as suggested, for example, by Jones and Pidduck (1950). Nevertheless, the dielectric wedge has been investigated by numerous authors in an attempt to arrive at approximate solutions either by analytical asymptotic methods, confining the solution to regions of interest like the apex of the wedge, a straight-out numerical solution, or a combined analytical-numerical approach. No solution to date can be said to be an unqualified success, especially in regard to GTD-type edge diffraction coefficients. In some cases authors have made rather immoderate claims for their solution only to find they do not stand the test of closer scrutiny. For this aspect, and to obtain a general overview of the various attempts at solving the dielectric wedge problem, the reader is referred to the papers of Bates (1973), Joo *et al.* (1984), and Yeo *et al.* (1985). These three papers include all of the relevant literature to date on the subject and we will not attempt an independent survey here. From the GTD point of view we have yet to see the availability of reliable dielectric edge diffraction coefficients, nevertheless ongoing work, such as that of Bates *et al.* (1985), offer promising means of eventually achieving this end.

5.11 Summary

Wedge diffraction
When a geometrical field $\boldsymbol{E}^i, \boldsymbol{H}^i$ given from eqn. 4.41 as

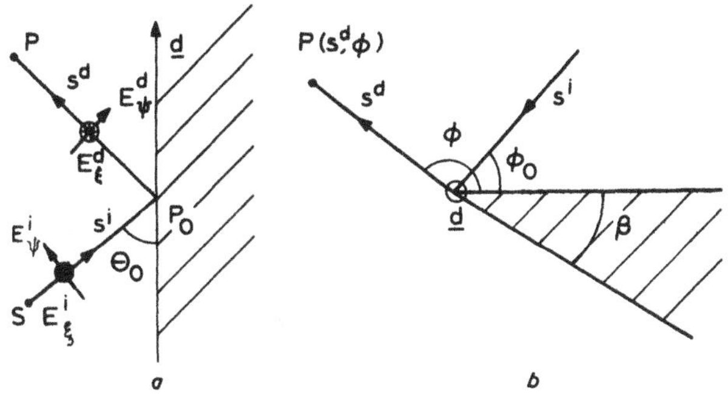

Fig. 5.21 Diffraction at a wedge

$$E^i(s^i) = E^i(q)\left[\frac{\rho_1^i\rho_2^i}{(\rho_1^i + s^i)(\rho_2^i + s^i)}\right]^{1/2} \exp(-jks^i);$$
$$H^i = \sqrt{\left(\frac{\hat{\epsilon}}{\mu}\right)}\hat{s}_0^i \times E^i \qquad (124a)$$

or alternatively from eqn. 4.42

$$E^i(s^i) = E^i(q)\left[\frac{\det \mathbf{Q}^i(s_0)}{\det \mathbf{Q}^i(0)}\right]^{1/2} \exp(-jks^i);$$
$$s^i = s_0^i + \tfrac{1}{2}\boldsymbol{b}^{iT}\mathbf{Q}^i\boldsymbol{b}^i \qquad (124b)$$

is incident upon a perfectly conducting wedge with a straight edge and planar surfaces (see Fig. 5.21) the resultant diffracted field may be written about the diffracted rays from the edge point P_0 as

$$E^d(s^d) = \mathbf{D}E^i(P_0)\left[\frac{\rho_3}{s^d(\rho_3 + s^d)}\right]^{1/2} \exp(-jks^d);$$
$$H^d = \sqrt{\left(\frac{\hat{\epsilon}}{\mu}\right)}\hat{s}^d \times E^d \qquad (125)$$

Points of edge diffraction, such as P_0, are determined by applying Fermat's principle for edge diffraction: edge diffracted rays from a point S to a point P are those rays for which the optical path length between S and P with one point on the diffracting edge is stationary with respect to infinitesimal variations in path. Mathematically this means that at the diffraction point

$$\hat{s}^i \cdot \hat{d} = \hat{s}^d \cdot \hat{d} \qquad (126)$$

where $\hat{d}$ is the unit vector in the direction of the tangent to the edge.

The incident field is expressed in terms of a ray based co-ordinate system (s_0^i, ψ, ξ) where E_ψ^i and E_ξ^i relate to electric and magnetic polarisation. Similarly the diffracted field is expressed in terms of a co-ordinate system (s^d, ψ, ξ) along the diffracted ray where E_ψ^d and E_ξ^d are associated with the electric and magnetic polarisation for the wedge problem. In eqn. 125 the quantity ρ_3 is given for a straight edge as

$$\frac{1}{\rho_3} = Q_{11}^i \tag{127}$$

and the diffraction matrix **D** is

$$\mathbf{D} = \begin{bmatrix} D^e & 0 \\ 0 & D^m \end{bmatrix}$$

where the components D^e, D^m of this matrix are given as

$$\begin{aligned} D^{e,m} &= \{h(\Phi^i) + h(-\Phi^i)\} \mp \{h(\Phi^r) + h(-\Phi^r)\}; \\ \Phi^{i,r} &= \phi \mp \phi_0, \quad 0 \leqslant \phi, \phi_0 \leqslant 2\pi \end{aligned} \tag{128a}$$

where

$$h(\Phi^{i,r}) = -\epsilon^{i,r}\sqrt{(\sigma)}K_-(v^{i,r})\Lambda^{i,r} \sim \frac{-\csc\theta_0 \cot\left(\dfrac{\pi + \Phi^{i,r}}{2N}\right)}{N\sqrt{(8j\pi k)}} \quad \text{for } v^{i,r} > 3 \tag{128b}$$

The angle θ_0 is the incident angle to the edge, ϕ_0 is the incident angle measured from one of the wedge faces, and ϕ is the angle of the diffracted ray measured from the same wedge face. If is essential to ensure that when evaluating ϕ and ϕ_0 they are within the range 0 to 2π.

The wedge angle is given by β from which we define

$$N = \frac{2\pi - \beta}{\pi}$$

With the superscript $i(r)$ relating to the incident (reflected) geometrical optics field, the quantities in eqn. 128b are given as follows:

$$\epsilon^{i,r} = \text{sgn}(a^{i,r}) = \begin{bmatrix} +1 & \text{for illuminated region} \\ -1 & \text{for shadow region} \end{bmatrix} \quad \text{with respect to the associated geometrical optics field.}$$

where

$$a^{i,r} = \sqrt{(2)}\cos\tfrac{1}{2}(\Phi^{i,r} + 2n\pi N)$$

The values of the integer n are chosen to satisfy the following two conditions on the optical boundaries:

$$|2n\pi N + \Phi^{i,r}| = \pi$$

$$\Lambda^{i,r} = 1$$

where

$$\Lambda^{i,r} = \frac{a^{i,r}}{\surd(2)N} \cot\left(\frac{\pi + \Phi^{i,r}}{2N}\right)$$

Removed from these boundaries the diffraction term $h(\pm\Phi^{i,r})$ is independent of n. Special care must be taken if a reflection boundary is near a wedge face and this is described in eqn. 25. Note that we may have either a reflection and shadow boundary, or two reflection boundaries with no shadow boundary. If we measure the angles ϕ, ϕ_0 from an illuminated wedge face the correct behaviour across the two possible reflection boundaries is ensured by putting $n = 0$ in $h(-\Phi^r)$, and $n = -1$ in $h(\Phi^r)$. When a shadow boundary exists in visible space we have $h(\pm\Phi^i)_{n=0}$ depending on whether $\Phi^i = \mp\pi$ on the shadow boundary with the remaining term given as $h(\mp\Phi^i)_{n=-1}$.

Since the incident field impinges upon flat surfaces, the reflected wavefront principal radii of curvature, ρ_1^r, ρ_2^r are the same as the incident radii of curvature ρ_1^i, ρ_2^i so that

$$\rho_1^i = \rho_1^r = \rho_1$$

$$\rho_2^i = \rho_2^r = \rho_2$$

(where ρ_1, ρ_2 are the values of the incident wavefront curvature measured at the point of diffraction P_0) and the value for σ is now

$$\sigma = \frac{\rho_1\rho_2(\rho_3 + s^d)s^d}{(\rho_1 + s^d)(\rho_2 + s^d)\rho_3}$$

When the incident field is a spherical wave, then $\rho_1 = \rho_2 = \rho_3 = \rho$ and σ simplifies to

$$\sigma = \frac{\rho s^d}{\rho + s^d}$$

This is also the value for σ when the incident field is from a line source parallel to the edge so that $\rho_1 = \rho_3 = \infty$. For plane wave incidence σ is simply given by s^d. The modified Fresnel integral K_- has been defined in Section 2.3.1. Its argument $v^{i,r}$ for edge diffraction is given as

$$v^{i,r} = \surd(k\sigma)|a^{i,r}|\sin\theta_0$$

When the incident field is at grazing incidence to the wedge (i.e., $\phi_0 = 0$) the diffracted field as given by eqn. 125 is divided by 2.

Only the diffracted field has been given in this section. The geometrical optics field is computed separately from the methods given in Chapter 4.

If a caustic surface is grazed and σ changes sign then the above results are not valid. This problem is considered in the following chapter.

Half-plane diffraction

The important special case of the half-plane when $N = 2$ gives substantially simpler expressions for the diffraction coefficients D^e and D^m. Thus putting $N = 2$ in eqn. 128 yields

$$D^{e,m} = -\sqrt{(\sigma)}\{\epsilon^i K_-(v^i) \mp \epsilon^r K_-(v^r)\} \tag{129a}$$

where

$$v^{i,r} = \sqrt{(2k\sigma)}\,|\cos\tfrac{1}{2}(\phi \mp \phi_0)|\sin\theta_0$$

Removed from optical boundaries

$$D^{e,m} \sim -\{\sec\tfrac{1}{2}(\phi - \phi_0) \mp \sec\tfrac{1}{2}(\phi + \phi_0)\}\frac{\csc\theta_0}{\sqrt{(8j\pi k)}} \tag{129b}$$

Slope-diffraction

When the incident field E^i is zero at the edge diffraction point P_0, the leading term in the asymptotic expansion of the diffracted field is directly proportional to the first derivative of the incident field at P_0 and is known as the slope-diffraction term. It is given as, with $\hat{n} = \hat{\phi}_0$,

$$E^{sd} = \frac{1}{jk\sin\theta_0}\frac{\partial}{\partial\phi_0}\mathbf{D}\frac{\partial}{\partial n}E^i\left[\frac{\rho_3}{s^d(\rho_3 + s^d)}\right]^{1/2}\exp(-jks^d) \tag{130}$$

where, *for the wedge*

$$\frac{\partial}{\partial\phi_0}D^{e,m} = -[h'(\Phi^i) - h'(-\Phi^i)] \mp [h'(\Phi^r) - h'(-\Phi^r)] \tag{131a}$$

where

$$h'(\Phi) = \frac{\partial}{\partial\Phi}h(\Phi) = \sqrt{(\sigma)}\left[\sqrt{\left(\frac{k\sigma}{2}\right)}\sin\tfrac{1}{2}(\Phi + 2n\pi N)\sin\theta_0\right.$$

$$\left\{2jvK_-(v) - \sqrt{\left(\frac{j}{\pi}\right)}\right\}\Lambda + \frac{\epsilon}{2N}K_-(v)\left\{\sin\tfrac{1}{2}(\Phi + 2n\pi N)\right.$$

$$\left.\left.\cot\left(\frac{\pi + \Phi}{2N}\right) + \frac{a}{\sqrt{(2)}N}\csc^2\left(\frac{\pi + \Phi}{2N}\right)\right\}\right] \tag{131b}$$

When v is large the first two terms in eqn. 131*b* cancel to give

$$\frac{\partial}{\partial\Phi}h(\Phi) \sim \frac{1}{2N^2}\frac{1}{\sqrt{(8j\pi k)}}\csc^2\left(\frac{\pi+\Phi}{2N}\right)\csc\theta_0 \qquad (131c)$$

For the *half-plane* the equations are considerably simpler, being

$$\frac{\partial}{\partial\phi_0}D^{e,m} = -\sqrt{\left(\frac{k}{2}\right)}\sigma\sin\theta_0\left[\sin\tfrac{1}{2}\Phi^i\left\{2jv^iK_-(v^i)-\sqrt{\left(\frac{j}{\pi}\right)}\right\}\right.$$
$$\left.\pm\sin\tfrac{1}{2}\Phi^r\left\{2jv^rK_-(v^r)-\sqrt{\left(\frac{j}{\pi}\right)}\right\}\right]$$
$$\sim\frac{\csc\theta_0}{2\sqrt{(8j\pi k)}}\{\sin\tfrac{1}{2}(\phi-\phi_0)\sec^2\tfrac{1}{2}(\phi-\phi_0)$$
$$\pm\sin\tfrac{1}{2}(\phi+\phi_0)\sec^2\tfrac{1}{2}(\phi+\phi_0)\} \qquad (131d)$$

It is important to note that eqn. 130 is only true when the direct incident field E^i is zero at the point of diffraction. With E^i non-zero, eqn. 130 does not constitute the complete second order term although it has been treated as such by some authors. This can be seen by referring to Section 5.5 or the physical optics higher order asymptotic expansion for the half-plane given by eqn. 82.

Double-edge diffraction

In diffraction between two wedges as shown in Fig. 5.22, a straightforward application of the above formulas is valid provided that the transition field of one edge is not incident on the adjacent edge. In the event of the latter occurring we initially formulate the problem with the edge diffracted field from each wedge considered in isolation. To account for the effect of transition field diffraction from edge A to edge B in Fig. 5.22, we replace the modified Fresnel integral function $K_-(x)$ with the generalised Fresnel integral function $L_-(x, y)$. The specific substitutions are

$$K_-(\nu_A^{i,r})\to L_-(x_1,\nu_A^{i,r})$$
$$K_-(\nu_B^{i,r})\to L_-(x_2,\nu_B^{i,r}) \qquad (132a)$$

where $\nu^{i,r}$ is given above for single edge diffraction, the subscripts A and B refer to the diffracting edges, and x_1 and x_2 are given by

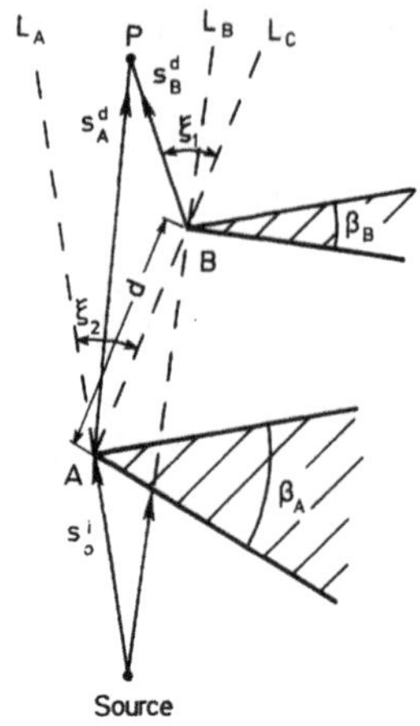

Fig. 5.22 Double-edge diffraction

$$x_1 = \epsilon_1 \left[\frac{2kds^d_B}{d + s^d_B}\right]^{1/2} |\sin \tfrac{1}{2}\xi_1|$$

$$x_2 = \epsilon_2 \left[\frac{2kds^i_0}{d + s^i_0}\right]^{1/2} |\sin \tfrac{1}{2}\xi_2| \qquad (132b)$$

where ϵ_1 accounts for the blockage effect of wedge B on the diffracted field from wedge A. Thus $\epsilon_1 = 1$ when no blockage takes place, otherwise $\epsilon_1 = -1$. The unit term $\epsilon_2 = 1$ if edge B is directly illuminated by the incident field and is -1 otherwise.

Impedance wedge

When the geometrical optics field of eqn. 124 is incident upon an impedance wedge, the resultant reflected field has reflection coefficients dependent on the face impedances given by eqn. 119. The edge diffracted field is given by eqn. 125 but with the components $D^{e,m}$ of the diffraction matrix given by

$$D^{e,m} = [\Psi^{e,m}(\phi_0)]^{-1}\{\Psi^{e,m}(\phi + \pi)[h(\Phi^i) - h(\Phi^r)] + \Psi^{e,m}(\phi - \pi)[h(-\Phi^i) - h(-\Phi^r)]\} \qquad (133)$$

where the function $\Psi^{e,m}(x)$ is described in Section 5.9. Note that this equation applies only for normal incidence ($\theta_0 = \pi/2$).

References

AAS, J.A. (1979): 'On the accuracy of the uniform geometrical theory of diffraction close to a 90° wedge', *IEEE Trans.*, **AP-27**, pp. 704–705.
AHLUWALIA, D.S. (1970): 'Uniform asymptotic theory of diffraction by the edge of a three-dimensional body', *SIAM J. Appl. Math.*, **18**, pp. 287–301.
AHLUWALIA, D.S., LEWIS, R.M., and BOERSMA, J. (1968): 'Uniform asymptotic theory of diffraction by a plane screen', *ibid.*, **16**, pp. 783–807.
BATES, R.H.T. (1973): 'Wavefunctions for prisms', *Int. J. Elect.*, **34**, pp. 81–95.
BATES, R.H.T., YEO, T.S., and WALL, D.J.N. (1985): 'Towards an algorithm for dielectric edge diffraction coefficients', *Proc. IEE,* **132**, pt. H, pp. 461–467.
BOERSMA, J. (1975): 'Diffraction by two parallel half-planes', *Quart. J. Mech. Appl. Math*, **28**, pp. 405–423.
BOERSMA, J., and LEE, S.W. (1977): 'High-frequency diffraction of a line-source field by a half-plane: solutions by ray techniques', *IEEE Trans.*, **AP-25**, pp. 171–179.
BOERSMA, J., and RAHMAT-SAMII, Y. (1980): 'Comparison of two leading uniform theories of edge diffraction with the exact uniform solution', *Radio Sci.*, **15**, pp. 1179–1199.
BOWMAN, J.J. (1967): 'High-frequency backscattering from an absorbing infinite strip with arbitrary face impedances', *Can. J. Phys.*, **45**, pp. 2409–2430.
BUCCI, O.M., and FRANCESCHETTI, G. (1976): 'Electromagnetic scattering by a half-plane with two face impedances', *Rad. Sci.*, **11**, pp. 49–59.
DESCHAMPS, G.A., BOERSMA, J., and LEE, S.W. (1984): 'Three-dimensional half-plane diffraction: exact solution and testing of uniform theories', *IEEE Trans.*, **AP-32**, pp. 264–271.
JAMES, G.L., and POULTON, G.T. (1973): 'Modified diffraction coefficients for focusing reflectors', *Electron. Lett.*, **9**, pp. 537–538.
JAMES, G.L. (1974): 'Edge diffraction at a curved screen', *ibid.*, **10**, pp. 167–168.
JAMES, G.L. (1977): 'Uniform diffraction coefficients for an impedance wedge', *ibid.*, **13**, pp. 403–404.
JAMES, G.L., and POULTON, G.T. (1979): 'Double knife-edge diffraction for curved screens', *IEE J. Microwave, Opt. and Acoust.*, **3**, pp. 221–223.
JAMES, G.L. (1980): 'Analysis of radiation pattern and G/T_A for shaped dual-reflector antennas', *IEE Proc.*, **127**, PtH, pp. 52–60.
JONES, D.S., and PIDDUCK, F.B. (1950): 'Diffraction by a metal wedge at large angles', *Quart. J. Math.*, **1**, pp. 229–237.
JONES, D.S. (1973): 'Double knife-edge diffraction and ray theory', *Quart. J. Mech. Appl. Math.*, **26**, pp. 1–18.
JOO, C.S., RA, J.W., and SHIN, S.Y. (1984): 'Scattering by right angle dielectric wedge', *IEEE Trans.*, **AP-32**, pp. 61–69.
KALOSHIN, V.A. (1982): 'Multiple diffraction on semiplanes', *Rad. Eng. El. Phy.*, **17**, pp. 56–61.
KELLER, J.B. (1962): 'Geometrical theory of diffraction', *J. Opt. Soc. Am.*, **52**, pp. 116–130.
KOUYOUMJIAN, R.G. and PATHAK, P.H. (1974): 'A uniform geometrical theory of diffraction for an edge in a perfectly conducting surface', *Proc. IEEE,* **62**, pp. 1448–1461.

LEE, S.W. and DESCHAMPS, G.A. (1976): 'A uniform asymptotic theory of electromagnetic diffraction by a curved wedge', *IEEE Trans.*, **AP-24**, pp. 25–34.

LEE, S.W. (1978): 'Path integrals for solving some electromagnetic edge diffraction problems', *J. Math. Phys.*, **19**, pp. 1414–1422.

LEWIS, R.M., and BOERSMA, J. (1969): 'Uniform asymptotic theory of edge diffraction', *ibid.*, **10**, pp. 2291–2305.

MALIUZHINETS, G.D. (1958): 'Excitation, reflection and emission of surface waves from a wedge with given face impedances', *Sov. Phys. Dokl.*, **3**, pp. 752–755.

MENENDEZ, R.C., and LEE, S.W. (1982): 'Analysis of rectangular horn antennas via uniform asymptotic theory', *IEEE Trans.*, **AP-30**, pp. 241–250.

MITSMAKHER, M.Yu. (1976): 'Generalized reflection coefficient in the diffraction of a plane electromagnetic wave at an impedance wedge', *Radiophys. Quantum. Electron.*, **19**, pp. 1063–1066.

PATHAK, P.H., and KOUYOUMJIAN, R.G. (1974): 'An analysis of the radiation from apertures in curved surfaces by the geometrical theory of diffraction', *Proc. IEEE*, **62**, pp. 1438–1447.

PAULI, W. (1938): 'On asymptotic series for functions in the theory of diffraction of light', *Phys. Rev.*, **54**, pp. 924–931.

RAHMAT-SAMII, Y., and MITTRA, R. (1977): 'On the investigation of diffracted fields at the shadow boundaries of staggered parallel plates – a spectral domain approach', *Rad. Sci.*, **12**, pp. 659–670.

RAHMAT-SAMII, Y., and MITTRA, R. (1978): 'Spectral analysis of high-frequency diffraction of an arbitrary incident field by a half plane-comparison with four asymptotic techniques', *ibid.*, **13**, pp. 31–48.

SANYAL, S., and BHATTACHARYYA, A.K. (1983): 'UAT analysis for *E*-plane near and far-field patterns of electromagnetic horn antennas', *IEEE Trans.*, **AP-31**, pp. 817–819.

SCHRETTER, S.J. and BOLLE, D.M. (1969): 'Surface currents induced on a wedge under plane wave illumination: an approximation', *ibid.*, **AP-17**, pp. 246–248.

TIBERIO, R., and KOUYOUMJIAN, R.G. (1982): 'An analysis of diffraction at edges illuminated by transition region fields', *Rad. Sci.*, **17**, pp. 323–336.

TIBERIO, R., BESSI, F., MANARA, G., and PELOSI, G. (1982): 'Scattering by a strip with two face impedances at edge-on incidence', *ibid.*, **17**, pp. 1199–1210.

TIBERIO, R. and PELOSI, G. (1983): 'High-frequency scattering from the edges of impedance discontinuities on a flat plate', *IEEE Trans.*, **AP-31**, pp. 590–596.

TIBERIO, R., and KOUYOUMJIAN, R.G. (1984): 'Calculation of the high-frequency diffraction by two nearby edges illuminated at grazing incidence', *ibid.*, **AP-32**, pp. 1186–1196.

TIBERIO, R., PELOSI, G., and MANARA, G. (1985): 'A uniform GTD formulation for the diffraction by a wedge with impedance faces', *ibid.*, **AP-33**, pp. 867–873.

TISHCHENKO, V.A., and KHESTANOV, R.Kh. (1974): 'Diffraction of a field with a light-shadow boundary at a half-plane', *Sov. Phys.-Dokl.*, **18**, pp. 644–646.

USLENGHI, P.L.E. (Ed.) (1978): 'Electromagnetic Scattering', (Academic Press).

VOGLER, L.E. (1982): 'An attenuation function for multiple knife-edge diffraction', *Rad. Sci.*, **17**, pp. 1541–1546.
VOLAKIS, J.L., and SENIOR, T.B.A. (1985): 'Simple expressions for a function occurring in diffraction theory', *IEEE Trans.*, **AP-33**, pp. 678–680.
WHITTEKER, J.H. (1984): 'Near-field ray calculation for multiple knife-edge diffraction', *Rad. Sci.*, **19**, pp. 975–986.
WOLF, P. (1966): 'Diffraction of a scalar wave by a plane screen', *SIAM J. Appl. Math.*, **14**, pp. 577–599.
YEO, T.S., WALL, D.J.N., and BATES, R.H.T. (1985): 'Diffraction by a prism', *J. Opt. Soc. Am.*, **2**, pp. 964–969.

Chapter 6

Diffraction by curved edges and surfaces

6.1 Plane wave diffraction around a circular cylinder

When a plane wave is normally incident upon a perfectly conducting circular cylinder in a lossless medium, as shown in Fig. 6.1, and has a field component V_z^i in the z-direction where

$$V_z^i = \exp(-jk\rho\cos\phi) \tag{1}$$

then for large cylinders where $ka \gg 1$ the exact solution in the *shadow region* for electric polarisation ($V_z^i = E_z^i$) is given from eqn. 3.131 as

$$E_z(\rho,\phi) = j\pi \sum_{n=1}^{\infty} \frac{H_{\nu_n}^{(1)}(ka) H_{\nu_n}^{(2)}(k\rho)}{\left.\frac{\partial}{\partial\nu} H_\nu^{(2)}(ka)\right|_{\nu=\nu_n}} \left[\exp\left\{-j\nu_n\left(\frac{\pi}{2}-\phi\right)\right\} + \exp\left\{-j\nu_n\left(\frac{\pi}{2}+\phi\right)\right\}\right] \tag{2a}$$

with the value of the complex variable ν_n determined from

$$H_{\nu_n}^{(2)}(ka) = 0 \tag{2b}$$

For the first few terms of this series we have

$$\nu_n = ka + \alpha_n \left(\frac{ka}{2}\right)^{1/3} \exp\left(-\frac{j\pi}{3}\right) \tag{3}$$

where $-\alpha_n$ are the zeros of the Airy function Ai, some of which are given in Table 2.1. Of particular interest to us is the field at points removed from the cylinder where $k\rho > |\nu_n|$, and eqn. 2 simplifies to (see eqn. 3.135)

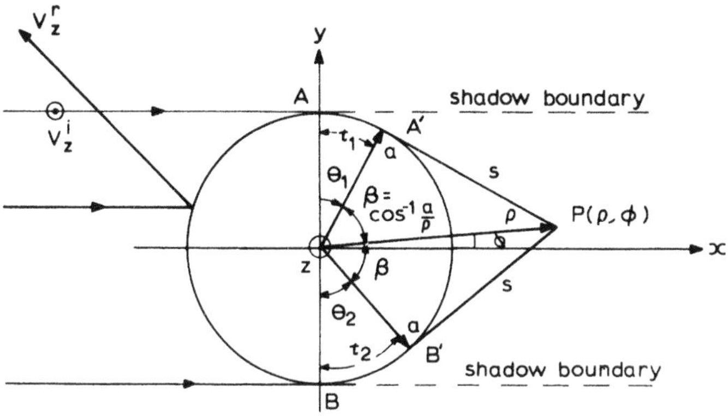

Fig. 6.1 Plane wave diffraction at a circular cylinder

$$E_z = \sum_{n=1}^{N} D_n^e \left[\exp\left\{-j\nu_n\left(\frac{\pi}{2} - \phi - \cos^{-1}\frac{a}{\rho}\right)\right\} + \exp\left\{-j\nu_n\left(\frac{\pi}{2} + \phi - \cos^{-1}\frac{a}{\rho}\right)\right\}\right] \frac{\exp(-jks)}{\sqrt{(8j\pi ks)}} \tag{4}$$

$$s = (\rho^2 - a^2)^{1/2}$$

$$D_n^e = \{\mathrm{Ai}'(-\alpha_n)\}^{-2}\, 2M \exp\left(\frac{j\pi}{6}\right); \qquad M = \left(\frac{ka}{2}\right)^{1/3}$$

Each term in this series is dependent on eqn. 3 being valid and hence this limits the upper value of N.

The corresponding expression for magnetic polarisation in the shadow region when $V_z^i = H_z^i$ is deduced from eqn. 3.148 as

$$H_z = \sum_{n=1}^{N} D_n^m \left[\exp\left\{-j\nu_n'\left(\frac{\pi}{2} - \phi - \cos^{-1}\frac{a}{\rho}\right)\right\} + \exp\left\{-j\nu_n'\left(\frac{\pi}{2} + \phi - \cos^{-1}\frac{a}{\rho}\right)\right\}\right] \frac{\exp(-jks)}{\sqrt{(8j\pi ks)}} \tag{5}$$

where

$$\nu_n' = ka + \alpha_n' M \exp\left(-\frac{j\pi}{3}\right)$$

$$D_n^m = \{\alpha_n'\{\mathrm{Ai}(-\alpha_n')\}^2\}^{-1}\, 2M \exp\left(\frac{j\pi}{6}\right)$$

and $-\alpha_n'$ are the zeros of the Airy function derivative Ai′, some of which are given in Table 2.1.

The equations in eqns. 4 and 5 can be readily given a ray optics interpretation. Referring to Fig. 6.1 it is seen that the angles θ_1, θ_2 are given by

$$\theta_1 = \frac{\pi}{2} - \phi - \cos^{-1}\frac{a}{\rho}; \qquad \theta_2 = \frac{\pi}{2} + \phi - \cos^{-1}\frac{a}{\rho}$$

and consequently

$$\tau_1 = \theta_1 a; \qquad \tau_2 = \theta_2 a$$

Substitution into eqns. 4 and 5 yields

$$E_z = \sum_{n=1}^{N} D_n^e \left[\exp\{-(jk + \Omega_n^e)\tau_1\} + \exp\{-(jk + \Omega_n^e)\tau_2\}\right] \frac{\exp(-jks)}{\sqrt{(8j\pi ks)}} \tag{6a}$$

$$H_z = \sum_{n=1}^{N} D_n^m \left[\exp\{-(jk + \Omega_n^m)\tau_1\} + \exp\{-(jk + \Omega_n^m)\tau_2\}\right] \frac{\exp(-jks)}{\sqrt{(8j\pi ks)}} \tag{6b}$$

where

$$\Omega_n^e = \frac{\alpha_n}{a} M \exp\left(\frac{j\pi}{6}\right) \tag{7a}$$

$$\Omega_n^m = \frac{\alpha_n'}{a} M \exp\left(\frac{j\pi}{6}\right) \tag{7b}$$

Each term in the series of eqn. 6 can now be interpreted in the following way. From the ***glancing points*** A and B on the cylinder in Fig. 6.1 the incident rays travel around the surface a distance τ_1 and τ_2 respectively. While on the surface the rays decay exponentially due to the ***attenuation constant*** Ω which is defined in eqn. 7 for both polarisations. As seen from eqn. 7 this constant is complex and thus gives an additional phase shift to the rays travelling around the surface. At the points A' and B' the rays leave the surface tangentially and continue unattenuated to the field point P as if they were emanating from line sources at A' and B'. The terms D_n^e and D_n^m simply act as amplitude weighting factors for the rays.

In fact surface rays continue beyond A' and B', encircling the cylinder an infinite number of times, and each time shedding a ray at A' and B' to the field point P. The summation of these multiple encircling rays modifies the field in eqn. 6 by the multiplication factor

$[1 - \exp\{-(jk + \Omega_n)2\pi a\}]^{-1}$. This term could have been retained in our development of the exact solution, but for large ka the exponential component, and hence the contribution of the encircling rays, is negligibly small and was rightly ignored.

The rays which propagate around the surface of the cylinder to give the diffracted field in the shadow are referred to as ***creeping rays.*** At each point on the surface a creeping ray is shed tangentially from the surface, as illustrated in Fig. 6.2. There are, in theory, an infinite number of these creeping rays which encircle the cylinder an infinite number of times. Fortunately, the attenuation constant takes care of most of these rays! As already mentioned the exponential attenuation of the rays as they encircle the cylinder ensures that they quickly reduce to a negligible level. Similarly, for the higher order creeping rays in the series given by eqn. 6. The Airy function roots $-\alpha_n, -\alpha_n'$ increase with n and only the first one or two terms are of importance provided that $\tau_{1,2}$ is non-zero. However, as $\tau_{1,2} \to 0$, an increasing number of terms in the series are required to maintain accuracy. Thus in the vicinity of the shadow boundaries where $\tau_{1,2} \simeq 0$, eqn. 6 is not a good representation of the field. We shall return to this problem shortly.

In the ***illuminated region*** it was shown in Section 3.4, from eqn. 3.138, that the field is given essentially by

$$V_z = V_z^i \mp V_z^r \tag{8}$$

where the upper (lower) sign applies for electric (magnetic) polarisation, and the reflected field component V_z^r is given in eqn. 3.138 as

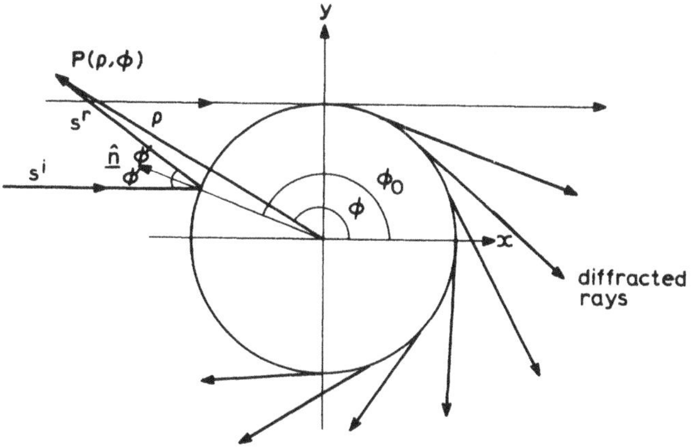

Fig. 6.2 Diffracted and reflected rays at a cylinder

$$V_z^r = \sqrt{\left(\frac{a}{2\rho}\sin\frac{|\phi|}{2}\right)}\exp\left\{-jk\left(\rho - 2a\sin\frac{|\phi|}{2}\right)\right\};$$
$$k\rho \to \infty, \quad 0 < |\phi| < \pi \tag{9}$$

Using the notation in Chapter 4, specifically eqn. 4.41, the geometrical optics reflected field from the cylinder at a point (a, ϕ_0) may be written as

$$\mp V_z^i(a, \phi_0)\left[\frac{\rho_2}{\rho_2 + s^r}\right]^{1/2}\exp(-jks^r) \xrightarrow[ks^r \to \infty]{} \mp V_z^i\sqrt{\left(\frac{\rho_2}{s^r}\right)}\exp(-jks^r)$$

where s^r is the distance along the reflected ray measured from the cylinder surface as shown in Fig. 6.2. Comparing this equation with eqn. 9 above gives

$$V_z^i(a, \phi_0) = \exp(jka\cos\phi') \sim \exp\left(jka\sin\frac{\phi}{2}\right)$$

$$s^r \sim \rho - a\sin\frac{\phi}{2}; \qquad \rho_2 \sim \frac{a}{2}\sin\frac{\phi}{2} = \frac{a}{2}\cos\phi'$$

and the reflected field for any field point in the illuminated region can now be given by

$$V_z^r(\rho, \phi) = \mp V_z^i(a, \phi_0)\left[\frac{a\cos\phi'}{a\cos\phi' + 2s^r}\right]^{1/2}\exp(-jks^r) \tag{10}$$

Note that as a shadow boundary is approached, eqn. 10 reduces to zero, leaving only the incident field in our above formulation. This, however, is not the true field behaviour and more about this will be mentioned later.

Since the scattered field is given by the geometrical optics reflected field in the illuminated region, we may use the methods given in Chapter 4 when we consider a general electromagnetic plane wave incident on the cylinder. In the shadow we may represent the electromagnetic diffraction by the same matrix notation using a ray-based co-ordinate system. The diffraction mechanism is identical at A and B in Fig. 6.1 so we need only consider the upper surface and by implication include diffraction from the lower surface of the cylinder.

Using the same ray-based co-ordinates (s, ψ, ξ) as before, an incident electromagnetic plane wave is decomposed into its electric and magnetic polarisation components, as shown in Fig. 6.3. The co-ordinates are oriented so that

$$\hat{\xi} = \hat{n}; \qquad \hat{s} = \hat{\psi} \times \hat{\xi} \tag{11}$$

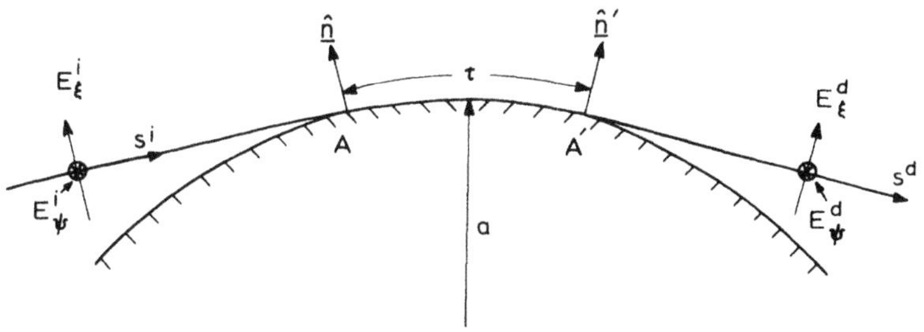

Fig. 6.3 Ray co-ordinates for diffraction around cylinder

where $\hat{n}$ is the outward normal from the surface. It is seen that E^i_ξ corresponds to magnetic polarisation and E^i_ψ to electric polarisation. The diffracted field from A' can now be written in the single equation

$$\begin{bmatrix} E^d_\psi(s^d) \\ E^d_\xi(s^d) \end{bmatrix} = \sum_{n=1}^{N} \frac{1}{\sqrt{(8j\pi k)}} \begin{bmatrix} D^e_n \exp(-\Omega^e_n \tau) & 0 \\ 0 & D^m_n \exp(-\Omega^m_n \tau) \end{bmatrix}$$

$$\begin{bmatrix} E^i_\psi(A) \\ E^i_\xi(A) \end{bmatrix} \frac{\exp\{-jk(\tau + s^d)\}}{\sqrt{(s^d)}} \qquad (12a)$$

which may be expressed in the compact form

$$E^d(s^d) = \sum_{n=1}^{N} \mathsf{D}_n E^i(A) \frac{\exp\{-jk(\tau + s^d)\}}{\sqrt{(s^d)}}; \quad H^d = \sqrt{\left(\frac{\epsilon}{\mu}\right)} \hat{s}^d \times E^d \qquad (12b)$$

We have considered so far only a normally incident plane wave. When the field is at oblique incidence, as shown in Fig. 6.4, one can use the technique given in Section 5.3 to derive the oblique incidence result from the normal incidence case. For the cylinder the only change in eqn. 12 is that the radius of the cylinder is replaced by the radius of curvature of the helical path the creeping ray now follows around the cylinder. This is illustrated in Fig. 6.4. The cross-section in Fig. 6.3 can be viewed as a cross-section of a part of this helical path and not of the cylinder.

We turn our attention now to the transition region about the shadow boundary. Our previous solutions have been adequate for the deep illuminated region, as illustrated in Fig. 6.5, where the field is determined from the geometrical optics field alone. Similarly, in the deep shadow region, the creeping wave formulation of eqn. 12 will predict the field accurately. In the intermediate transition region we must use

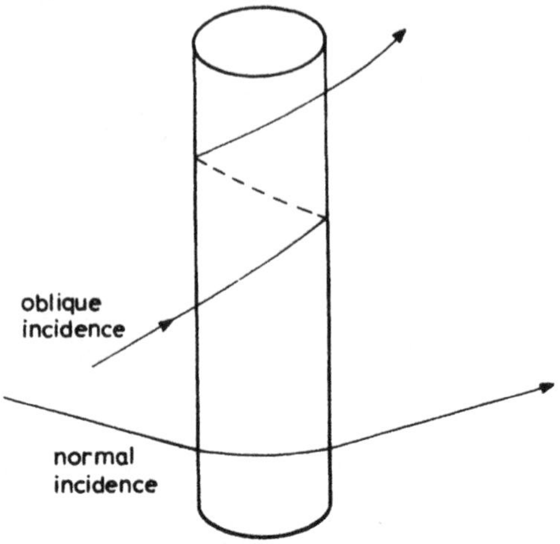

Fig. 6.4 Oblique and normal incidence to the cylinder

the formulation given in Section 3.4.3 for the total field. The extent of this transition region is determined by the value of δ. Since we shall use the geometrical optics reflected field only in the deep illuminated region we can define an effective 'reflection' boundary as in Fig. 6.5. Thus in region 1 the *total* field is given by the sum of the geometrical optics incident and reflected fields. In region 2 the field is given by the incident field plus a 'diffracted' field which will be determined from the results in Section 3.4.3. For the remaining region (3 + 4) the field is given entirely by the diffracted field. Borrowing the notation from edge diffraction we may write the total electric field in the form

$$E = U(\epsilon^i)E^i + U(\epsilon^r)E^r + U(-\epsilon^r)E^d \tag{13}$$

where ϵ^i, ϵ^r change sign at their respective boundaries.

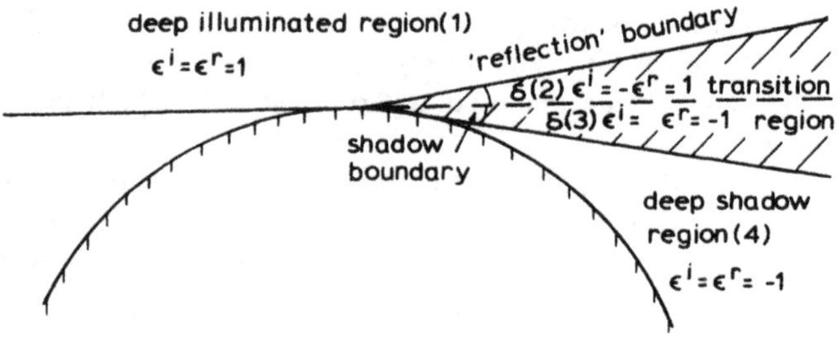

Fig. 6.5 Various regions about a convex surface

The diffracted field $\boldsymbol{E}^d$ for region 4 has been given above by eqn. 12. To extend this result to the transition region we use without further modification the first term (since we are only considering the upper surface of the cylinder) in the equations of eqn. 3.159. Consequently,

$$E^d(s^d) = \mathbf{D}E^i(A)\frac{\exp\{-jk(\tau + s^d)\}}{\sqrt{(s^d)}} \tag{14}$$

where

$$\mathbf{D} = -\sqrt{\left(\frac{2}{jk}\right)}M\begin{bmatrix} \tilde{p}_-\left(M\frac{\tau}{a}, y\right) & 0 \\ 0 & \tilde{q}_-\left(M\frac{\tau}{a}, y\right)\end{bmatrix}; \quad y = M^{-1}\sqrt{\left(\frac{ks^d}{2}\right)}$$

The functions within the diffraction matrix are the modified Pekeris functions defined by

$$\begin{aligned}\tilde{p}_\pm(x, y) &= p(x) - y\,\mathrm{sgn}\,(x)\,K_\pm(y|x|)\exp\left(\mp\frac{j\pi}{4}\right)\\ \tilde{q}_\pm(x, y) &= q(x) - y\,\mathrm{sgn}\,(x)\,K_\pm(y|x|)\exp\left(\mp\frac{j\pi}{4}\right)\end{aligned} \tag{15}$$

where $p(x)$, $q(x)$ are the Pekeris functions defined in Section 2.2.3. On the shadow boundary $\tau = 0$ and eqn. 14 becomes

$$E^d_\psi = -E^i_\psi(A)\exp(-jks^d)\left\{\tfrac{1}{2}\epsilon^i + p(0)\frac{\exp\left(-\frac{j\pi}{4}\right)}{y}\right\}$$

$$E^d_\xi = -E^i_\xi(A)\exp(-jks^d)\left\{\tfrac{1}{2}\epsilon^i + q(0)\frac{\exp\left(-\frac{j\pi}{4}\right)}{y}\right\}$$

and is seen to reduce to half the incident field, as for edge diffraction, *plus* an addition term proportional to the Pekeris function with zero argument. These particular values are given by equation 2.59*b* so that

$$E^d_\psi = -E^i_\psi(A)\exp(-jks^d)\left\{\tfrac{1}{2}\epsilon^i + \frac{0{\cdot}354}{y}\exp\left(-\frac{j\pi}{12}\right)\right\}$$

(on the shadow boundary)(16)

$$E^d_\xi = -E^i_\xi(A)\exp(-jks^d)\left\{\tfrac{1}{2}\epsilon^i - \frac{0{\cdot}307}{y}\exp\left(-\frac{j\pi}{12}\right)\right\}$$

These equations show that for E^d_ψ, corresponding to electric polarisation, the field along the shadow boundary is increased above that for edge diffraction, and the converse is true for magnetic polarisation. The value of y increases with distance from the cylinder causing the electric and magnetic polarisation components to converge towards the edge diffraction value on the shadow boundary.

As we move away from the shadow boundary, the modified Fresnel integral argument will rapidly increase, so that within the transition region it can be replaced by its asymptotic value giving the modified Pekeris functions as

$$\begin{aligned} \tilde{p}_{\pm}(x, y) &\sim p(x) - \frac{1}{2\sqrt{(\pi)}x} = \hat{p}(x) \\ & \qquad\qquad ; \qquad y|x| > 3{\cdot}0 \qquad (17) \\ \tilde{q}_{\pm}(x, y) &\sim q(x) - \frac{1}{2\sqrt{(\pi)}x} = \hat{q}(x) \end{aligned}$$

which now are the Pekeris carot functions $\hat{p}(x), \hat{q}(x)$ defined in Section 2.2.3. In the shadow region where $\tau > 0$, the functions as given in eqn. 17 used in the diffracted field formulation of eqn. 14 will retrieve the deep shadow result of eqn. 12. This follows directly by using eqn. 2.58*c*. Thus the extent of the transition region in the shadow must include $y|x| > 3{\cdot}0$. Similarly, for the illuminated region when $\tau < 0$ it was shown in Section 3.4.3 that provided we can approximate sin $\frac{\tau}{a} \simeq \frac{\tau}{a}$ with $M\frac{\tau}{a}$ a large negative number the geometrical optics reflected wave is obtained. Thus if the reflection boundary is chosen to satisfy these conditions and $y|x| > 3{\cdot}0$, then the discontinuity in the reflected and diffracted field at this boundary as formulated in eqn. 13 will fully compensate so as to give continuity of the total field. Typically a value of $\delta \simeq M^{-1}$ has been used in an attempt to achieve this objective.

In the illuminated region when $\tau < 0$ the diffraction point A' in Fig. 6.6 has apparently gone into the illuminated region and the path $OAA'P$ to the field point P is as shown in the figure. One would hesitate to call this a diffracted ray path as it does not obey the modified Fermat's principle to be given below. It should be viewed simply as a geometric path to assist in the calculation of the modified Pekeris functions for negative argument.

It is interesting to compare diffraction by a cylinder with edge diffraction. We have already noted the similarity of behaviour along the shadow boundary. In the shadow of a sharp edge the field was seen to decay algebraically while the creeping ray decays exponentially. Thus a convex surface casts a deeper shadow than that given by an edge. Also note that the surface is a caustic of the diffracted rays as was the edge for edge diffraction.

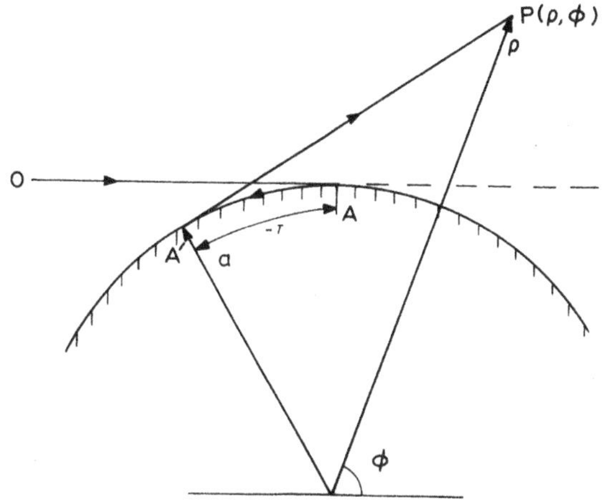

Fig. 6.6 Apparent diffraction point for the illuminated region

6.2 GTD formulation for smooth convex surface diffraction

In the previous section it was seen that an asymptotic evaluation of the exact solution for plane wave diffraction, at an infinitely long straight cylinder, gave the field in the shadow as being due to creeping rays which shed diffracted rays tangentially from the surface as they propagate. By analogy with geometrical optics and the GTD formulation for edge diffraction, we use this rigorous result to obtain approximate solutions for more complicated problems involving diffraction around a smooth convex surface. If an arbitrary field is incident upon a body having on some part of it a smooth, convex, perfectly conducting surface, then, as before, the GTD method offers an approximate solution to the diffracted field, by assuming that each point on the convex surface behaves locally as if it were part of an infinite straight circular cylinder, and that the incident field behaves locally as a plane wave. In diffraction around the cylinder the creeping wave travelled over the surface by the shortest possible route from A to A' in Fig. 6.3. For the cylinder this was along a helical path, and it implies that for a spherical surface it would be along a great circle. That this is indeed the case can be demonstrated by studying the high-frequency behaviour of the exact solution to scattering by a perfectly conducting sphere. (For example see Levy and Keller (1959), and for a good reference, Chapter 10 of Bowman *et al.* (1969)). Extending this concept to more general surfaces we make the assumption that the *creeping ray on a smooth convex surface follows a geodesic path.*

This leads to a statement of Fermat's principle applicable to creeping rays. Thus, *Fermat's principle for diffraction around a smooth convex surface states that diffracted rays from a point S to a point P are those continuous rays for which the optical path length between S and P where part of the path must lie on the surface is stationary with respect to infinitesimal variations in path.*

In general, the geodesics are not easily described except for the special cases of the cylindrical, spherical, and conical surfaces, and it is necessary to solve numerically the differential equations for the geodesic paths.

From our assumptions above we now formulate the diffracted field around an arbitrary smooth convex surface due to an incident pencil. For an astigmatic incident field at A in Fig. 6.7 each ray on the surface within the pencil will follow its own geodesic path until the diffraction point at A'. This will cause the cross section of the pencil at A' to differ, in general, from that at the incident point A. From the concepts discussed in Chapter 4, it can be appreciated that this alone will cause the field on the pencil to be altered by the factor

$$\sqrt{\left(\frac{d\eta}{d\eta'}\right)}$$

As noted earlier the surface is a caustic of the diffracted rays. This caustic is associated with the principal radius of curvature in the $\hat{\xi} - s^d$ plane. The other principal radius of curvature is given by ρ_3 as shown in

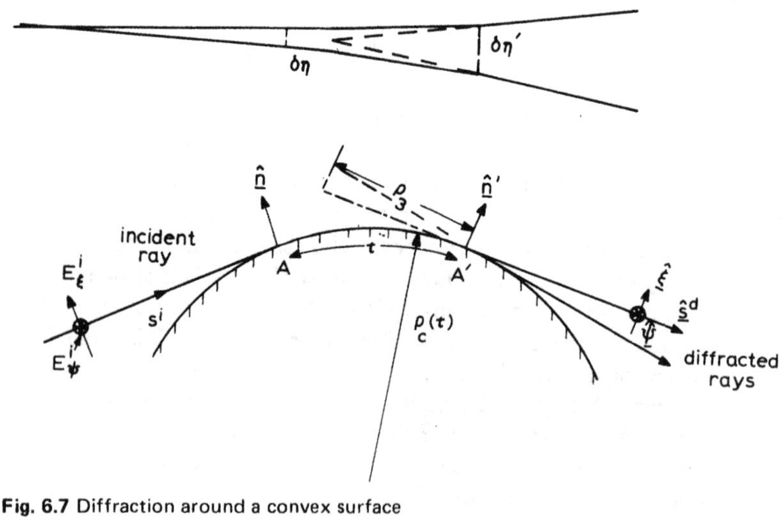

Fig. 6.7 Diffraction around a convex surface

Fig. 6.7. Unlike the corresponding factor for edge diffraction derived in Section 5.4 we cannot define ρ_3 further, and along with the value for $d\eta'$ we must evaluate it numerically for a given surface.

As for edge diffraction, ρ_3 modifies the field along the diffracted ray by

$$\left[\frac{\rho_3}{\rho_3 + s^d}\right]^{1/2}$$

In using the result in eqn. 12 for the deep shadow region of an arbitrary smooth convex surface, the radius of curvature of the surface ρ_c need not be a constant but may be a function of the geodesic path position τ. The attenuation constants in eqn. 12 are therefore a function of τ and the quantity $\Omega_n^{e,m}\tau$ becomes

$$\int_0^T \Omega_n^{e,m}(\tau')d\tau'$$

In the amplitude term $D_n^{e,m}$ the radius of curvature is replaced by its mean over the geodesic path. For most applications it is sufficient to take the mean of the curvatures at the incident point and at the point of diffraction, or alternatively, the square root of their product.

Collecting the various components, the GTD formulation for the diffracted field in the deep shadow of an arbitrary smooth convex surface can be written as

$$E^d(s^d) = \sum_{n=1}^{N} \mathbf{D}_n E^i(A) \left[\frac{\rho_3}{s^d(\rho_3 + s^d)}\right]^{1/2} \exp\{-jk(\tau + s^d)\}$$
$$H^d(s^d) = \sqrt{\left(\frac{\hat{\epsilon}}{\mu}\right)}\hat{s}^d \times E^d \tag{18}$$

where

$$\mathbf{D}_n = \sqrt{\left(\frac{d\eta}{d\eta'}\right)}\frac{1}{\sqrt{(8j\pi k)}}\begin{bmatrix} D_n^e \exp\left\{-\int_0^T \Omega_n^e(\tau')d\tau'\right\} & 0 \\ 0 & D_n^m \exp\left\{-\int_0^T \Omega_n^m(\tau')d\tau'\right\} \end{bmatrix}$$

$$D_n^e = \{\mathrm{Ai}'(-\alpha_n)\}^{-2}2\bar{M}\exp\left(\frac{j\pi}{6}\right); \quad \bar{M} = \left[\frac{k}{2}\sqrt{\{\rho_c(0)\rho_c(\tau)\}}\right]^{1/3}$$

$$D_n^m = \{\alpha_n'\{\mathrm{Ai}(-\alpha_n')\}^2\}^{-1} 2\bar{M} \exp\left(\frac{j\pi}{6}\right)$$

$$\Omega_n^e(\tau') = \frac{\alpha_n}{\rho_c(\tau')} M(\tau') \exp\left(\frac{j\pi}{6}\right); \quad M(\tau') = \left[\frac{k\rho_c(\tau')}{2}\right]^{1/3}$$

$$\Omega_n^m(\tau') = \frac{\alpha_n'}{\rho_c(\tau')} M(\tau') \exp\left(\frac{j\pi}{6}\right)$$

If the incident pencil is convergent in nature it is possible that $\dfrac{\rho_3}{\rho_3 + s^d}$ will go negative. The square root sign introduces a possible phase shift of $\exp(\pm j\frac{\pi}{2})$. The correct choice is $\exp(j\frac{\pi}{2})$ as will be discussed in Section 6.6.

The diffraction matrix as given in eqn. 18 is suitable only for the deep shadow region. For the transition region we must use a modified version of the diffraction matrix given in eqn. 14. In the region of the shadow boundary when $|\tau| \to 0$, the following approximations for the quantities in eqn. 18 are valid:

$$\begin{aligned} \int_0^\tau \Omega_n^{e,m}(\tau')\,d\tau' &\simeq \Omega_n^{e,m}(0)\,\tau \\ \sqrt{\left(\frac{d\eta}{d\eta'}\right)} &= 1 \\ M(\tau') \simeq \bar{M} &\simeq M(0) \end{aligned} \tag{19}$$

From eqn. 14 the diffracted field in the transition region for an astigmatic incident pencil can be written as

$$E^d(s^d) = \mathbf{D} E^i(A) \left[\frac{\rho_3}{s^d(\rho_3 + s^d)}\right]^{1/2} \exp\{-jk(\tau + s^d)\} \tag{20a}$$

where

$$\mathbf{D} = -\sqrt{\left(\frac{2}{jk}\right)} M(0) \begin{bmatrix} \tilde{p}_-(\Delta, y) & 0 \\ 0 & \tilde{q}_-(\Delta, y) \end{bmatrix};$$

$$\Delta = \int_0^\tau \frac{M}{\rho_c}\,d\tau' \simeq \frac{M(0)}{\rho_c(0)}\,\tau$$

The value of y must now be chosen to ensure continuity of field across the shadow boundary. Using the UTD approach as outlined in Section 5.4 for edge diffraction, we deduce the appropriate value for y to be

$$y = M^{-1}\left[\frac{ks^d \rho_1 \rho_2(\rho_3 + s^d)}{2(\rho_1 + s^d)(\rho_2 + s^d)\rho_3}\right]^{1/2} \tag{20b}$$

where ρ_1 and ρ_2 are the principal radii of curvature of the incident pencil at the glancing point A on the convex surface. At the boundary between the transition and the deep shadow regions the formulation for the field changes from that of eqn. 20 to eqn. 18. Provided that the assumptions in eqn. 19 remain valid up to this point, the changeover will be smooth. The quantity $\sqrt{(d\eta/d\eta')}$ may be retained throughout the entire shadow region.

For a converging pencil, y, as defined in eqn. 20*b*, can go negative. We shall consider this case along with that for edge diffraction in Section 6.6. In the meantime we shall assume that the field does not traverse any caustic points in the transition region.

We must now determine the behaviour in the illuminated region. For plane wave incidence on a circular cylinder it was shown earlier that the diffracted field formulation for the transition region (i.e. in eqn. 14) yielded the geometrical optics reflected field at the 'reflection' boundary. Let us consider this reflected field for the general case of an astigmatic incident field upon a curved surface. Using the methods given in Chapter 4 the problem is set up in Fig. 6.8 as in Fig. 4.5. The phase along the axial ray will be unchanged and we need only concern ourselves with the amplitude of the field about this ray. This will be determined by the curvature matrix $\mathbf{Q}^r$ of the reflected wavefront at P_0 given in eqn. 4.55*a* as

$$\mathbf{Q}^r = \begin{bmatrix} 2C_{11}\cos\nu_i + Q^i_{11} & 2C_{12} - Q^i_{12} \\ 2C_{12} - Q^i_{12} & 2C_{22}\sec\nu_i + Q^i_{22} \end{bmatrix} \tag{21}$$

assuming that the components are known for the incident wavefront

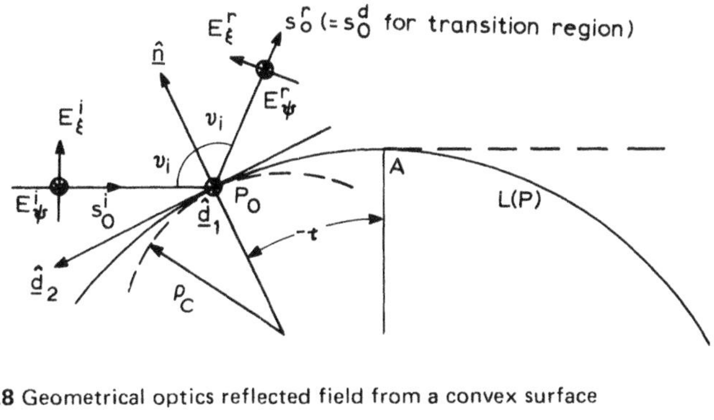

Fig. 6.8 Geometrical optics reflected field from a convex surface

curvature matrix $\mathbf{Q}^i$ and the curvature matrix $\mathbf{C}$ for the convex surface. The angle ν_i is the incident angle to the outward normal at the point of reflection, and may be written as

$$\nu_i = \frac{\pi}{2} - \gamma \tag{22}$$

From the components of $\mathbf{Q}^r$, the principal radii of curvature of the reflected wavefront, measured from the reflection point P_0, are given from eqn. 4.46 as

$$\frac{1}{\rho^r_{1,2}} = \tfrac{1}{2}[Q^r_{11} + Q^r_{22} \pm \sqrt{\{(Q^r_{11} - Q^r_{22})^2 + 4Q^{r2}_{12}\}}]$$

For field points within the transition region the value of γ in eqn. 22 will be small and tend to zero as the shadow boundary is approached. In this region, therefore,

$$\frac{1}{\rho^r_1} \simeq Q^r_{11} \simeq 2C_{11}\gamma + Q^i_{11} \tag{23a}$$

$$\frac{1}{\rho^r_2} \simeq Q^r_{22} \simeq \frac{2C_{11}}{\gamma} = \frac{2}{\rho_c\gamma} \tag{23b}$$

This result tells us several important features about the reflected field within the transition region. For example, the principal radii of curvature are in the principal planes $\hat{\boldsymbol{\xi}} - \hat{s}$, $\hat{\boldsymbol{\psi}} - \hat{s}$ used to analyse the diffraction process. The curvature ρ^r_2 in the $\hat{\boldsymbol{\xi}} - \hat{s}$ plane is seen to be unaffected by the curvature of the incident wavefront and is therefore identical to the value given in eqn. 10. In the other principal plane the curvature ρ^r_1 can be directly associated with the value of ρ_3 in the illuminated region. The quantity C_{11} is identified as the inverse of the *traverse* (to ρ_c) *curvature* ρ_{tc} of the surface. Thus

$$\frac{1}{\rho_3} = \frac{2\gamma}{\rho_{tc}} + Q^i_{11} \quad \text{for} \quad \tau < 0 \tag{24}$$

We conclude, therefore, that if an astigmatic pencil is incident on an arbitrary convex surface, the diffracted field given by eqn. 20 will give continuity of field at the 'reflection' boundary. Note that in the illuminated region we put $\sqrt{(d\eta/d\eta')} = 1$ since the diffracted field is not considered to be given by creeping rays emanating from the glancing point in this region.

In the deep illuminated region we calculate the geometrical optics reflected field directly (by the methods discussed in Chapter 4) with the total field given by the construction of eqn. 13. With eqn. 18 or 20 used for the diffracted field component in eqn. 13 we obtain an essentially uniform solution for the total field.

The solution just described has been subsequently refined by Pathak *et al.* (1980). In the shadow region their solution is in essence given by eqn. 20 above but without incorporating the approximations of eqn. 19. For the illuminated region their solution differs in that the modified Pekeris functions have arguments different to that used in the shadow region. This is done to retrieve the reflected geometrical optics field uniformly, without resorting to the construction of eqn. 13 above, while still retaining continuity of the field across the shadow boundary. We will now give details of this alternative uniform solution for the total field in the illuminated region.

For a field point P in the illuminated region, and denoting P_0 as the geometrical optics reflection point from the convex surface as in Fig. 6.8, we can write the solution for the total field from Pathak *et al.* (1980) as

$$E(P) = E^i(P) + \mathbf{R}E^i(P_0)\left[\frac{\rho_1^r\rho_2^r}{(\rho_1^r + s_0^r)(\rho_2^r + s_0^r)}\right]^{1/2} \exp(-jks_0^r) \tag{25a}$$

where

$$\mathbf{R} = \begin{bmatrix} R^e & 0 \\ 0 & R^m \end{bmatrix}$$

and s_0^r is the distance from P_0 to P.

This equation has the appearance of the usual geometrical optics field with the exception of $R^{e,m}$ which are given by

$$\begin{Bmatrix} R^e \\ R^m \end{Bmatrix} = -\frac{2}{\sqrt{-\Delta_L}} \exp\left[-j\left(\frac{\Delta_L^3}{12} + \frac{\pi}{4}\right)\right] \begin{Bmatrix} \tilde{p}_-(\Delta_L, y_L) \\ \tilde{q}_-(\Delta_L, y_L) \end{Bmatrix} \tag{25b}$$

where

$$\Delta_L = -2M\cos\nu_i \tag{25c}$$

$$y_L = M^{-1}\left[\frac{ks_0^r\rho_1^i\rho_2^i(\rho_1^r + s_0^r)}{2(\rho_1^i + s_0^r)(\rho_2^i + s_0^r)\rho_1^r}\right]^{1/2} \tag{25d}$$

All of the quantities in Δ_L, y_L are evaluated at the reflection point P_0.

In the deep illuminated region $\Delta_L \to -\infty$ and it follows from the asymptotic expansion of the modified Pekeris functions that $R^{e,m} \to \mp 1$

as required. On the shadow boundary $\Delta_L \to 0$ and it is readily shown that eqn. 25*a* is continuous with eqn. 20 as the shadow boundary is crossed. (Note in eqn. 25*a* as the shadow boundary is approached the infinity in the term $-2/\sqrt{-\Delta_L}$ is compensated by ρ_2^r – see eqn. 23*b* – to produce a finite field).

For the illuminated region, both uniform solutions yield results that, in practical applications, differ very little from each other. The uniform solution given by eqn. 25 is more convenient to implement in practice and for this reason is the preferred method here.

We have so far considered perfectly conducting convex surfaces. The extension of the above solution to smooth convex impedance surfaces is, unlike the case of edges in impedance surfaces discussed in Section 5.9, easily obtained. The problem has been considered by James (1980) who based his solution on the work of Wait and Conda (1959). Utilising the present formulation of eqns. 20 and 25 the only change necessary for impedance surfaces is to replace the modified Pekeris functions $\tilde{p}_-(x, y)$, $\tilde{q}_-(x, y)$ with a new modified Pekeris function $\tilde{Q}_-(x, y, z)$ ($Q(x, y, z)$ in James (1980)) defined by

$$\tilde{Q}_\pm(x, y, z) = V(x, z) - y \operatorname{sgn}(x) K_\pm(y|x|) \exp\left(\mp j\frac{\pi}{4}\right) \tag{26}$$

where the two-argument Pekeris function $V(x, z)$ is described in Section 2.2.3. For a given surface impedance Z_s we can define (as was done earlier for the impedance wedge) complex Brewster angles $\nu^e(\nu^m)$ for the case when the electric (magnetic) field of the incident wave is parallel to the surface. These angles are defined by

$$\nu^e = \cos^{-1}(Z_0/Z_s); \qquad \nu^m = \cos^{-1}(Z_s/Z_0)$$

where Z_0 is the impedance of the medium surrounding the wedge. The diffracted field in the shadow region is given by eqn. 20 with the following replacements for the transition functions:

$$\tilde{p}_-(\Delta, y) \to \tilde{Q}_-(\Delta, y, \gamma^e)$$
$$\tilde{q}_-(\Delta, y) \to \tilde{Q}_-(\Delta, y, \gamma^m) \tag{27a}$$

where

$$\gamma^{e,m} = -jM \cos \nu^{e,m} \tag{27b}$$

In the illuminated region we use the formulation given by eqn. 25*a* in preference to that in James (1980). The reflection coefficients $R^{e,m}$ are given by eqn. 25 with the following replacements for the transition functions:

$$\tilde{p}_-(\Delta_L, y_L) \rightarrow \tilde{Q}_-(\Delta_L, y_L, \gamma^e)$$
$$\tilde{q}_-(\Delta_L, y_L) \rightarrow \tilde{Q}_-(\Delta_L, y_L, \gamma^m) \tag{27c}$$

In the deep illuminated region $\Delta_L \rightarrow -\infty$ and it follows that the function $\tilde{Q}_-(\Delta_L, y_L, \gamma^{e,m}) \sim \hat{V}(\Delta_L, \gamma^{e,m})$, where this Pekeris carot function is defined in eqn. 2.60. The asymptotic expansion of $\hat{V}(x, y)$ (given by eqn. 2.60*d* for large negative values of x) when substituted into the expressions for the reflection coefficients yields

$$R^{e,m} = \frac{\cos \nu_i - \cos \nu^{e,m}}{\cos \nu_i + \cos \nu^{e,m}}$$

which are the Fresnel reflection coefficients for the impedance surface as needed to represent the reflected field in the deep illuminated region.

In the deep shadow region $\Delta \rightarrow \infty$, $\tilde{Q}_-(\Delta, y, \gamma) \sim \hat{V}(\Delta, \gamma)$, and the residue series solution given by eqn. 2.60*c* can be used to evaluate the field. The solution is given by eqn. 18 where the quantities $D_n^{e,m}$, $\Omega_n^{e,m}$ are now given by

$$D_n^{e,m} = [\bar{\alpha}_n^{e,m} \mathrm{Ai}^2(-\bar{\alpha}_n^{e,m}) + \mathrm{Ai}'^2(-\bar{\alpha}_n^{e,m})]^{-1} 2\bar{M} \exp\left(\frac{j\pi}{6}\right)$$

$$\Omega_n^{e,m} = \frac{\bar{\alpha}_n^{e,m}}{\rho_c(\tau')} M(\tau') \exp\left(\frac{j\pi}{6}\right)$$

where $\bar{\alpha}_n^{e,m}$ are the roots of the equation

$$\mathrm{Ai}'(-\bar{\alpha}_n^{e,m}) + \gamma^{e,m} \exp\left(-\frac{j\pi}{3}\right) \mathrm{Ai}(-\bar{\alpha}_n^{e,m}) = 0.$$

Finally, it is important to note that the impedance surface solution applies only to fields that are locally normally incident to the surface. As noted earlier in Section 5.9, the extension to oblique incidence for impedance surfaces is difficult since, unlike the perfectly conducting case, the field components are coupled.

6.3 Radiation from sources on a smooth convex surface

The solution in the previous section was for an incoming field from a source situated at some point away from the surface. When the sources creating the field are situated on the surface, such as a radiating aperture or a surface mounted antenna, appropriate GTD formulations have been suggested by a number of authors. The most notable contributions from which we develop a solution here are to be found in Pathak and

Kouyoumjian (1974), Mittra and Safavi-Naini (1979) and Pathak *et al.* (1981). We begin with the canonical solution for magnetic line sources given in Section 3.4.4. As before we have a deep illuminated region, transition and deep shadow regions, as illustrated in Fig. 6.9. The scattered field is related to an axial magnetic current m_a and a transverse magnetic current m_t. For magnetic line sources on a circular cylinder (where the elemental currents are directed axially or transversely) the diffracted field in the transition region is given by eqn. 3.164. Using the ray-based co-ordinates for the electric field it follows from the first term in the equations in eqn. 3.164 that along ray 1 in Fig. 6.9

$$\begin{bmatrix} E_\psi^d(s^d) \\ \\ E_\xi^d(s^d) \end{bmatrix} = -\sqrt{\left(\frac{jk}{8\pi}\right)} \begin{bmatrix} \dfrac{1}{jM} f\left(M\dfrac{\tau}{a}\right) & 0 \\ \\ 0 & g\left(M\dfrac{\tau}{a}\right) \end{bmatrix} \begin{bmatrix} m_t \\ \\ m_a \end{bmatrix} \cdot \frac{\exp\{-jk(\tau + s^d)\}}{\sqrt{(s^d)}} \tag{28a}$$

where the functions f and g are the Fock functions described in Section 2.2.3.

Eqn. 28*a* may be written in the compact form

$$\boldsymbol{E}^d(s^d) = \mathbf{D}\boldsymbol{m}\frac{\exp\{-jk(\tau + s^d)\}}{\sqrt{(s^d)}} \tag{28b}$$

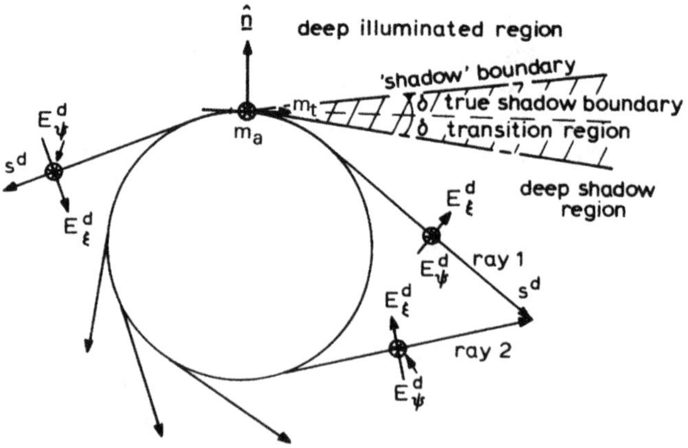

Fig. 6.9 Line source excitation on the convex surface

The remaining terms in eqn. 3.164 account for the ray that creeps around the cylinder in the opposite direction, as shown by ray 2 in Fig. 6.9. This can be described by applying eqn. 28 to this ray path.

In the shadow region when $\tau > 0$ we may use the residue series solution for the Fock functions given in Section 2.2.3. Consequently, eqn. 28a becomes

$$\begin{bmatrix} E^d_\psi(s^d) \\ E^d_\xi(s^d) \end{bmatrix} = \sqrt{\left(\frac{jk}{8\pi}\right)} \sum_{n=1}^{N} \begin{bmatrix} D^e_n \exp(-\Omega^e_n \tau) & 0 \\ 0 & D^m_n \exp(-\Omega^m_n \tau) \end{bmatrix} \cdot$$

$$\cdot \begin{bmatrix} m_t \\ m_a \end{bmatrix} \frac{\exp\{-jk(\tau + s^d)\}}{\sqrt{(s^d)}}; \qquad \tau > 0 \qquad (28c)$$

where

$$D^e_n = \left\{\exp\left(\frac{j\pi}{6}\right) M \,\mathrm{Ai}'(-\alpha_n)\right\}^{-1}$$

$$D^m_n = \{\alpha'_n \mathrm{Ai}(-\alpha'_n)\}^{-1}$$

$$\Omega^e_n = \frac{\alpha_n}{a} M \exp\left(\frac{j\pi}{6}\right)$$

$$\Omega^m_n = \frac{\alpha'_n}{a} M \exp\left(\frac{j\pi}{6}\right)$$

This equation is similar to the equivalent expression given by eqn. 12 for an incident plane wave on the cylinder. Note that the attenuation coefficients are identical. As before the number of terms that can be taken in the series is limited by the inherent assumption in the Fock function representation that ν_n and ν'_n are in the region of ka.

In the deep illuminated region the field is given by twice the direct radiation from the sources due to the image effect. It was shown in Section 3.4.4 that provided $\sin \tau/a \simeq \tau/a$ with $M(\tau/a)$ a large number, the Fock functions retrieve the geometrical optics field when $\tau < 0$. If we choose a 'shadow' boundary as shown in Fig. 6.9 where $\delta \cong M^{-1}$, and we express the total field as

$$E^t = U(\epsilon^i)E^i + U(-\epsilon^i)E^d$$

then the discontinuity in the geometrical optics and diffracted field at this boundary will compensate to ensure continuity of the total field.

An alternative solution for the illuminated region is to redefine the Fock function arguments in the same manner as the previous section for the case of the source removed from the surface. The electric field $\boldsymbol{E}$ in the illuminated region is now written as

$$\boldsymbol{E} = \mathbf{D}_L \boldsymbol{m} \frac{\exp(-jks)}{\sqrt{(s)}} \tag{29a}$$

where s is the distance of the source to the field point, $\mathbf{D}_L$ given by

$$\mathbf{D}_L = -\sqrt{\left(\frac{jk}{8\pi}\right)} \begin{bmatrix} \frac{1}{jM} f(\Delta_L) & 0 \\ 0 & g(\Delta_L) \end{bmatrix} \exp\left(\frac{-j\Delta_L^3}{3}\right), \tag{29b}$$

$\Delta_L = -M \cos \nu_i$ and ν_i is the radiation angle measured from the outward normal $\hat{\boldsymbol{n}}$ from the source (viz, $\nu_i = \cos^{-1} \hat{\boldsymbol{s}} \cdot \hat{\boldsymbol{n}}$). It is easily shown that eqn. 29 goes uniformly into the geometrical optics field in the deep illuminated region and provides continuity with eqn. 28 at the shadow boundary.

If we make the usual high-frequency assumption that the field given by eqns. 28 and 29 expresses the local behaviour of the field, we may extend this result to give a solution for point sources. With the magnetic current term $\boldsymbol{m}$ now taken to be a point source the radial dependence of $\boldsymbol{m}$ is changed to

$$\frac{\exp(-jks)}{\sqrt{(8j\pi ks)}} \longrightarrow \frac{\exp(-jks)}{4\pi s}$$

This follows directly by comparing the 2-dimensional and 3-dimensional Green's function representations as given in Section 2.1.2.

Radiation from the source along the surface follows the geodesic path. If the radius of curvature is infinite then the divergence of nearby rays from the source along the surface will follow straight paths and cause the field to decay by $\rho^{-1/2}$. In terms of the quantities defined in Fig. 6.10 we may write

$$\rho^{-1/2} = \sqrt{\left(\frac{d\beta}{d\eta'}\right)}$$

When the surface is curved the rays emanating from A' will have a radius of curvature ρ_3 in the tangent plane to the surface at A'. This curvature will only be measured from the source at A if the surface is flat, and generally we have $d\eta' = \rho_3 d\beta'$ where $d\beta'$ is the angle between adjacent rays in the surface tangent plane at A'. The divergence factor of the field along the surface is, therefore,

$$\sqrt{\left(\frac{d\beta}{\rho_3 d\beta'}\right)}$$

As before, the field along the diffracted ray from A' will be modified by the factor

$$\left[\frac{\rho_3}{\rho_3 + s^d}\right]^{1/2}$$

If the point source is situated on a convex surface with a varying radius of curvature $\rho_c(\tau')$ as in Fig. 6.11a, then we can apply the methods of the previous section to obtain a uniform GTD solution. The procedure is very similar and we need only quote the result here. Thus in the shadow region the field along a diffracted ray s^d is given by

$$\boldsymbol{E}^d(s^d) = \mathbf{D}\boldsymbol{m}\frac{\exp\{-jk(\tau + s^d)\}}{\sqrt{\{s^d(\rho_3 + s^d)\}}}; \qquad \tau > 0 \qquad (30a)$$

where

$$\mathbf{D} = -\frac{jk}{4\pi}\sqrt{\left(\frac{d\beta}{d\beta'}\right)}\begin{bmatrix}\frac{1}{j\bar{M}}f(\Delta) & 0 \\ 0 & g(\Delta)\end{bmatrix}$$

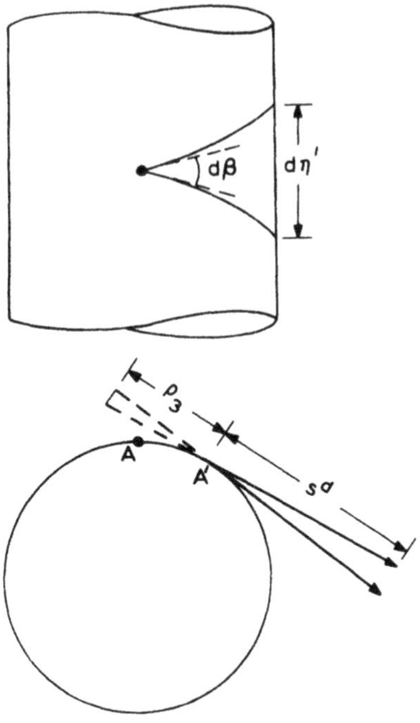

Fig. 6.10 Point source on the cylinder

$$\Delta = \int_0^{\tau} \frac{M(\tau')}{\rho_c(\tau')} d\tau'$$

$$\bar{M} = \left[\frac{k}{2} \sqrt{\{\rho_c(0)\rho_c(\tau)\}}\right]^{1/3}$$

In the illuminated region we derive the expression for the field from eqn. 29 as

$$E(s) = \mathbf{D}_L \boldsymbol{m} \frac{\exp(-jks)}{s} \tag{30b}$$

$$\mathbf{D}_L = \frac{-jk}{4\pi} \begin{bmatrix} \frac{1}{jM(0)} f(\Delta_L) & 0 \\ 0 & g(\Delta_L) \end{bmatrix} \exp\left(-\frac{j\Delta_L^3}{3}\right)$$

$$\Delta_L = -M(0) \cos \nu_i$$

Equation 30 provides a uniform solution for the total field for a magnetic point source on a convex perfectly conducting smooth surface. In this solution the geometrical optics field is retrieved in the deep illuminated region, the field is continuous through the shadow boundary, and in the deep shadow region the residue series for the Fock functions yields the creeping wave solution.

For an arbitrary convex surface it is, in general, necessary to determine the geodesic path and the quantity $\sqrt{d\beta/d\beta'}$ numerically. Furthermore, the geodesic path leaving the source will not necessarily be in the direction of a principal radii of curvature of the surface. In this case surface ray torsion can occur and this gives a contribution to the leading term for the field in addition to that given by eqn. 30.

To account for the torsional effect we adapt the results of Pathak *et al.* (1981) to suitably modify eqn. 30. We first define the directions $\hat{C}_1$, $\hat{C}_2$ of the principal surface curvatures ρ_1^c, ρ_2^c at the source point A in Fig. 6.11(*a*). These are chosen such that we ensure $\rho_1^c \geqslant \rho_2^c$ and are rotated through an angle γ relative to the source surface coordinates $\hat{\psi}_0$, $\hat{s}_0$ as illustrated in Fig. 6.11(*b*). The effect of torsion is given by modifying $\boldsymbol{m}$ in eqn. 30 as follows

$$\boldsymbol{m} = \begin{bmatrix} m_t \\ m_a \end{bmatrix} \longrightarrow \begin{bmatrix} m_t + Tm_a \\ m_a \end{bmatrix} \tag{31a}$$

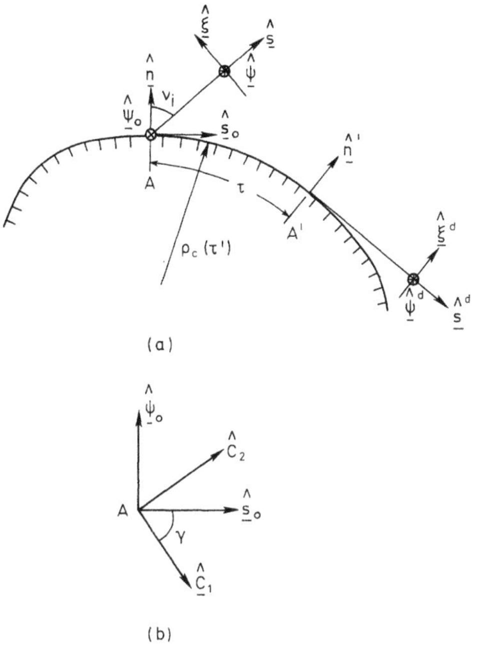

Fig. 6.11 Point source on a convex surface

where in the shadow region

$$T = T_s = \tfrac{1}{2}\sin 2\gamma \left[\frac{\rho_1^c - \rho_2^c}{\rho_2^c \cos^2\gamma + \rho_1^c \sin^2\gamma}\right] \tag{31b}$$

and in the illuminated region, the dominant effect is given by

$$T \approx T_s\left[\frac{1}{jM(0)} f(\Delta_L) - g(\Delta_L)\cos\nu_i\right]\exp\left(\frac{-j\Delta_L^3}{3}\right) \tag{31c}$$

The solution for the field given by eqn. 30 remains uniform throughout with the inclusion of torsion.

The above solution for a magnetic point source on a convex surface is commonly used to solve for the electromagnetic field radiated by an aperture in a smooth perfectly conducting convex surface. If $\boldsymbol{E}(A)$ is the electric field at a point A in the aperture and $\hat{\boldsymbol{n}}_A$ is the local outward normal to the surface, then the elemental magnetic source $\boldsymbol{m}$ at A is given by $\boldsymbol{E}(A) \times \hat{\boldsymbol{n}}_A da$ where da is the elemental area at A. Substituting for $\boldsymbol{m}$ in eqn. 30 and integrating over the aperture yields the total radiated field. Applications of this technique are given, for example, in Pathak *et al.* (1981).

Equation 30 can also be used for a short monopole situated at A in Fig. 6.11. If the electric point source $\boldsymbol{e}$ is given by $\hat{\boldsymbol{n}}e_n$, then the resultant radiated field may be determined from eqn. 30 with the following replacements. In the shadow region

$$m \longrightarrow \begin{bmatrix} T_s \\ 1 \end{bmatrix} e_n Z_0$$

where Z_0 is the free space impedance of the medium surrounding the monopole and T_s is given by eqn. 31b. In the illuminated region

$$m \longrightarrow \begin{bmatrix} T \\ 1 \end{bmatrix} \sin \nu_i e_n Z_0$$

where T is given by eqn. 31c.

The uniform GTD solution given by eqn. 30 is for the radiated field at points well-removed from the surface. In some cases it is of interest to have expression for the field on the surface. This has particular application in evaluating the mutual coupling between apertures on convex surfaces. A number of asymptotic solutions based on the surface Fock functions (not given here) have been developed for the surface field representation. We will not consider these solutions in detail here, but a comprehensive critical survey is to be found in Bird (1984) on the leading techniques proposed to date.

6.4 Higher order terms

The asymptotic solutions developed so far for diffraction around a smooth convex surface have not included the effect of the rate of change in the surface curvature. Such effects are contained within the higher order terms in the asymptotic expansion of the diffracted field. For a slowly varying curvature where GTD can be applied, the correction to the amplitude terms will not be particularly significant. The corrections can be important in the sensitive attenuation constants $\Omega_n^{e,m}$ which, it will be recalled, are applicable for the source on, or removed from, the surface.

Higher order terms for a cylinder of variable curvature were studied by Franz and Klante (1959) who obtained a correction to the attenuation constants dependent on the curvature and its first two derivatives.

Their results were obtained from an asymptotic expansion of the integral equation for the creeping wave. Beginning with the same integral equation, Keller and Levy (1959) derived similar results which they showed to be in agreement with the asymptotic solution for diffraction by an elliptic and a parabolic cylinder. Both of these papers were for cylindrical bodies only. For a general convex surface the effect of the transverse curvature ρ_{tc} to the geodesic arc was given by Hong (1967) for magnetic polarisation and by Voltmer (1970) for electric polarisation, although in the latter case the transverse curvature only appears in the amplitude term.

The attenuation constants, therefore, correct to second order, are

$$\Omega_n^e = \frac{\alpha_n}{\rho_c} M \exp\left(\frac{j\pi}{6}\right)\left\{1 + \frac{\alpha_n}{M^2}\left(\frac{1}{60} - \frac{2}{45}\rho_c \frac{\partial^2 \rho_c}{\partial \tau^2} + \frac{4}{135}\left(\frac{\partial \rho_c}{\partial \tau}\right)^2\right)\exp\left(\frac{-j\pi}{3}\right)\right\}$$

$$\Omega_n^m = \frac{\alpha_n'}{\rho_c} M \exp\left(\frac{j\pi}{6}\right)\left\{1 + \frac{\alpha_n'}{M^2}\left(\frac{1}{60} - \frac{2}{45}\rho_c \frac{\partial^2 \rho_c}{\partial \tau^2} + \frac{4}{135}\left(\frac{\partial \rho_c}{\partial \tau}\right)^2\right)\exp\left(\frac{-j\pi}{3}\right) + \frac{1}{(M\alpha_n')^2}\left(\frac{1}{10} - \frac{\rho_c}{4\rho_{tc}} - \frac{1}{60}\rho_c \frac{\partial^2 \rho_c}{\partial \tau^2} + \frac{1}{90}\left(\frac{\partial \rho_c}{\partial \tau}\right)^2\right)\exp\left(\frac{-j\pi}{3}\right);\right\}$$

$$\rho_c < \rho_{tc}$$

For details of the derivation of these results the reader is referred to the original papers.

6.5 Diffraction at a discontinuity in curvature

The diffracted field resulting from a discontinuity in curvature has been studied by Weston (1962), Senior (1972), Kaminetsky and Keller (1972). In this last paper an asymptotic series solution was constructed which was similar to that used in Section 5.5 to obtain higher order edge diffraction terms. They considered an impedance boundary condition on a surface where the discontinuity could occur in any higher order derivative of the surface, whether along a curve or at an isolated point. For the special case of a perfectly conducting 2-dimensional surface with a discontinuity in curvature their results agreed with those obtained by Senior (1972) using a different approach. This was based on the earlier work of Weston (1962) who modelled the discontinuity in curvature by the junction of two parabolic cylinders. This allowed the field on the surface in the vicinity of the junction to be formulated rigorously and an asymptotic evaluation of the subsequent integrals yielded the diffraction coefficients.

All these results are non-uniform in the sense that they are invalid in the vicinity of the reflection boundary. We will now derive a uniform solution based on the physical optics approximation which can be corrected to obtain the rigorous result at field points removed from the reflection boundary. We begin by considering a magnetically polarised plane wave incident upon a truncated perfectly conducting circular cylinder such that we have straight edges at A and B as shown in Fig. 6.12. The scattered magnetic field $\boldsymbol{H}^s$ due to the electric surface current $\boldsymbol{J}_s$ flowing on the surface is given from eqn. 2.16 as

$$\boldsymbol{H}^s = jk \int_0^{\phi_1} \boldsymbol{J}_s \times \hat{\boldsymbol{P}} \frac{\exp(-jkP)}{\sqrt{(8j\pi kP)}} a d\phi$$

With the incident field given by

$$\boldsymbol{H}^i = \hat{z} \exp\{jk\rho \cos(\phi - \phi_0)\}$$

we use the physical optics approximation to the currents

$$\boldsymbol{J}_s = 2\hat{\boldsymbol{n}} \times \boldsymbol{H}^i$$

to yield the scattered magnetic field as

$$\boldsymbol{H}^s = \hat{z} \int_0^{\phi_1} f(\phi') \exp\{jkg(\phi')\} d\phi' \tag{32}$$

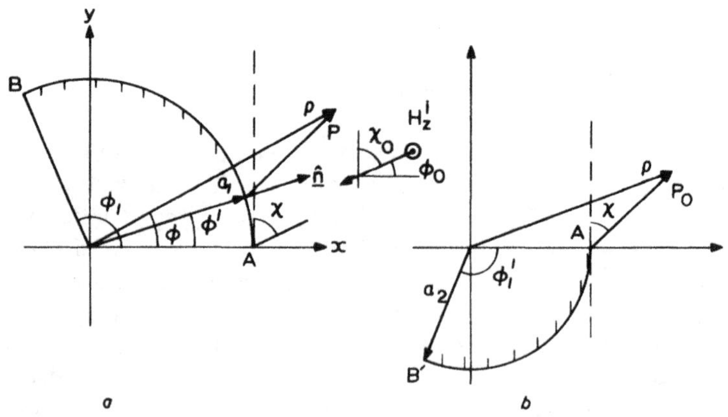

Fig. 6.12 Construction for the discontinuity in curvature problem

where

$$f(\phi') = \frac{2jka_1}{P\sqrt{(8j\pi kP)}}\{\rho\cos(\phi-\phi')-a_1\}$$

$$g(\phi') = a_1\cos(\phi_0-\phi')-P$$

$$P = \{\rho^2+a_1^2-2a_1\rho\cos(\phi-\phi')\}^{1/2}$$

This equation can be evaluated asymptotically using the method of stationary phase given in Section 2.3.1. In particular we will concentrate on the contribution from the endpoint at $\phi'=0$. To evaluate the first two terms in the asymptotic expansion about this endpoint as given by eqn. 2.80 requires the first derivative of the amplitude function $f(\phi')$ and the first two derivatives of the phase function $g(\phi')$.

Relating these various quantities in terms of distance P_0 from the edge at A and angle χ to the tangent plane at the edge, as in Fig. 6.12, so that

$$\sin\chi = \frac{\rho\cos\phi-a_1}{P_0}, \qquad \cos\chi = \frac{\rho\sin\phi}{P_0}$$

we get

$$\left.\begin{aligned}
f(0) &= \frac{2jka_1}{\sqrt{(8j\pi kP_0)}}\sin\chi \\
f'(0) &= \frac{2jka_1}{\sqrt{(8j\pi kP_0)}}\cos\chi\left(1+\frac{3}{2}\frac{a_1}{P_0}\sin\chi\right) \\
g(0) &= a_1\cos\phi_0-P_0 \\
g'(0) &= a_1(\cos\chi_0+\cos\chi); \qquad \chi_0 = \frac{\pi}{2}-\phi_0 \\
g''(0) &= -a_1\left\{\sin\chi_0+\sin\chi\left(1+\frac{a_1}{P_0}\sin\chi\right)\right\}
\end{aligned}\right] \tag{33}$$

Thus for the diffracted field H_z^d from the edge at $\phi'=0$, the first two terms in the asymptotic series are from eqn. 2.80

$$H_z^d \sim -\exp(jkg)\frac{1}{jkg'}\left(f-\frac{1}{jk}\frac{f'g'-fg''}{(g')^2}\right)\Bigg|_{\phi'=0} \tag{34a}$$

When $g'\rightarrow 0$, an optical boundary is being approached, and for a valid solution we can use the Fresnel integral formulation of eqn. 2.86, viz.,

$$H_z^d \sim \pm\frac{|g'|}{g'}f\exp(jkg)\exp(\mp jv^2)\sqrt{\left(\frac{2}{k|g''|}\right)}F_\pm(v)\Bigg|_{\phi'=0};$$

$$g''\gtrless 0, \quad v = \sqrt{\left(\frac{k}{2|g''|}\right)}|g'| \tag{34b}$$

where the first term in eqn. 34*a* is retrieved for large v. The complete second term, however, is not recovered. To be specific, from eqn. 2.36

$$F_{+}(x) \sim \frac{1}{2x} \exp\left\{\pm j\left(x^2 + \frac{\pi}{2}\right)\right\}\left(1 \pm \frac{1}{2jx^2} + \ldots\right)$$

so that substitution into eqn. 34*b* when v is large yields

$$H_z^d \sim -\exp(jkg)\frac{1}{jkg'}\left\{f + \frac{fg''}{jk(g')^2}\right\}\Bigg|_{\phi'=0} \qquad (34c)$$

Compared to eqn. 34*a* it will be seen that the term in f' is absent. This is because the Fresnel integral formulation used in eqn. 34*b* was derived for only the first order term. The complete Fresnel integral formulation to second order is, in fact, given by eqn. 34*b* with f replaced by $f - \dfrac{f'g'}{g''}$ and an additional term of

$$-\frac{1}{jk}\frac{f'}{g''}\exp(jkg)$$

Note that on the reflection boundary, eqn. 34*b* gives the complete leading term for the diffracted field. As for edge diffraction, our uniform solution will be required to produce the correct leading term along the reflection boundary and retrieve the rigorous solution at field points removed from this boundary. To this end simpler expressions result if we begin with the formulation of eqn. 34*b*. For the cylinder in Fig. 6.12*b* the diffraction components for the edge at A are as in eqn. 33 but with a_1 replaced by a_2. If these two cylinders are now connected at A as in Fig. 6.13*a* then the scattered field will be formulated as

$$H_z^s = \int_{B'}^{B} dl = \int_{A}^{B} dl - \int_{A}^{B'} dl$$

and hence the diffracted field from the discontinuity in curvature at A is given by the difference between the endpoint contributions for the two cylinders at A. Only terms dependent on a_1 or a_2 will remain and these occur only in the second term of eqn. 34*a*. Thus from eqn. 34*a* the diffracted field H_z^d for a discontinuity in curvature, with the phase referred to A, is

$$H_z^d \sim \frac{2\exp(-jkP_0)}{jk\sqrt{(8j\pi kP_0)}}\frac{1+\cos(\chi-\chi_0)}{(\cos\chi_0+\cos\chi)^3}\left(\frac{1}{a_1}-\frac{1}{a_2}\right) \qquad (35a)$$

This is invalid in the vicinity of the reflection boundary when $\chi_0 \simeq \pi - \chi$ and to obtain a useful solution the Fresnel integral formulation

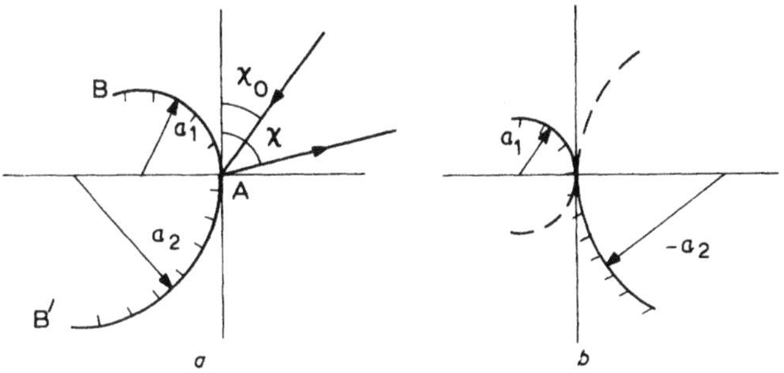

Fig. 6.13 Examples of discontinuity in curvature

of eqn. 34*b* must be used. Assuming initially that both a_1 and a_2 are positive, then the second derivative of the phase function *g* is seen from eqn. 33 to be always negative. Therefore from eqn. 34*b*, with the phase referred to *A* as before, we get a uniform solution to the diffracted field in the form

$$H_z^d \sim -\frac{|\cos\chi_0+\cos\chi|}{\cos\chi_0+\cos\chi}\,\frac{2jk\sin\chi}{\sqrt{(8j\pi kP_0)}}\sqrt{\left(\frac{2}{k}\right)}\exp(-jkP_0)\left[\frac{a_1\exp(jv_1^2)}{\sqrt{(|g_1''|)}}F_-(v_1)-\frac{a_2\exp(jv_2^2)}{\sqrt{(|g_2''|)}}F_-(v_2)\right] \qquad (35b)$$

where

$$v_{1,2}=\sqrt{\left(\frac{k}{2|g_{1,2}''|}\right)}\,|g_{1,2}'|$$

and the subscripts refer to either a_1 or a_2 used in the quantities given in eqn. 33. Removed from the reflection boundary the asymptotic solution to the Fresnel integral gives, from eqns. 34*c* and 35*b*, the diffracted field as

$$H_z^d \sim \frac{2\exp(-jkP_0)}{jk\sqrt{(8j\pi kP_0)}}\,\frac{\sin\chi(\sin\chi_0+\sin\chi)}{(\cos\chi_0+\cos\chi)^3}\left[\frac{1}{a_1}-\frac{1}{a_2}\right] \qquad (35c)$$

To obtain the true asymptotic solution of eqn. 35*a* for field points removed from the reflection boundary it is necessary to multiply eqn. 35*c* by the correction factor

$$C^m = \frac{1+\cos(\chi-\chi_0)}{\sin\chi(\sin\chi_0+\sin\chi)} \qquad (36a)$$

Note that on the reflection boundary when $\chi_0=\pi-\chi$ this correction factor is equal to unity and multiplying eqn. 35*b* by C^m effects a uniform solution which smoothly reduces to eqn. 35*a* away from this boundary.

For electric polarisation where the incident field is given by

$$E^i = \hat{z}\exp\{jk\rho\cos(\phi-\phi_0)\}$$

the scattered field is derived from eqn. 2.16 and is in the same form as eqn. 32 with H^s replaced by E^s and

$$f(\phi') = -\frac{2jka_1}{\sqrt{(8j\pi kP)}}\cos(\phi'-\phi_0)$$

The phase function $g(\phi')$ is the same as before. Proceeding with the analysis the diffracted field removed from the reflection boundary is given as

$$E_z^d \sim -H_z^d \qquad \text{(as in eqn. 35}a\text{)}$$

The corresponding uniform solution is as in eqn. 35*b* with a change in sign, and $\sin\chi$ replaced by $\sin\chi_0$. Also, the correction factor C^e for electric polarisation is

$$C^e = \frac{1+\cos(\chi-\chi_0)}{\sin\chi_0(\sin\chi_0+\sin\chi)} \tag{36b}$$

which goes to unity on the reflection boundary.

We can now make use of the results in Senior (1972) or Kaminetsky and Keller (1972) to improve our diffracted field still further. In their results expressions similar to eqn. 35*a* are given but with the additional term

$$\frac{2\exp(-jkP_0)}{jk\sqrt{(8j\pi kP_0)}}\frac{1+\cos(\chi+\chi_0)}{(\cos\chi_0+\cos\chi)^3}\left[\frac{1}{a_1}-\frac{1}{a_2}\right] \tag{37}$$

added to both the magnetic and electric polarised diffracted fields. Further correction (multiplication) factors C_1^e, C_1^m to correct eqn. 35*a*, and its equivalent for electric polarisation, by using eqn. 37 are deduced as

$$C_1^e = \frac{2\sin\chi_0\sin\chi}{1+\cos(\chi-\chi_0)}, \qquad C_1^m = \frac{2(1+\cos\chi\cos\chi_0)}{1+\cos(\chi-\chi_0)}$$

Our composite correction factors C^e and C^m for the uniform solution now become

$$C^e = \frac{2\sin\chi}{\sin\chi_0+\sin\chi}; \qquad C^m = \frac{2(1+\cos\chi\cos\chi_0)}{\sin\chi(\sin\chi_0+\sin\chi)} \tag{38a}$$

and the diffracted field is given by

$$\begin{bmatrix} E_z^d \\ H_z^d \end{bmatrix} = \begin{bmatrix} C^e \sin\chi_0 \\ -C^m \sin\chi \end{bmatrix} \frac{|\cos\chi_0 + \cos\chi|}{\cos\chi_0 + \cos\chi} 2j\sqrt{(2k)} \left[\frac{a_1 \exp(jv_1^2)}{\sqrt{|g_1''|}} F_-(v_1) - \frac{a_2 \exp(jv_2^2)}{\sqrt{(|g_2''|)}} F_-(v_2) \right] \frac{\exp(-jkP_0)}{\sqrt{(8j\pi k P_0)}} \tag{38b}$$

where

$$v_{1,2} = \sqrt{\left(\frac{k}{2|g_{1,2}''|}\right)} |g_{1,2}'|$$

$$g_{1,2}' = a_{1,2}(\cos\chi_0 + \cos\chi)$$

$$g_{1,2}'' = -a_{1,2}\left\{\sin\chi_0 + \sin\chi\left(1 + \frac{a_{1,2}}{P_0}\sin\chi\right)\right\}$$

When the Fresnel integral arguments are large eqn. 38*b* reduces to

$$\begin{aligned} E_z^d &\sim -\frac{4}{jk}\left(\frac{1}{a_1} - \frac{1}{a_2}\right) \frac{\sin\chi_0 \sin\chi}{(\cos\chi_0 + \cos\chi)^3} \frac{\exp(-jkP_0)}{\sqrt{(8j\pi kP_0)}} \\ H_z^d &\sim \frac{4}{jk}\left(\frac{1}{a_1} - \frac{1}{a_2}\right) \frac{1 + \cos\chi\cos\chi_0}{(\cos\chi_0 + \cos\chi)^3} \frac{\exp(-jkP_0)}{\sqrt{(8j\pi kP_0)}} \end{aligned} \tag{38c}$$

Eqn. 38 is valid for the diffracted field in the region of space where $0 < \chi < \pi$. For the remaining portion of real space the diffracted field is determined by the creeping waves which arise for the source generated by the discontinuity in curvature. These creeping waves will be generally weak since the change in curvature has resulted in a term which, although behaving as an edge diffracted wave, is of higher order. (The same order, in fact, as the slope-diffraction term discussed in Section 5.5). If it is desired to include these creeping waves then they can be computed using the formulation given in Section 6.3. There may also be creeping waves created directly from the incident field and these may be included by using the approach given in Section 6.2. The scattered field is given by the inclusion of the geometrical optics field reflected from the surface which can be evaluated by employing the methods given in Chapter 4.

The solution so far has been for a change in curvature in a convex surface. If the surface is concave then the problem is of little practical interest since the geometrical optics field will dominate the solution. Eqn. 38*c* is valid, however, if the radii of curvature a_1, a_2 are taken as negative. (Note that eqn. 38*b* is not applicable, as will become clear in the following discussion.) Of interest is the case illustrated in Fig. 6.13*b* where one radius, a_2, is negative. This may be, for example, an approximation to the cross-section of a reflector antenna with a rolled edge. Returning to the diffracted field formulation of eqn. 34*b* we see that the choice of Fresnel integral is determined by the sign of g'', and from eqn. 33 we have

$$g''_{1,2} = -a_{1,2}\left\{\sin\chi_0 + \sin\chi\left(1 + \frac{a_{1,2}}{P_0}\sin\chi\right)\right\}$$

When $a_{1,2}$ was positive this term always remained negative. For the situation in Fig. 6.13*b*, however, a_2 is negative and the sign of g''_2 changes from negative to positive when

$$\sin\chi_0 > -\sin\chi\left(1 + \frac{a_2}{P_0}\sin\chi\right)$$

In the region of the reflection boundary $\chi_0 \simeq \pi - \chi$ so that for a_2 negative we get

$$g''_2 \gtrless 0 \quad \text{for} \quad P_0 \gtrless \frac{|a_2|}{2}\sin\chi_0$$

Following through the analysis the diffracted field is given by eqn. 38*b* with the replacement

$$-a_2\exp(jv_2^2)F_-(v_2) \rightarrow \pm|a_2|\exp(\mp jv_2^2)F_\pm(v_2); \quad g''_2 \gtrless 0 \quad (39)$$

When g''_2 changes sign it indicates that the ray has passed through a caustic. (This problem as it applies to edge diffraction and diffraction around a smooth convex surface is to be considered in the next section).

Eqn. 38 is for a plane wave at normal incidence to a cylindrical surface having a discontinuity in curvature. When the incident field is from a line source at a distance ρ from the discontinuity, one finds by performing the same analysis as above that provided $k\rho \gg 1$, the asymptotic solution yields eqn. 38 with P_0 replaced in $g''_{1,2}$ by σ where

$$\sigma = \frac{\rho P_0}{P_0 + \rho} \quad (40)$$

For the general case of an astigmatic ray at an oblique incidence of angle θ_0 on an arbitrary curved surface having a discontinuity in curvature, we may use the methods used in Sections 5.3 and 5.4 for edge diffraction to extend eqn. 38 to include this situation. Thus for field points removed from the reflection boundary eqn. 38*c* becomes

$$E_z^d \sim -\frac{4}{jk}\left(\frac{1}{a_1} - \frac{1}{a_2}\right)\frac{\sin\chi_0\sin\chi\csc^2\theta_0}{(\cos\chi_0 + \cos\chi)^3}\left[\frac{\rho_3}{\rho_3 + s^d}\right]^{1/2} \frac{\exp(-jks^d)}{\sqrt{(8j\pi ks^d)}} \quad (41a)$$

$$H_z^d \sim \frac{4}{jk}\left(\frac{1}{a_1}-\frac{1}{a_2}\right)\frac{(1+\cos\chi\cos\chi_0)\csc^2\theta_0}{(\cos\chi_0+\cos\chi)^3}\left[\frac{\rho_3}{\rho_3+s^d}\right]^{1/2} \frac{\exp(-jks^d)}{\sqrt{(8j\pi ks^d)}} \tag{41b}$$

The diffracted rays along s^d now form a cone of semi-angle θ_0 emanating from the discontinuity in curvature, and the value of ρ_3 is given, as for edge diffraction, by eqn. 5.39.

A uniform solution for oblique incidence can be obtained from eqn. 38 together with eqn. 40 to include a spherical incident wave. This follows since knowledge of the 2-dimensional uniform solution with the incident field given by a line source leads to a GTD uniform solution for spherical wave incidence.

A uniform solution, however, is not possible from the above equations for a general astigmatic incidence field to be included.

Collecting our results and using the ray-based vector diffraction formulation given earlier (i.e., in eqn. 5.125) the diffracted field can be expressed as

$$\begin{aligned} E^d(s^d) &= \mathbf{D}E^i(A)\left[\frac{\rho_3}{s^d(\rho_3+s^d)}\right]^{1/2}\exp(-jks^d); \\ H^d &= \sqrt{\left(\frac{\hat{\epsilon}}{\mu}\right)}\,\hat{s}^d \times E^d \end{aligned} \tag{42}$$

The diffraction matrix is given as before, by

$$\mathbf{D} = \begin{bmatrix} D^e & 0 \\ 0 & D^m \end{bmatrix}$$

where the components D^e, D^m for oblique incidence applicable to a plane wave, a cylindrical wave, or a spherical wave incidence, are obtained from eqns. 38 to 41 as

$$\begin{aligned} D^e &= \sin\chi\sin\chi_0\,\Xi \\ D^m &= -(1+\cos\chi\cos\chi_0)\,\Xi \end{aligned} \tag{43a}$$

where

$$\Xi = \frac{2}{\sin\chi_0+\sin\chi}\,\frac{|\cos\chi_0+\cos\chi|}{\cos\chi_0+\cos\chi}\sqrt{\left(\frac{j}{\pi\sin\theta_0}\right)}\left[\frac{a_1\exp(jv_1^2)}{\sqrt{(|g_1''|)}}F_-(v_1) \pm \frac{|a_2|\exp(\mp jv_2^2)}{\sqrt{(|g_2''|)}}F_\pm(v_2)\right]; \qquad g_2'' \gtrless 0$$

$$v_{1,2} = \sqrt{\left(\frac{k \sin\theta_0}{2|g''_{1,2}|}\right)}\,|g'_{1,2}|; \quad g'_{1,2} = a_{1,2}(\cos\chi_0 + \cos\chi)$$

$$g''_{1,2} = -a_{1,2}\left\{\sin\chi_0 + \sin\chi\left(1 + \frac{a_{1,2}}{\sigma}\sin\chi\right)\right\}; \quad \sigma = \frac{s^d \rho}{\rho + s^d}$$

and ρ is the distance from the source to the discontinuity. The value of a_1 is always taken to be positive whereas a_2 can be positive or negative depending on whether the surface associated with a_2 is convex or concave.

Removed from the reflection boundary the asymptotic value for the Fresnel integrals reduces Ξ to

$$\Xi \sim -\frac{4}{jk}\left(\frac{1}{a_1} - \frac{1}{a_2}\right)\frac{\csc^2\theta_0}{(\cos\chi_0 + \cos\chi)^3}\frac{1}{\sqrt{(8j\pi k)}} \tag{43c}$$

Some examples of discontinuity in curvature diffraction are shown in Fig. 6.14 for a magnetically polarised incident plane wave. In Fig. 6.14*a*

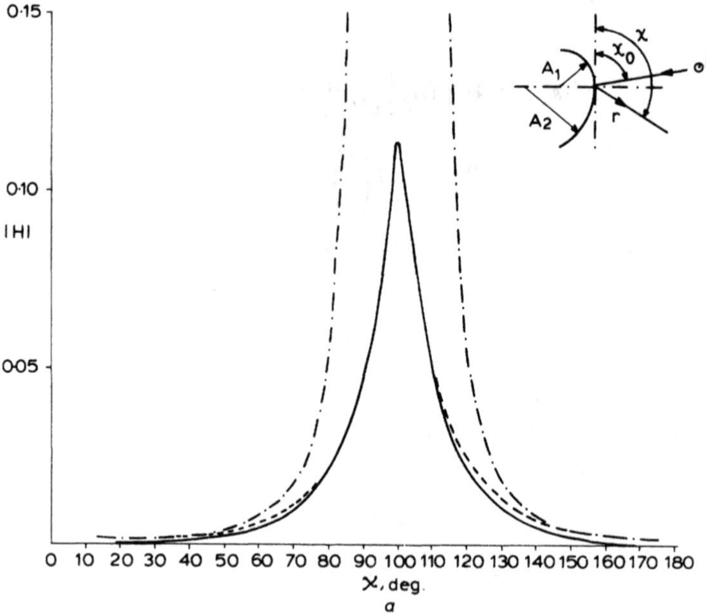

Fig. 6.14 Diffracted field from a discontinuity in curvature
a Physical optics only
$A_1 = 10\lambda$
$A_2 = 100\lambda$
— · — non-uniform solution
– – – – Fresnel integral value
——— Fresnel integral with correction factor
(the last two: uniform solution)

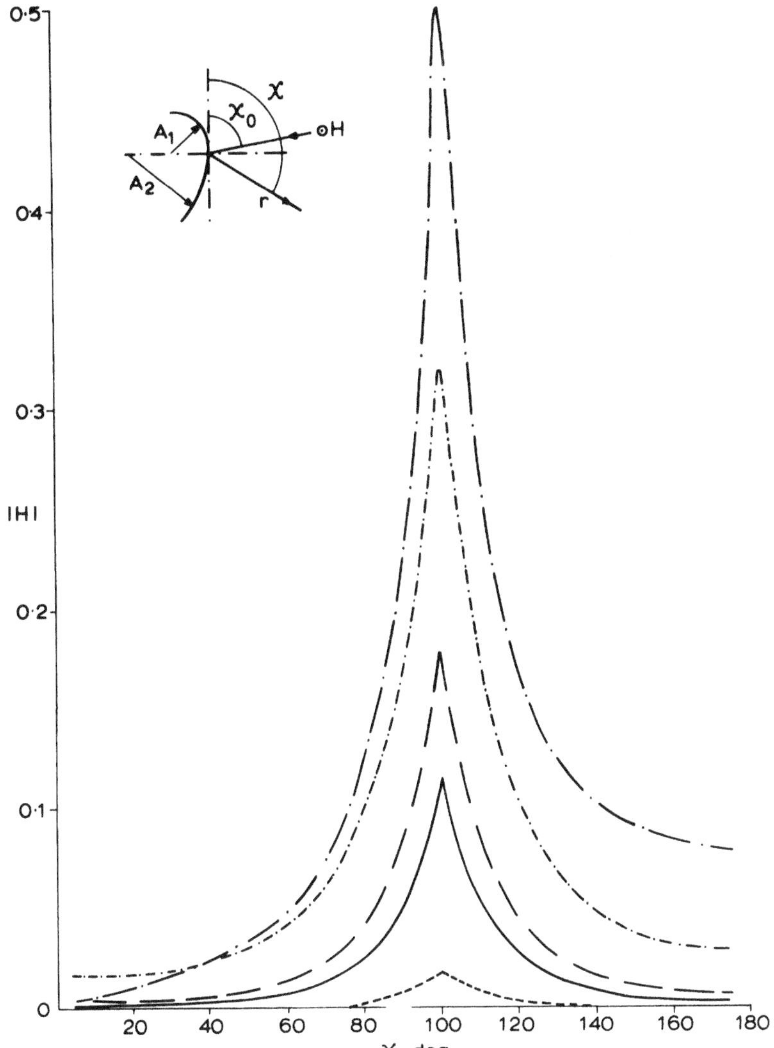

b

Fig. 6.14 Diffracted field from a discontinuity in curvature

b Physical optics solution with correction factor from the rigorous nonuniform solution

—·— exact half-plane

$$\left.\begin{array}{l} \text{—·—}\ A_1 = 1\,\lambda \\ \text{— —}\ A_1 = 5\,\lambda \\ \text{——}\ A_1 = 10\,\lambda \\ \text{---------}\ A_1 = 50\,\lambda \end{array}\right\}\ A_2 = 100\,\lambda$$

the diffracted field computed from the physical optics approximation is given. It is seen in the example that the non-uniform result is invalid over a large sector and illustrates the need for a uniform solution as developed in this section. Also the correction factor of eqn. 36*a* smoothly connects the Fresnel integral formulation into the true asymptotic solution for field points removed from the reflection boundary. In Fig. 6.14*b* the complete result fo eqns. 42 and 43 is shown for magnetically polarised incident plane wave on surfaces with various discontinuities in curvature.

6.6 Curved edge diffraction and the field beyond a caustic

The edge diffracted field was given about the diffracted rays in Section 5.1.1, as

$$E^d(s^d) = \mathbf{D}E^i(\mathrm{P}_0)\left[\frac{\rho_3}{s^d(\rho_3+s^d)}\right]^{1/2}\exp(-jks^d); \quad \mathbf{D} = \begin{bmatrix} D^e & 0 \\ 0 & D^m \end{bmatrix} \tag{44}$$

where

$$D^{e,m} = \{h(\Phi^i)+h(-\Phi^i)\} \mp \{h(\Phi^r)+h(-\Phi^r)\}$$

$$h(\Phi^{i,r}) = -\epsilon^{i,r}\{\sigma^{i,r}\}^{1/2}K_-(v^{i,r})\,\Lambda^{i,r} \sim \frac{-\csc\theta_0}{N\sqrt{(8j\pi k)}}\cot\left(\frac{\pi+\Phi^{i,r}}{2N}\right);$$

$$\sigma^{i,r} > 0$$

$$\sigma^{i,r} = \frac{\rho_1^{i,r}\rho_2^{i,r}(\rho_3+s^d)s^d}{(\rho_1^{i,r}+s^d)(\rho_2^{i,r}+s^d)\rho_3}$$

$$v^{i,r} = \sqrt{(k\sigma^{i,r})}\,|a^{i,r}|\sin\theta_0; \quad a^{i,r} = \sqrt{(2)}\cos\tfrac{1}{2}(\Phi^{i,r}+2n\pi N)$$

$$\Lambda^{i,r} = \frac{a^{i,r}}{\sqrt{2N}}\cot\left(\frac{\pi+\Phi^{i,r}}{2N}\right)$$

and ρ_3 is defined by eqn. 5.39.

In Section 5.4 only straight edges and surfaces were considered so that $\sigma^i = \sigma^r$ and this quantity was restricted to being positive. We will now attempt to lift these restrictions to include more general edge diffraction phenomena. A similar problem exists for diffraction by a convex surface where y, given by eqn. 20*b*, was restricted to being positive. This will be discussed at the end of this section.

Consider first an incident ray on a general wedge, as in Fig. 6.15, where the edge and both wedge faces will, in general, be curved. Removed from optical boundaries the asymptotic form of $h(\Phi^{i,r})$ is the same as before and there is little difficulty in computing the edge diffracted field. The effective source along the edge will give rise to creeping rays around a convex wedge face as in Fig. 6.15*b*. This phenomenon may be included by using the creeping ray formulation given in Section 6.3. For a concave wedge face the edge diffracted rays will reflect from the surface as shown in Fig. 6.15*b*. These reflected rays along the wedge face can be evaluated, using the methods given in Chapter 4, until the edge is approached where this simple ray picture

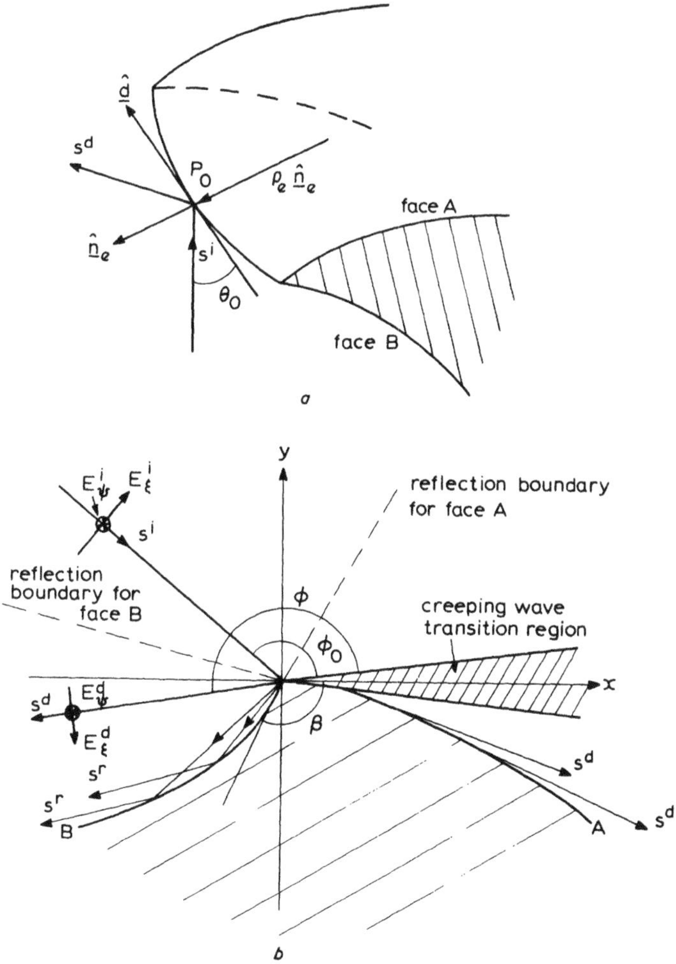

Fig. 6.15 Diffraction at a curved wedge

is invalid. Further investigation of this problem can be found in Ishihara *et al.* (1978). The general problem of diffraction by edges in concave surfaces has to date received limited attention. For some recent work see Idemen and Felsen (1981) and Serbest (1984).

As an optical boundary is approached we must use the modified Fresnel integral formulation for $h(\Phi)$. In this term the radii of curvature $\rho_1^{i,r}\,\rho_2^{i,r}$ of the corresponding geometrical optics field will not be the same as for straight wedge faces since the reflected field is modified by the surface curvature. Thus for the reflection boundary appropriate to face A in Fig. 6.15 the radii of curvature ρ_1^r, ρ_2^r will be dependent on the reflected ray from this wedge face. and similarly for face B. For example, a plane wave incident on the wedge given in Fig. 6.15 will give, along reflection boundary A, values for ρ_1^r, ρ_2^r to produce a *diverging* pencil, while $\rho_1^r\rho_2^r$ along reflection boundary B will produce a *converging* pencil. If a shadow boundary exists, then $\rho_1^i\rho_2^i$ will remain as the values given by the incident field.

The formulation for transition field diffraction between two planar wedges developed in Section 5.8 and the planar impedance wedge solution given in Section 5.9 both involve the function $h(\Phi^{i,r})$. It follows from the above discussion that both these formulations can be applied directly to general curved wedges provided we use the appropriate radii of curvature $\rho_1^{i,r}$, $\rho_2^{i,r}$ for the incident field and each wedge face in the term $\sigma^{i,r}$ where it appears within the function $h(\Phi^{i,r})$. No further details need be given here.

The value of the integer n in the diffraction term was discussed in Section 5.2 (see eqn. 5.25) and these apply to the curved wedge problems. For the shadow boundary related terms, the correct choice for n in $h(\pm\Phi^i)$, is $n = 0$ for a shadow boundary and $n = -1$ for the remaining term. This latter choice for n is only important when a reflection boundary is close to a wedge face. When this situation doesn't arise the asymptotic value of $h(\pm\Phi^i)_{n=-1}$ applies which is independent of n. We can then let $n = 0$ for both terms to give the Pauli expression (see eqn. 5.29)

$$h(-\Phi^i)_{n=0} + h(\Phi^i)_{n=0} = -\epsilon^i\sqrt{(\sigma^i)}K_-(v^i)\left[\frac{\dfrac{2}{N}\sin\dfrac{\pi}{N}\cos\tfrac{1}{2}\Phi^i}{\cos\dfrac{\Phi^i}{N} - \cos\dfrac{\pi}{N}}\right] \tag{45}$$

The quantity in square brackets goes to unity on a shadow boundary. Using this equation in the edge diffraction formulation of eqn. 44

means that we have only to consider three modified Fresnel integral terms corresponding to the three possible optical boundaries.

The remaining problem is when σ goes negative, which can occur as the field passes through a caustic. To determine the nature of the diffracted field for this situation we must appeal to the asymptotic formulations of the field integrals given in Section 2.3. Consider first the simple case of a magnetically polarised plane wave incident on a cylindrical surface as shown in Fig. 6.16. The resultant scattered field $\boldsymbol{H}^s$ was given in the previous section by eqns. 32 and 33 using the physical optics approximation to the currents. We can use these results directly to study the field as it passes through a caustic. From eqn. 32 $\boldsymbol{H}^s$ is given as

$$\boldsymbol{H}^s = \hat{z}\int_0^{\phi_1} f(\phi')\exp\{jkg(\phi')\}d\phi'$$

The contribution from the endpoint at $\phi' = 0$ is the edge diffraction component $\boldsymbol{H}_z^d$ from A and is given, from eqn. 2.86, as

$$H_z^d = \pm\frac{|g'|}{g'} f\exp(jkg)\exp(\mp jv^2)\sqrt{\left(\frac{2}{k|g''|}\right)}F_\pm(v)\bigg|_{\phi'=0};$$

$$g''(0) \gtrless 0 \qquad (46a)$$

where

$$v = \sqrt{\left(\frac{k}{2|g''|}\right)}|g'|$$

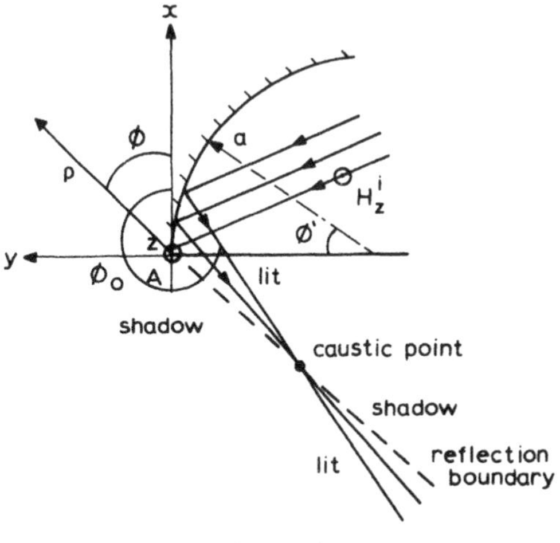

Fig. 6.16 Behaviour of the field beyond a caustic

When $v > 3{\cdot}0$ then

$$H_z^d \sim -\frac{1}{jk}\frac{f(0)}{g'(0)}\exp\{jkg(0)\} \tag{46b}$$

On an optical boundary $v = 0$ and H_z^d becomes

$$H_z^d = \pm\tfrac{1}{2}\frac{|g'|}{g'} f \exp(jkg)\sqrt{\left(\frac{2\pi}{k|g''|}\right)}\exp\left(\pm\frac{j\pi}{4}\right)\bigg|_{\phi'=0}; \quad g''(0) \gtrless 0 \tag{47}$$

which is simply half of the geometrical optics field associated with that boundary. Referring the phase to the edge the quantities in eqn. 47 are given from eqn. 33 as

$$\begin{aligned} f &= \frac{2jka}{\sqrt{(8j\pi k\rho)}}\sin\phi; \qquad g = -\rho \\ g' &= a(\cos\phi_0 + \cos\phi) \\ g'' &= -a\left\{\sin\phi_0 + \sin\phi\left(1 + \frac{a}{\rho}\sin\phi\right)\right\} \end{aligned} \tag{48}$$

Along the shadow boundary where $\phi_0 = \pi + \phi$ the field propagates as a plane wave. No caustic exists and g'' is seen to remain negative. On the reflection boundary where $\phi_0 = \pi - \phi$ a caustic point occurs when $g'' = 0$, i.e. where

$$\rho = -\frac{a}{2}\sin\phi_0 = -\rho_2^r \tag{49}$$

The sign of g'' changes from negative to positive as ρ passes (as it increases) through this point. Substituting eqn. 48 into eqn. 47 yields for $\pi < \phi_0 < 2\pi$

$$\begin{aligned} H_z^d &= -\tfrac{1}{2}\frac{|\cos\phi_0 + \cos\phi|}{\cos\phi_0 + \cos\phi}\left[\frac{|\rho_2^r|}{|\rho_2^r| - \rho}\right]^{1/2}\exp(-jk\rho); \quad \rho < |\rho_2^r| \\ &= \tfrac{1}{2}\frac{|\cos\phi_0 + \cos\phi|}{\cos\phi_0 + \cos\phi}\left|\frac{|\rho_2^r|}{|\rho_2^r| - \rho}\right|^{1/2}\exp\left(\frac{j\pi}{2}\right)\exp(-jk\rho); \quad \rho > |\rho_2^r| \end{aligned} \tag{50}$$

Two important consequences result from this equation. The first is that the phase shift of $\exp\left(\frac{j\pi}{2}\right)$ as the field passes through the caustic has been shown. The second is the division of the illuminated and shadow regions for the reflected field defined by the reflection boundary. Previously we defined ϵ^r as equal to $+1$ on the illuminated side and -1 on the shadow side of the reflection boundary and this could be defined as

$$\epsilon^r = \frac{|\cos\frac{1}{2}(\phi+\phi_0)|}{\cos\frac{1}{2}(\phi+\phi_0)}$$

When the caustic is passed, however, the illumination and shadow regions change sides as illustrated in Fig. 6.16 and our definition of ϵ^r must now be

$$\epsilon^r = \pm\frac{|\cos\frac{1}{2}(\phi+\phi_0)|}{\cos\frac{1}{2}(\phi+\phi_0)} = \pm\frac{|a^r|}{a^r}; \quad \rho \lessgtr |\rho_2^r| \tag{51}$$

so that eqn. 50 becomes (for $\pi < \phi_0 < 2\pi$)

$$\begin{aligned} H_z^d &= -\tfrac{1}{2}\epsilon^r \left|\frac{|\rho_2^r|}{|\rho_2^r|-\rho}\right|^{1/2} \exp(-jk\rho); \qquad \rho < |\rho_2^r| \\ &= -\tfrac{1}{2}\epsilon^r \exp\left(\frac{j\pi}{2}\right)\left|\frac{|\rho_2^r|}{|\rho_2^r|-\rho}\right|^{1/2} \exp(-jk\rho); \qquad \rho > |\rho_2^r| \end{aligned} \tag{52}$$

A similar expression exists if there is a caustic on the shadow boundary (i.e. ρ_2^i finite and negative).

Returning to eqn. 46 we note that as the caustic is passed, the diffracted field representation changes sign in the Fresnel integral, i.e. from F_- to F_+ in this instance. Relating to eqn. 44 with $\rho_1 = \rho_3 = \infty$, the diffraction term $h(\Phi^{i,r})$ can now be defined as

$$h(\Phi^{i,r}) = -\epsilon^{i,r}\{j\sigma^{i,r}\}^{1/2}K_\pm\{\sqrt{(k\sigma^{i,r})}\,|a^{i,r}|\sin\theta_0\}\Lambda^{i,r}\exp\left(\pm\frac{j\pi}{4}\right);$$

$$\frac{\rho_2^{i,r}}{\rho_2^{i,r}+s^d} \lessgtr 0 \tag{53}$$

where

$$\sigma^{i,r} = \left|\frac{\rho_2^{i,r}s^d}{\rho_2^{i,r}+s^d}\right|$$

Either side of the caustic this will provide a smooth transition across the optical boundaries. It is important to note that the changeover of the regions across an optical boundary only occurs for the caustic in the cross-sectional plane at the edge, as in Fig. 6.16. If the surface is also curved in the plane perpendicular to the figure then the appropriate asymptotic formulation is that given by eqn. 2.102, i.e.,

$$\begin{aligned} H_z^d &= \pm\frac{|g'_{u_1}|}{g'_{u_1}} f\exp(jkg)\exp(\mp jv^2)\sqrt{\left(\frac{2}{k|g''_{u_1u_1}|}\right)} \\ &\quad F_\pm(v)P_{u_2}\Big|_{\substack{u_1=\alpha(u_{e2})\\ u_2=u_{e2}}}; \quad \begin{array}{l} g''_{u_1u_1} \gtrless 0 \\ g'_{u_2}(\alpha(u_{e2}),u_{e2})=0 \end{array} \\ &\sim \frac{f\exp(jkg)}{-jkg'_{u_1}}P_{u_2}\Big|_{\substack{u_1=\alpha(u_{e2})\\ u_2=u_{e2}}} \end{aligned} \tag{54}$$

where

$$P_{u_2} = \sqrt{\left(\frac{2\pi}{k|g''_{u_2u_2}|}\right)} \exp\left(\pm\frac{j\pi}{4}\right); \quad g''_{u_2u_2} \gtrless 0,$$

$$v = \sqrt{\left(\frac{2}{k|g''_{u_1u_1}|}\right)} |g'_{u_1}|$$

In this equation, u_1 is the variable in which the integral terminates, and corresponds to ϕ' above. The quantity u_2 is then the variable in the perpendicular plane and does not influence the behaviour of the Fresnel integral. In this plane along the optical boundaries lie the other principal radii of curvature $\rho_1^{i,r}$ and if a caustic is traversed the $\exp\dfrac{j\pi}{2}$ phase shift is provided in the term P_{u_2}. Away from the optical boundaries the radius of curvature associated with P_{u_2} is that for the edge rays, ρ_3. Relating eqn. 54 to the GTD formulation in eqn. 44 we see that if $\rho_1^{i,r} = \rho_3$ on the optical boundaries, then by using eqn. 53 for $h(\Phi^{i,r})$, we obtain the same behaviour for the field as given in eqn. 54.

A difficulty arises when $\rho_1^{i,r}$ does not equal ρ_3 on the optical boundaries and $\rho_2^{i,r}$ is not in the cross-sectional plane to the edge. For this more general situation the GTD formulation of eqn. 44 will only be valid before and after the three caustics are traversed. Specifically we have $h(\Phi^{i,r})$ as defined in eqn. 53 but with $\sigma^{i,r}$ given as

$$\sigma^{i,r} = \left|\frac{\rho_2^{i,r}\rho_1^{i,r}(\rho_3 + s^d)s^d}{(\rho_2^{i,r} + s^d)(\rho_1^{i,r} + s^d)\rho_3}\right| \tag{55}$$

and the upper (lower) sign in eqn. 53 applying to field points before (after) the three caustics are traversed. The asymptotic formulations of the field integrals given in Section 2.3 remain valid in all regions removed from caustics, however, and can be used in the transition regions when the GTD formulation fails. If the currents are approximated by physical optics then, as shown earlier, the leading term of the asymptotic solution will be obtained in the transition regions. Outside these regions one can either use the GTD formulation directly or correct the field by using appropriate correction factors as discussed in Section 5.6. The correction factors given by eqns. 5.78 and 5.81 are applicable only for the half-plane but terms appropriate for a wedge can be easily derived.

In a similar way we can treat the transition region representation for a converging incident field diffraction around a smooth convex surface. As an example consider an astigmatic pencil incident on such a surface with the diffracted field valid though the transition region given from eqn. 20 as

$$E^d(s^d) = \mathbf{D}E^i(A)\left[\frac{\rho_3}{s^d(\rho_3+s^d)}\right]^{1/2}\exp\{-jk(\tau+s^d)\} \qquad (56)$$

where

$$\mathbf{D} = -\sqrt{\left(\frac{2}{jk}\right)}M\begin{bmatrix}\tilde{p}_-(\Delta,y) & 0\\ 0 & \tilde{q}_-(\Delta,y)\end{bmatrix}$$

$$\Delta = \int_0^\tau \frac{M}{\rho_c}\,d\tau', \quad y = M^{-1}\left[\frac{ks^d\rho_1\rho_2(\rho_3+s^d)}{2(\rho_1+s^d)(\rho_2+s^d)\rho_3}\right]^{1/2}$$

The definitions of the various quantities in this equation are to be found in Section 6.2.

Eqn. 56 is restricted to field points prior to any caustic of ρ_1, ρ_2 or ρ_3 being traversed in the transition region. Outside this region the only caustic for the diffracted field is that associated with ρ_3 and this presents no difficulty. In the transition region, however, we have the same problem as for edge diffraction. Redefining y in eqn. 56 as

$$y = M^{-1}\sqrt{\left(\frac{k}{2}\sigma^i\right)} \qquad (57)$$

with σ^i as given by eqn. 55, then by analogy with the edge diffraction formulation, one finds that the GTD formulation of eqn. 56 for a convex surface is valid in the general case, once all three possible caustics have been passed, by replacing $\tilde{p}_-$ and $\tilde{q}_-$ with $\tilde{p}_+$ and $\tilde{q}_+$ where from eqn. 15

$$\tilde{p}_\pm(x,y) = p(x) - y\,\mathrm{sgn}(x)K_\pm(y|x|)\exp\left(\mp\frac{j\pi}{4}\right) \sim p(x) - \frac{1}{2\sqrt{(\pi)}x}$$

$$\tilde{q}_\pm(x,y) = q(x) - y\,\mathrm{sgn}(x)K_\pm(y|x|)\exp\left(\mp\frac{j\pi}{4}\right) \sim q(x) - \frac{1}{2\sqrt{(\pi)}x} \qquad (58)$$

As for edge diffraction the illuminated and shadow regions have changed sides across the shadow boundary. For the special case when ρ_2^i is in the cross-sectional plane at glancing incidence this change will occur. as before, at $\rho_2/(\rho_2+s^d) = 0$. If $\rho_1^i = \rho_3$ on the shadow boundary then eqn. 56. in conjunction with eqns. 57 and 58, will be valid between the caustics ρ_1^i and ρ_2^i in the transition region. As an example, Fig. 6.17 shows a converging field incident on a convex surface which illustrates the behaviour across the shadow boundary.

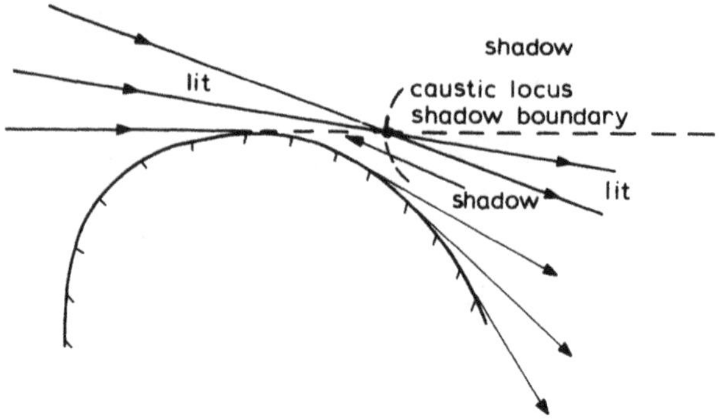

Fig. 6.17 Converging field incident on a curved surface

6.7 Evaluating the field at caustics

All ray-tracing methods predict infinities for the field at caustics and an alternative approach must be used to evaluate the field in caustic regions. For the geometrical optics field, as a *surface caustic* is grazed, one may use asymptotic formulations such as given in Section 2.3 provided that the field is initially expressed in terms of an integral equation. A surface caustic implies that the third derivative of the phase function in the field integrals is non-zero. When this is not the case, such as at a point caustic, these asymptotic formulations fail and it is necessary to evaluate the field integrals numerically. Two simple examples are a spherical reflector and a parabolic cylindrical reflector with a plane wave incident on axis. Using the physical optics approximation for the induced currents on the reflector surfaces, an asymptotic evaluation of the resultant field integral yields the leading term at points removed from caustics as the geometrical optics component. In the vicinity of the surface caustic generated by the spherical surface, we may use the Airy function representation for the field given by eqn. 2.76, but this formulation is not applicable for the line caustic at the focus of the parabolic cylinder. It is not difficult physically to see why this should be the case. For the spherical reflector each point along the caustic surface still relates to a specific localised region on the reflector surface, as for the geometrical optics field. This is not true for the parabolic cylinder where we must consider the total current distribution in order to evaluate the field at the caustic.

Other approaches to evaluating the field in caustic regions can be found, for example, in the work of Kay and Keller (1954) and Ludwig (1966). Also Yu.I. Orlov has made several contributions to evaluating the field at caustics (for example see Orlov, 1976).

A commonly occurring caustic of edge diffracted rays is when a circular edge is illuminated symmetrically. This circular edge may be formed by a disc, the base of a cone, or the rim of a circularly symmetric reflector. In all these cases if the edge is illuminated symmetrically, either by a source situated along the axis or an axially incident plane wave, then the axis is a caustic of the edge diffracted rays. In other words, diffracted rays from all points around the edge have a common intersection along the axis. A finite value for the field at such a caustic can be obtained by using *equivalent edge currents* derived in the following way. We consider initially the far field from an infinitely long z-directed electric current J_z (magnetic current M_z) given from eqn. 2.17 as

$$\begin{aligned} E_z &= -j\omega\mu J_z \frac{\exp(-jk\rho)}{\sqrt{(8j\pi k\rho)}} \\ E_\phi &= -jk M_z \frac{\exp(-jk\rho)}{\sqrt{(8j\pi k\rho)}} \end{aligned} \tag{59}$$

It was noted earlier that, removed from transition regions about the optical boundaries, the edge diffracted field from the wedge appeared as a line source situated along the edge. Thus to derive this *equivalent* line source simply means equating eqn. 59 with the appropriate expression for the diffracted field. The simplest case is that of a plane wave at normal incidence to the wedge, for which from eqns. 5.19 and 5.28

$$\begin{bmatrix} E_z^d \\ E_\phi^d \end{bmatrix} = \begin{bmatrix} E_z^i \\ E_\phi^i \end{bmatrix} (u_d^i \mp u_d^r)$$

where

$$u_d^{i,r} \sim \frac{2}{N}\sin\frac{\pi}{N}\left[\cos\frac{\pi}{N} - \cos\frac{\phi \mp \phi_0}{N}\right]^{-1} \frac{\exp(-jk\rho)}{\sqrt{(8j\pi k\rho)}}$$

and E_z^i, E_ϕ^i are the values of the incident field components at the edge. Equating this equation with eqn. 59 yields the equivalent edge currents as

$$J_z = -\frac{E_z^i}{j\omega\mu}\frac{2}{N}\sin\frac{\pi}{N}\left\{\left[\cos\frac{\pi}{N} - \cos\frac{\phi-\phi_0}{N}\right]^{-1} - \left[\cos\frac{\pi}{N} - \cos\frac{\phi+\phi_0}{N}\right]^{-1}\right\} \tag{60a}$$

$$M_z = -\frac{E_\phi^i}{jk}\frac{2}{N}\sin\frac{\pi}{N}\left\{\left[\cos\frac{\pi}{N} - \cos\frac{\phi-\phi_0}{N}\right]^{-1} + \left[\cos\frac{\pi}{N} - \cos\frac{\phi+\phi_0}{N}\right]^{-1}\right\}$$

and for the special case of the half-plane when $N = 2$ this reduces to

$$\left.\begin{aligned} J_z &= -\frac{E_z^i}{j\omega\mu}\,\frac{4\sin\frac{1}{2}\phi\sin\frac{1}{2}\phi_0}{\cos\phi_0+\cos\phi} \\ M_z &= \frac{E_\phi^i}{jk}\,\frac{4\cos\frac{1}{2}\phi\cos\frac{1}{2}\phi_0}{\cos\phi_0+\cos\phi} \end{aligned}\right\} \quad \text{(for the half-plane)} \qquad (60b)$$

It is now assumed that in an element of a curved edge, equivalent edge currents are given by eqn. 60 for a correspondingly oriented wedge or half-plane. These fictitious currents are then used in the usual integral solutions for the electromagnetic diffracted field. In this way we obtain a finite value for the field at an edge diffracted caustic. This technique can be reconciled with the asymptotic methods given in Section 2.3, and this is discussed in the example of Section 7.4.

Unlike real currents, the equivalent currents given in eqn. 60 depend upon the direction of observation. This means that in each integration step we must evaluate the appropriate value for the angle ϕ. The consequent numerical integration is considerably slowed down by this calculation. Since the equivalent current formulation is usually required only in the region of the axial caustic we can approximate J_z, M_z by their values on the axis in the region where the field is being evaluated. This means that J_z, M_z can be taken outside the integration and the procedure is considerably simplified.

It must be noted that equivalent currents are rigorous only in the direction of the diffracted ray and should not be expected to yield reliable results in directions removed from the diffracted ray. This point is discussed by Burnside and Peters (1974) and Knott and Senior (1974).

The equivalent current concept was first used by Millar (1956, 1957*a*, 1957*b*) to solve for the axial and aperture fields in the case of an aperture in a plane screen. It was revived by Ryan and Peters (1969) for back scatter from a right circular cone and has since been applied to various problems by many authors. Our solution above was for a normally incident plane wave, but other types of equivalent currents can be readily derived. For example, James and Kerdemelidis (1973) obtained equivalent currents for a line source near the edge of a half-plane. They also obtained slope-diffraction equivalent currents applicable for a non uniform line source near the edge. Both these formulations are valid through the optical boundaries and were used to analyse reflector antenna radiation patterns. The slope-diffraction equivalent current was subsequently rederived by Mentzer *et al.* (1975) and used to predict radiation from pyramidal horn antennas. In this application the edges at the aperture are of finite length and the equivalent current method takes this into account. Equivalent edge currents valid for all angles of observation have been derived by Michaeli (1984). (See also comments by Knott, 1985). While these currents are non-uniform and considerably more complicated to those given by eqn. 60, they are nevertheless readily amenable to numerical calculation.

6.8 Summary

Curved edge diffraction

When a geometrical optics field given by

$$E^i(s^i) = E^i(q)\left[\frac{\rho_1^i\rho_2^i}{(\rho_1^i+s^i)(\rho_2^i+s^i)}\right]^{1/2} \exp(-jks^i);$$
$$H^i = \sqrt{\left(\frac{\epsilon}{\mu}\right)}\,\hat{s}_0^i \times E^i \tag{61}$$

is incident upon a curved wedge, as in Fig. 6.15, then the resultant diffracted field may be written about the diffracted rays from the edge point P_0 as

$$E^d(s^d) = \mathsf{D}E^i(P_0)\left[\frac{\rho_3}{s^d(\rho_3+s^d)}\right]^{1/2} \exp(-jks^d);$$
$$H^d = \sqrt{\left(\frac{\epsilon}{\mu}\right)}\,\hat{s}^d \times E^d \tag{62}$$

where

$$\sqrt{\left(\frac{\rho_3}{\rho_3+s^d}\right)} = \left|\frac{\rho_3}{\rho_3+s^d}\right|^{1/2}; \qquad \frac{\rho_3}{\rho_3+s^d} > 0$$
$$= \exp\left(\frac{j\pi}{2}\right)\left|\frac{\rho_3}{\rho_3+s^d}\right|^{1/2}; \qquad \frac{\rho_3}{\rho_3+s^d} < 0$$

Points of edge diffraction are determined by applying Fermat's principle for edge diffraction as given in Sections 5.4 and 5.8. As before. the incident and diffracted fields are expressed in terms of a ray-based co-ordinate system $(s_0^{i,d}, \psi, \xi)$, where E_ψ (E_ξ) is associated with electric (magnetic) polarisation. The quantity ρ_3 in eqn. 62 is given as

$$\frac{1}{\rho_3} = Q_{11}^i - \frac{1}{\rho_e \sin^2\theta_0}(\hat{s}^i\cdot\hat{n}_e - \hat{s}^d\cdot\hat{n}_e) \tag{63}$$

where $\hat{n}_e$ is the outward normal from the edge along the direction of the edge curvature ρ_e at P_0, θ_0 is the incident angle to the edge, and

$Q_{11}^{i'}$ is the appropriate component in the incident field curvature matrix.

In the diffraction matrix

$$\mathbf{D} = \begin{bmatrix} D^e & 0 \\ 0 & D^m \end{bmatrix}$$

the components are

$$D^{e,m} = \{h(\Phi^i)_{n=0} + h(-\Phi^i)_{n=0}\} \mp \{h(\Phi^r)_{n=-1} + h(-\Phi^r)_{n=0}\};$$

$$\Phi^{i,r} = \phi \mp \phi_0, 0 \leqslant \phi, \phi_0 \leqslant 2\pi \tag{64}$$

provided that the incident and diffraction angles ϕ_0 and ϕ are measured from an illuminated wedge face, and a reflection boundary is not close to a wedge face.

The diffraction term $h(\Phi^{i,r})$ is given by

$$h(\Phi^{i,r}) = -\epsilon^{i,r}\{j|\sigma^{i,r}|\}^{1/2} K_\pm(v^{i,r})\Lambda^{i,r} \exp\left(\pm\frac{j\pi}{4}\right); \quad \sigma^{i,r} \lessgtr 0 \tag{65}$$

where

$$\epsilon^{i,r} = \mp \operatorname{sgn}(a^{i,r}); \quad \sigma^{i,r} \lessgtr 0$$ (= 1 for illuminated region, and −1 for the shadow.)

$$\sigma^{i,r} = \frac{\rho_1^{i,r}\rho_2^{i,r}(\rho_3 + s^d)s^d}{(\rho_1^{i,r} + s^d)(\rho_2^{i,r} + s^d)\rho_3}$$

$$v^{i,r} = \sqrt{(k|\sigma^{i,r}|)}|a^{i,r}|\sin\theta_0$$

$$a^{i,r} = \sqrt{(2)}\cos\tfrac{1}{2}(\Phi^{i,r} + 2n\pi N); \qquad N = \frac{2\pi - \beta}{\pi}$$

$$\Lambda^{i,r} = \frac{a^{i,r}}{\sqrt{(2)}N}\cot\left(\frac{\pi + \Phi^{i,r}}{2N}\right); \qquad \Lambda^{i,r} = 1$$ on optical boundaries.

The modified Fresnel integral $K_\pm$ is defined in Section 2.2.1. We may combine the incident field diffraction terms to give

$$h(\Phi^i)_{n=0} + h(-\Phi^i)_{n=0} = -\epsilon^i\{j|\sigma^i|\}^{1/2}K_\pm(v^i)\left[\frac{\dfrac{2}{N}\sin\dfrac{\pi}{N}\cos\dfrac{\Phi^i}{2}}{\cos\dfrac{\Phi^i}{N} - \cos\dfrac{\pi}{N}}\right]$$

$$\exp\left(\pm\frac{j\pi}{4}\right); \qquad \sigma^i \lessgtr 0 \tag{66}$$

Combined with eqn. 64 we get the diffracted field as the sum of three terms associated with the three possible optical boundaries. In each term we will have, in general, different values for $\rho_1^{i,r}, \rho_2^{i,r}$. If a shadow boundary exists, then ρ_1^i, ρ_2^i will remain as the values given by the incident field at the point of diffraction. The values for ρ_1^r, ρ_2^r must be associated with the appropriate wedge face, and are determined when evaluating the geometrical optics reflected field using the techniques given in Chapter 4.

The condition $\sigma^{i,r} \lessgtr 0$ as given in eqns. 65 and 66, needs qualification in that these equations are only generally valid for field points situated before and after the three caustics are traversed. This problem is discussed in more detail in Section 6.6. Removed from the transition regions about the optical boundaries when $v^{i,r} > 3{\cdot}0$ no such difficulties arise, since

$$h(\Phi^{i,r}) \sim \frac{-\csc\theta_0}{\sqrt{(8j\pi k)}}\frac{1}{N}\cot\left(\frac{\pi+\Phi^{i,r}}{2N}\right) \quad \text{for} \quad v^{i,r} > 3{\cdot}0 \qquad (67)$$

and this is independent of $\sigma^{i,r}$.

The special case of the half-plane when $N = 2$ yields $D^{e,m}$ as

$$\begin{aligned} D^{e,m} &= -\left[\epsilon^i\{j|\sigma^i|\}^{1/2}K_\pm(v^i)\exp\left(\pm\frac{j\pi}{4}\right)\Bigg|_{\sigma^i \lessgtr 0}\right] \\ &\quad \pm\left[\epsilon^r\{j|\sigma^r|\}^{1/2}K_\pm(v^r)\exp\left(\pm\frac{j\pi}{4}\right)\Bigg|_{\sigma^r \lessgtr 0}\right] \end{aligned} \qquad (68a)$$

Removed from optical boundaries

$$D^{e,m} \sim -\{\sec\tfrac{1}{2}(\phi-\phi_0) \mp \sec\tfrac{1}{2}(\phi+\phi_0)\}\frac{\csc\theta_0}{\sqrt{(8j\pi k)}} \qquad (68b)$$

Diffraction at a discontinuity in curvature

When an incident plane wave, cylindrical wave, or sperical wave is incident upon a discontinuity in curvature, as in Fig. 6.13, the resultant diffracted field in the region $0 < \chi < \pi$ is given by eqn. 62 where the diffraction matrix components $D^{e,m}$ are now

$$\begin{aligned} D^e &= \sin\chi\sin\chi_0\,\Xi \\ D^m &= -(1+\cos\chi\cos\chi_0)\,\Xi \end{aligned} \qquad (69a)$$

with

$$\Xi = \frac{2}{\sin\chi_0+\sin\chi}\,\frac{|\cos\chi_0+\cos\chi|}{\cos\chi_0+\cos\chi}\sqrt{\left(\frac{j}{\pi\sin\theta_0}\right)}\left[\frac{a_1\exp(jv_1^2)}{\sqrt{|g_1''|}}F_-(v_1)\right.$$

$$\pm \frac{|a_2| \exp(\mp j v_2^2)}{\sqrt{(|g_2''|)}} F_\pm(v_2) \Bigg] ; \qquad g_2'' \gtrless 0$$

$$v_{1,2} = \sqrt{\left(\frac{k \sin\theta_0}{2|g_{1,2}''|}\right)} |g_{1,2}'|; \quad g_{1,2}' = a_{1,2}(\cos\chi_0 + \cos\chi);$$

$$g_{1,2}'' = -a_{1,2}\left\{\sin\chi_0 + \sin\chi\left(1 + \frac{a_{1,2}}{\sigma}\sin\chi\right)\right\}; \quad \sigma = \frac{s^d \rho}{\rho + s^d}$$

and ρ is the distance from the source to the discontinuity. Removed from the reflection boundary

$$\Xi \sim -\frac{4}{jk}\left(\frac{1}{a_1} - \frac{1}{a_2}\right) \frac{\csc^2\theta_0}{(\cos\chi_0 + \cos\chi)^3} \frac{1}{\sqrt{(8j\pi k)}} \tag{69b}$$

The value of ρ_3 is given by eqn. 63 where ρ_e is interpreted as the curvature of the line of discontinuity.

Diffraction around a smooth convex surface

(a) Source removed from the surface

When the geometrical optics field of eqn. 61 is incident upon a smooth convex surface as in Fig. 6.7. the diffracted field in the deep shadow region can be expressed about the diffracted rays as

$$E^d(s^d) = \sum_{n=1}^{N} \mathbf{D}_n E^i(A) \left[\frac{\rho_3}{s^d(\rho_3 + s^d)}\right]^{1/2} \exp\{-jk(\tau + s^d)\};$$

$$H^d(s^d) = \sqrt{\left(\frac{\hat{\epsilon}}{\mu}\right)} \hat{s}^d \times E^d \tag{70}$$

where

$$\mathbf{D}_n = \sqrt{\left(\frac{\partial\eta}{\partial\eta'}\right)} \frac{1}{\sqrt{(8j\pi k)}}$$

$$\begin{bmatrix} D_n^e \exp\left[-\int_0^\tau \Omega_n^e(\tau')d\tau'\right] & 0 \\ 0 & D_n^m \exp\left[-\int_0^\tau \Omega_n^m(\tau')d\tau'\right] \end{bmatrix}$$

$$D_n^e = 2\bar{M}\exp\left(\frac{j\pi}{6}\right)\{\mathrm{Ai}'(-\alpha_n)\}^{-2}; \quad \bar{M} = \left(\frac{k}{2}\left[\sqrt{\{\rho_c(0)\rho_c(\tau)\}}\right]\right)^{1/3}$$

$$D_n^m = 2\bar{M}\exp\left(\frac{j\pi}{6}\right)\{\alpha_n'\{\mathrm{Ai}(-\alpha_n')\}^2\}^{-1}$$

$$\Omega_n^e(\tau') = \frac{\alpha_n}{\rho_c(\tau')} M(\tau') \exp\left(\frac{j\pi}{6}\right); \quad \Omega_n^m(\tau') = \frac{\alpha_n'}{\rho_c(\tau')} M(\tau') \exp\left(\frac{j\pi}{6}\right);$$

$$M(\tau') = \{\tfrac{1}{2} k \rho_c(\tau')\}^{1/3}$$

Some of the Airy functions $\mathrm{Ai}(-\alpha_n')$, $\mathrm{Ai}'(-\alpha_n)$ and their roots α_n, α_n' are to be found in Table 2.1. The values of ρ_3 and $\sqrt{(d\eta/d\eta')}$ must, in general, be determined numerically as the incident pencil creeps around the convex surface. This creeping ray along τ follows the geodesic path around the surface determined from Fermat's principle for diffraction around a smooth convex surface: diffracted rays from a point S to a point P are those continuous rays for which the optical length between S and P, where part of the path must lie on the surface, is stationary with respect to infinitesimal variations in path.

The radius of curvature along the geodesic path is given by ρ_c. In the deep shadow, one term in the series given by eqn. 70 usually suffices in describing the diffracted field. As the transition region is approached it is necessary to take further terms to maintain accuracy. In the transition region itself the diffracted electric field within the shadow region is given by

$$E^d(s^d) = \mathbf{D} E^i(A) \left[\frac{\rho_3}{s^d(\rho_3 + s^d)}\right]^{1/2} \exp\{-jk(\tau + s^d)\} \tag{71}$$

where

$$\mathbf{D} = -\sqrt{\left(\frac{2}{jk}\right)} M(0) \begin{bmatrix} \tilde{p}_\pm(\Delta, y) & 0 \\ 0 & \tilde{q}_\pm(\Delta, y) \end{bmatrix}; \qquad \sigma^i \lessgtr 0$$

$$\Delta = \int_0^\tau \frac{M}{\rho_c} d\tau' \simeq \frac{M(0)\tau}{\rho_c(0)}; \quad y = M^{-1}\sqrt{(\tfrac{1}{2} k \sigma^i)}; \quad M = \{\tfrac{1}{2} k \rho_c(\tau')\}^{1/3}$$

and $\tilde{p}_\pm, \tilde{q}_\pm$ are the modified Pekeris functions

$$\tilde{p}_\pm(x, y) = p(x) - y \,\mathrm{sgn}(x) K_\pm(y|x|) \exp\left(\mp\frac{j\pi}{4}\right) \sim p(x) - \frac{1}{2\sqrt{(\pi)}x}$$

$$\tilde{q}_\pm(x, y) = q(x) - y \,\mathrm{sgn}(x) K_\pm(y|x|) \exp\left(\mp\frac{j\pi}{4}\right) \sim q(x) - \frac{1}{2\sqrt{(\pi)}x}$$

The value of σ^i is the same as for edge diffraction, i.e.,

$$\sigma^i = \frac{\rho_1^i \rho_2^i (\rho_3 + s^d) s^d}{(\rho_1^i + s^d)(\rho_2^i + s^d)\rho_3}$$

and the transition region expression of eqn. 71 is (as for edge diffraction) only universally valid for field points situated before and after the three caustics are traversed.

The extent of the transition region must be sufficient to allow the modified Fresnel integral argument, $y\,|x\,|$, to exceed a value of 3·0 so that its asymptotic expansion may be used. (see Section 2.2.1). A smooth transition should then be affected between the formulations of eqns. 71 and 70 as one progresses from the transition region to the deep shadow region. (For more details see Section 6.2).

In the illuminated region we write the solution for the total field at P as

$$E(P) = E^i(P) + \mathbf{R}E^i(P_0)\left[\frac{\rho_1^r\rho_2^r}{(\rho_1^r + s_0^r)(\rho_2^r + s_0^r)}\right]^{1/2} \exp(-jks_0^r) \tag{72}$$

$\mathbf{R} = \begin{bmatrix} R^e & 0 \\ 0 & R^m \end{bmatrix}$, P_0 is the geometrical optics reflection point, s_0^r is the distance from P_0 to the field point at P,

$$\begin{Bmatrix} R^e \\ R^m \end{Bmatrix} = -\frac{2}{\sqrt{-\Delta_L}} \exp\left[-j\left(\frac{\Delta_L^3}{12} + \frac{\pi}{4}\right)\right]\begin{Bmatrix} \tilde{p}_\pm(\Delta_L, y_L) \\ \tilde{q}_\pm(\Delta_L, y_L) \end{Bmatrix}; \quad \sigma_L \lessgtr 0$$

The values for Δ_L, y_L, σ_L are evaluated at the reflection point P_0 (see Fig. 6.8) and are given as

$$\begin{aligned} \Delta_L &= -2M\cos\nu_i \\ y_L &= M^{-1}\sqrt{(\tfrac{1}{2}k\sigma_L)} \\ \sigma_L &= \frac{s_0^r\rho_1^i\rho_2^i(\rho_1^r + s_0^r)}{(\rho_1^i + s_0^r)(\rho_2^i + s_0^r)\rho_1^r} \end{aligned}$$

Equations 71 and 72 are continuous through the shadow boundary and eqn. 72 goes uniformly into the geometrical optics field in the deep illuminated region.

If the convex surface has an impedance surface the above equations apply but with the transition functions $\tilde{p}_\pm(x, y)$, $\tilde{q}_\pm(x, y)$ replaced by the transition functions $\tilde{Q}_\pm(x, y, \gamma^e)$, $\tilde{Q}_\pm(x, y, \gamma^m)$; see Section 6.2.

(b) Source on the surface

When a magnetic *point source* $\boldsymbol{m}$ is situated on the surface we have, as before, a deep illuminated region, transition region and a deep shadow region. In the last region the diffracted field is very similar to eqn. 70 being given about the diffracted rays as

$$E^d(s^d) = \sum_{n=1}^{N} \mathbf{D}_n\boldsymbol{m}\,\frac{\exp\{-jk(\tau + s^d)\}}{\sqrt{\{s^d(s^d + \rho_3)\}}} \tag{73}$$

where

$$\mathbf{D}_n = -\frac{jk}{4\pi}\sqrt{\left(\frac{d\beta\rho_3}{d\eta'}\right)}\cdot$$

$$\cdot\begin{bmatrix} D_n^e \exp\left[-\int_0^\tau \Omega_n^e(\tau')d\tau'\right] & 0 \\ 0 & D_n^m \exp\left[-\int_0^\tau \Omega_n^m(\tau')d\tau'\right] \end{bmatrix}$$

$$\cdot D_n^e = \left\{\exp\left(\frac{j\pi}{6}\right)\bar{M}\,\mathrm{Ai}'(-\alpha_n)\right\}^{-1}$$

$$D_n^m = \{\alpha_n'\,\mathrm{Ai}(-\alpha_n')\}^{-1}$$

and $d\beta$ is the angle between adjacent rays as shown in Fig. 6.10. The magnetic force $\boldsymbol{m}$ is resolved into the components m_t (transversely directed to ρ_c) and m_a (axially directed to ρ_c) as in Fig. 6.9. In the transition region within the shadow region we have

$$E^d(s^d) = \mathbf{D}\boldsymbol{m}\frac{\exp\{-jk(\tau+s^d)\}}{\sqrt{\{s^d(s^d+\rho_3)\}}} \tag{74}$$

where

$$\mathbf{D} = -\frac{jk}{4\pi}\sqrt{\left(\frac{d\beta\rho_3}{d\eta'}\right)}\begin{bmatrix}\frac{1}{j\bar{M}}f(\Delta) & 0 \\ 0 & g(\Delta)\end{bmatrix} \qquad \Delta \text{ as above}$$

and f, g are the Fock functions defined in Section 2.2.3.

In the illuminated region we express the total field as

$$E(s) = \mathbf{D}_L\boldsymbol{m}\frac{\exp(-jks)}{s} \tag{75}$$

where s is the distance of the source to the field point, $\mathbf{D}_L$ given by

$$\mathbf{D}_L = -\frac{jk}{4\pi}\begin{bmatrix}\frac{1}{jM(0)}f(\Delta_L) & 0 \\ 0 & g(\Delta_L)\end{bmatrix}\exp\left(-j\frac{\Delta_L^3}{3}\right)$$

$$\Delta_L = -M(0)\cos\nu_i$$

and ν_i is the radiation angle measured from the outward normal $\hat{\boldsymbol{n}}$ from the source.

Equation 74 and 75 are continuous through the shadow boundary and eqn. 75 goes uniformly into the geometrical optics field in the deep illuminated region.

References

BIRD, T.S. (1984): 'Comparison of asymptotic solutions for the surface field excited by a magnetic dipole on a cylinder', *IEEE Trans.*, **AP-32**, pp. 1237–1244.

BOWMAN, J.J., SENIOR, T.B.A., and USLENGHI, P.L.E. (1969): 'Electromagnetic and acoustic scattering by simple shapes', (North-Holland Publishing Company).

BURNSIDE, W.D., and PETERS, L. (1974): 'Edge diffracted caustic fields', *IEEE Trans.*, **AP-22**, pp. 620–623.

FRANZ, W., and KLANTE, K. (1959): 'Diffraction by surfaces of variable curvature', *ibid.*, **AP-7**, pp. S68–S70.

HONG, S. (1967): 'Asymptotic theory of electromagnetic and acoustic diffraction by smooth convex surfaces of variable curvature', *J. Math. Phys.*, **8**, pp. 1223–1232.

IDEMEN, M., and FELSEN, L.B. (1981): 'Diffraction of a whispering gallery mode by the edge of a thin concave cylindrical curved surface', *IEEE Trans.* **AP-29**, pp. 571–579.

ISHIHARA, T., FELSEN, L.B., and GREEN, A. (1978): 'High-frequency fields excited by a line source located on a perfectly conducting concave cylindrical surface', *ibid.*, **AP-26**, pp. 757–767.

JAMES, G.L., and KERDEMELIDIS, V. (1973): 'Reflector antenna radiation pattern analysis by equivalent edge currents', *ibid.*, **AP-21**, pp. 19–24.

JAMES, G.L. (1980): 'GTD solution for diffraction by convex corrugated surfaces', *IEE Proc.*, **127**, pt. H, pp. 257–262.

KAMINETSKY, L., and KELLER, J.B. (1972): 'Diffraction coefficients for higher order edges and vertices', *SIAM J. Appl. Math.*, **22**, pp. 109–134.

KAY, I., and KELLER, J.B. (1954): 'Asymptotic evaluation of the field at a caustic', *J. Appl. Phys.*, **25**, pp. 876–883.

KELLER, J.B., and LEVY, B.R. (1959): 'Decay exponents and diffraction coefficients for surface waves on surfaces of non-constant curvature', *IRE Trans.*, **AP-7**, pp. S52–S61.

KNOTT, E.F., and SENIOR, T.B.A. (1974): 'Comparison of three high-frequency diffraction techniques', *Proc. IEEE*, **62**, pp. 1468–1474.

KNOTT, E.F. (1985): 'The relationship between Mitzner's ILDC and Michaeli's equivalent currents', *IEEE Trans.*, **AP-33**, pp. 112–114.

LEVY, B.R., and KELLER, J.B. (1959): 'Diffraction by a smooth object', *Commun. Pure Appl. Math.*, **12**, pp. 159–209.

LUDWIG, D. (1966): 'Uniform asymptotic expansions at a caustic', *ibid.*, **19**, pp. 215–250.

MENTZER, C.A., PETERS, L., and RUDDUCK, R.C. (1975): 'Slope diffraction and its application to horns', *IEEE Trans.*, **AP-23**, pp. 153–159.

MICHAELI, A. (1984): 'Equivalent edge currents for arbitrary aspects of observation', *ibid.*, **AP-32**, pp. 252–258.

MILLER, R.F. (1956): 'An approximate theory of the diffraction of an electromagnetic wave by an aperture in a plane screen', *Proc. IEE*, **103C**, pp. 177–185.
MILLAR, R.F. (1957*a*): 'The diffraction of an electromagnetic wave by a circular aperture', *ibid.*, **104C**, pp. 87–95.
MILLER, R.F. (1957*b*): 'The diffraction of an electromagnetic wave by a large aperture', *ibid.*, **104C**, pp. 240–250.
MITTRA, R., and SAFAVI-NAINI, S. (1979): 'Source radiation in the presence of smooth convex bodies', *Rad. Sci.*, **14**, pp. 217–237.
ORLOV, Yu.I. (1976): 'Modification of the geometrical theory of diffraction of waves in the vicinity of the caustic of the boundary wave', *Rad. Eng. El. Phys.*, **21**, pp. 50–58.
PATHAK, P.H., and KOUYOUMJIAN, R.G. (1974): 'An analysis of the radiation from apertures in curved surfaces by the geometrical theory of diffraction', *Proc. IEEE*, **62**, pp. 1438–1447.
PATHAK, P.H., BURNSIDE, W.D., and MARHEFKA, R.J. (1980): 'A uniform GTD analysis of the diffraction of electromagnetic waves by a smooth convex surface', *IEEE Trans.*, **AP-28**, pp. 631–642.
PATHAK, P.H., WANG, N., BURNSIDE, W.D., and KOUYOUMJIAN, R.G. (1981): 'A uniform GTD solution for the radiation from sources on a convex surface', *ibid.*, **AP-29**, pp. 609–622.
RYAN, C.E., and PETERS, L. (1969): 'Evaluation of edge-diffracted fields including equivalent currents for the caustic regions', *ibid.*, **AP-17**, pp. 292–299.
SERBEST, A.H. (1984): 'Diffraction coefficients for a curved edge with soft and hard boundary conditions', *IEE Proc.*, **131**, pt. H, pp. 383–389.
SENIOR, T.B.A. (1972): 'The diffraction matrix for a discontinuity in curvature', *IEEE Trans.*, **AP-20**, pp. 326–333.
VOLTMER, D.R. (1970): 'Diffraction by doubly curved convex surfaces', Ph.D. dissertation, Ohio State University, Columbus, Ohio, USA.
WAIT, J.R., and CONDA, A.M. (1959): 'Diffraction of electromagnetic waves by smooth obstacles for grazing angles', *J. Res. Nat. Bur. Stand.*, **63D**, pp. 181–197.
WESTON, V.H. (1962): 'The effect of a discontinuity in curvature in high-frequency scattering', *IRE Trans.*, **AP-10**, pp. 775–780.

Chapter 7

Application to some radiation and scattering problems

To illustrate some of the applications of GTD a few worked examples are given in this Chapter. The first example involves evaluation of the geometrical optics field only using the methods given in Chapter 4. This is followed by the 2-dimensional edge diffraction problem of radiation from a parallel-plate waveguide. A comparison is made between GTD, the aperture field method, and the exact solution. Higher order diffraction effects are considered which include use of the slope-diffraction term and coupling effects between edges. The third example studies the effect of placing a reflecting plate in front of the parallel-plate waveguide aperture. This plate assumes various profiles in order to include wedge diffraction as well as half-plane diffraction in the analysis. The problem is considered initially as a 2-dimensional one and the extension to 3-dimensions is given. Both theoretical and experimental results are presented. The fourth example evaluates the edge diffracted field from a reflector antenna. It includes Fermat's principle for edge diffraction (to determine the edge diffraction points) and the use of the equivalent edge current method for the axial caustic region. A comparison is given with the physical optics approach. The final example analyses radiation from a flanged circular waveguide using a uniform equivalent edge current solution. Measured results are also presented.

7.1 Geometrical optics field reflected from a reflector antenna

This example will illustrate the use of the geometrical optics ray tracing techniques given in Chapter 4. The problem is shown in Fig. 7.1 where a point source at $S(r_1, \theta_1, \phi_1)$ is radiating in the presence of a circularly symmetric reflector antenna, and we are required to determine the reflected geometrical optics field at the point $P(r_2, \theta_2, \phi_2)$. In practice,

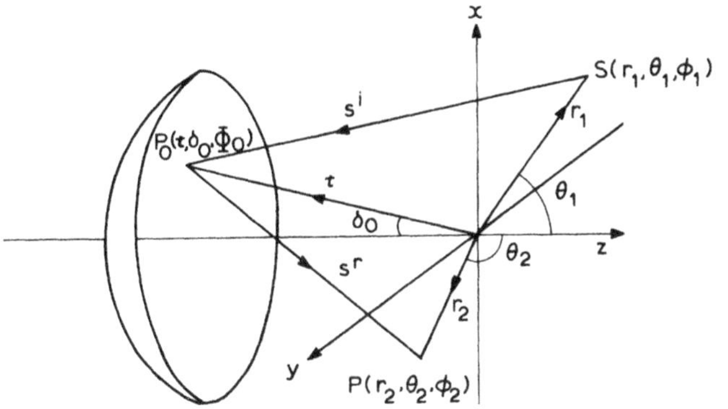

Fig. 7.1 Co-ordinates for the reflector

the reflector may be a paraboloid where the source S is situated at or near the focus, and we wish to evaluate the fields in the aperture plane. Another case is when the source is taken to infinity and the incident field is now a plane wave. The field point P may then be a position on a subreflector from which we can evaluate the current distribution using the physical optics appromixation.

To evaluate the reflected field we first establish the point, or points, $P_0(\tau, \delta_0, \Phi_0)$ on the reflector such that the ray path $s^i + s^r$ is stationary (Fermat's principle). The profile of the reflector is given by

$$\tau = \tau(\delta) \tag{1}$$

and we write s^i, s^r as

$$\begin{aligned} s^{i,r}(\delta, \Phi) &= \{\tau^2 + r_{1,2}^2 - 2r_{1,2}\chi_{1,2}\}^{1/2}; \\ \chi_{1,2} &= \sin\theta_{1,2}\sin\delta\cos(\phi_{1,2} - \Phi) - \cos\theta_{1,2}\cos\delta \end{aligned} \tag{2}$$

For the ray path $s^i + s^r$ to be stationary, we must solve the simultaneous equations

$$\frac{\partial s^i}{\partial \delta} + \frac{\partial s^r}{\partial \delta} = 0; \qquad \frac{\partial s^i}{\partial \Phi} + \frac{\partial s^r}{\partial \Phi} = 0 \tag{3}$$

to obtain the reflection point at $P_0(\tau, \delta_0, \Phi_0)$. Solution of these equations invariably involves numerical search procedures such as the various hill-climbing techniques that are available.

Once points such as P_0 have been established, we proceed to evaluate the incident curvature matrix $\mathbf{Q}^i$ and the interface curvature matrix $\mathbf{C}$ about these points. The incident field is produced from a point source at S and the principal radii of curvature of the incident wavefront at P_0 are given as $\rho_1^i = \rho_2^i = s^i$. When the principal radii of curvature are equal, then from eqn. 4.45 we see that the components of the curvature matrix are independent of any ray-based co-ordinate system i.e.,

$$Q^i_{11} = Q^i_{22}, \quad Q^i_{12} = 0 \quad \text{when} \quad \rho^i_1 = \rho^i_2$$

and the incident curvature matrix for this case becomes

$$\mathbf{Q}^i = \begin{bmatrix} \frac{1}{s^i} & 0 \\ 0 & \frac{1}{s^i} \end{bmatrix} \tag{4}$$

For the interface curvature matrix **C** we first determine the principal radii of curvature ρ^c_1, ρ^c_2 of the surface at A and then make the necessary co-ordinate transformation into the plane of reflection. The profile of the reflector surface is illustrated in Fig. 7.2. With the profile given by $\tau(\delta)$ we can write the ρ, z components as

$$\rho = \tau(\delta)\sin\delta; \qquad z = \tau(0) - \tau(\delta)\cos\delta \tag{5}$$

At any point, such as A, on Γ the tangent plane through that point, an angle β is made with the z-axis as shown in Fig. 7.2. Thus we can write the normal at that point as

$$\hat{\boldsymbol{n}} = \hat{z}\sin\beta - \hat{\boldsymbol{\rho}}\cos\beta \tag{6}$$

Furthermore, we have $\cos\beta = \dfrac{dz}{d\Gamma}$ and $\sin\beta = \dfrac{d\rho}{d\Gamma}$. From eqn 5 we get

$$dz = (\tau\sin\delta - \tau'\cos\delta)d\delta$$

$$d\rho = (\tau\cos\delta + \tau'\sin\delta)d\delta$$

$$d\Gamma = \sqrt{(dz^2 + d\rho^2)} = \sqrt{(\tau^2 + \tau'^2)}d\delta$$

and the values for $\cos\beta$, $\sin\beta$ now become

$$\left.\begin{aligned} \cos\beta &= \frac{dz}{d\Gamma} = \frac{dz}{d\delta}\frac{d\delta}{d\Gamma} = \frac{\tau\sin\delta - \tau'\cos\delta}{\sqrt{(\tau^2 + \tau'^2)}} \\ \sin\beta &= \frac{d\rho}{d\Gamma} = \frac{d\rho}{d\delta}\frac{d\delta}{d\Gamma} = \frac{\tau\cos\delta + \tau'\sin\delta}{\sqrt{(\tau^2 + \tau'^2)}} \end{aligned}\right\} \tag{7}$$

The equation of the straight line, L, along the normal $\hat{\boldsymbol{n}}$ is

$$\rho = a - \cot\beta z \tag{8}$$

For the radius of curvature, ρ^c_2, of the interface in the $d_2 - n$ plane

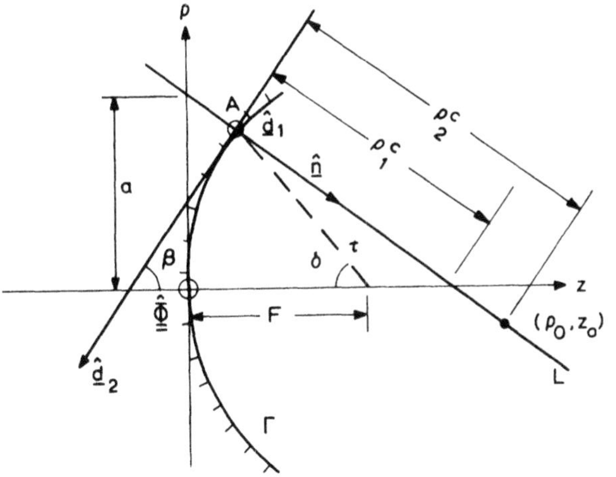

Fig. 7.2 Reflector profile

about the line L, we have the condition that $\dfrac{d\rho}{d\delta} = \dfrac{dz}{d\delta} = 0$ at the local centre of curvature (ρ_0, z_0). Therefore, from eqn. 8

$$z_0 = \frac{\dfrac{da}{d\delta}}{\dfrac{d}{d\delta}\cot\beta}$$

From eqn. 7 we derive $\cot\beta$ as

$$\cot\beta = \frac{\tau\sin\delta - \tau'\cos\delta}{\tau\cos\delta + \tau'\sin\delta}$$

and from Fig. 7.2 we see that

$$a = \tau\sin\delta + (F - \tau\cos\delta)\cot\beta$$

By straightforward differentiation the value for z_0 becomes

$$z_0 = \frac{\tau^2(\tau'\sin\delta + F) - \tau\tau''(F - \tau\cos\delta) + 2\tau'^2\{F + \frac{1}{2}(\tau'\sin\delta - \tau\cos\delta)\}}{\tau^2 - \tau\tau'' + 2\tau'^2}$$

and the corresponding value for ρ_0 is determined via eqn. 8. The calculation of ρ_2^c follows directly. From Fig. 7.2 we have

$$\rho_2^c = \{(z_0 - F + \tau\cos\delta)^2 + (\rho_0 - \tau\sin\delta)^2\}^{1/2}$$

which finally reduces to

$$\rho_2^c = \frac{\{\tau^2 + \tau'^2\}^{3/2}}{\tau^2 - \tau\tau'' + 2\tau'^2} \tag{9}$$

The other principal radius of curvature ρ_1^c in the $d_1 - n$ plane for a circularly symmetrical interface is given simply by the distance along L between the intersection of the z-axis and the interface at A:

$$\rho_1^c = \frac{\tau \sin\delta}{\cos\beta} = \frac{\tau \sin\delta\,(\tau^2 + \tau'^2)^{1/2}}{\tau\sin\delta - \tau'\cos\delta} \tag{10}$$

With the principal radii of curvature for Γ given by eqns. 9 and 10, the curvature matrix for reflection in the $\rho - z$ plane can be written as

$$\mathbf{C}_0 = \begin{bmatrix} -\rho_1^{c^{-1}} & 0 \\ 0 & -\rho_2^{c^{-1}} \end{bmatrix} \tag{11}$$

Note that the radii of curvature are taken to be negative since the interface is locally *concave* to the chosen normal.

Reflection will not always take place in the $d_2 - n$ plane and eqn. 11 must be modified by the appropriate co-ordinate transformation into the reflection plane. This reflection plane contains the unit vectors, $\hat{s}^i, \hat{n}$ and $\hat{s}^r$, where

$$\begin{aligned} s^i\hat{s}^i &= -\mathbf{r}_1 + \boldsymbol{\tau} \\ &= \hat{x}(\tau\sin\delta_0\cos\Phi_0 - r_1\sin\theta_1\cos\phi_1) \\ &\quad + \hat{y}(\tau\sin\delta_0\sin\Phi_0 - r_1\sin\theta_1\sin\phi_1) - \hat{z}(\tau\cos\delta_0 + r_1\cos\theta_1); \end{aligned} \tag{12}$$

$$\begin{aligned} s^r\hat{s}^r &= \mathbf{r}_2 - \boldsymbol{\tau} \\ &= -\hat{x}(\tau\sin\delta_0\cos\Phi_0 - r_2\sin\theta_2\cos\phi_2) - \hat{y}(\tau\sin\delta_0\sin\Phi_0 \\ &\quad - r_2\sin\theta_2\sin\phi_2) + \hat{z}(\tau\cos\delta_0 + r_2\cos\theta_2); \end{aligned} \tag{13}$$

and from eqn. 6

$$\hat{n} = -\hat{x}\cos\beta\cos\Phi_0 - \hat{y}\cos\beta\sin\Phi_0 + \hat{z}\sin\beta \tag{14}$$

In the reflection plane we resolve the incident electric field as in Fig. 7.3 into an electric polarisation component E^i_ψ and a magnetic polarisation component E^i_ξ, where

$$\hat{s}^i = \hat{\boldsymbol{\psi}} \times \hat{\boldsymbol{\xi}} \tag{15}$$

The unit vector $\hat{\boldsymbol{\psi}}$ is normal to the reflection plane and we have

$$\hat{s}^i \times \hat{n} = \sin\nu_i\,\hat{\boldsymbol{\psi}} \tag{16a}$$

$$\hat{\boldsymbol{\psi}}\cdot\hat{n} = \hat{\boldsymbol{\psi}}\cdot\hat{s}^{i,r} = 0 \tag{16b}$$

These equations can be solved to yield $\hat{\boldsymbol{\psi}}$ and ν_i in terms of the variables τ, δ_0, Φ_0, $r_{1,2}$, $\theta_{1,2}$, $\phi_{1,2}$, β, $s^{i,r}$. The derivation, although straightforward, yields expressions which are rather lengthy for the general case. At this point we will consider only the special case where the source recedes to infinity and we have an incident plane wave upon the reflector. We also take $\phi_1 = 0$ and assume an incident electric field

$$E^i = \hat{x}E_x^i \cos\theta_1 - \hat{z}E_z^i \sin\theta_1 \tag{17}$$

With $\tau \ll r_1$ and $\phi_1 = 0$ we derive from eqns 12 to 16

$$\sin\nu_i = \{\cos^2\beta\sin^2\Phi_0 + \cos^2\theta_1(\tan\theta_1\sin\beta + \cos\beta\cos\Phi_0)^2\}^{1/2} \tag{18}$$

and for $\hat{\boldsymbol{\psi}}$

$$\hat{\boldsymbol{\psi}} = -\csc\nu_i\{\hat{x}\cos\theta_1\cos\beta\sin\Phi_0 - \hat{y}(\sin\theta_1\sin\beta + \cos\theta_1\cos\beta\cos\Phi_0)$$
$$-\hat{z}\sin\theta_1\cos\beta\sin\Phi_0 \tag{19}$$

For the remaining component of the ray-based co-ordinate system we have, from eqn. 15

$$\hat{\boldsymbol{\xi}}^{i,r} = \hat{s}^{i,r} \times \hat{\boldsymbol{\psi}}$$

to yield

$$\hat{\boldsymbol{\xi}}^i = \hat{x}\cos\theta_1 A + \hat{y}B - \hat{z}\sin\Phi_0 A \tag{20a}$$

$$\begin{aligned}\boldsymbol{\xi}^r = -\hat{x}&\left\{\frac{A}{s^r}(r_2\cos\theta_2 + \tau\cos\delta_0) - \frac{B\sin\theta_1}{s^r}\right.\\&\left.(r_2\sin\theta_2\sin\phi_2 - \tau\sin\delta_0\sin\Phi_0)\right\}\\&-\hat{y}\left\{\frac{B\cos\theta_1}{s^r}(r_2\cos\theta_2 + \tau\cos\delta_0) + \frac{B\sin\theta_1}{s^r}\right.\\&\left.(r_2\sin\theta_2\cos\phi_2 - \tau\sin\delta_0\cos\Phi_0)\right\}\\&+\hat{z}\left\{\frac{B\cos\theta_1}{s^r}(r_2\sin\theta_2\sin\phi_2 - \tau\sin\delta_0\sin\Phi_0) + \frac{A}{s^r}\right.\\&\left.(r_2\sin\theta_2\cos\phi_2 - \tau\sin\delta_0\cos\Phi_0)\right\}\end{aligned} \tag{20b}$$

where

$$A = \csc\nu_i(\sin\theta_1\sin\beta + \cos\theta_1\cos\beta\cos\Phi_0)$$

$$B = \csc\nu_i\cos\beta\sin\Phi_0$$

The curvature matrix for the surface as given by eqn. 11 is only valid for the $d_1 - d_2$ co-ordinate system as in Fig. 7.2, where

$$\hat{d}_1 = \hat{\boldsymbol{\Phi}}$$

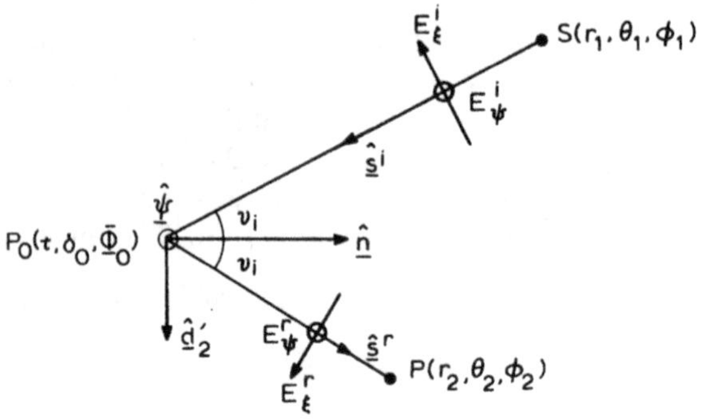

Fig. 7.3 Plane of reflection

We require, in general, the surface curvature matrix to relate to the $d'_1 - d'_2$ co-ordinate system, where

$$\hat{d}'_1 = \hat{\psi}$$

so that it is necessary to evaluate the angle α in Fig. 7.4, where

$$\cos\alpha = \hat{\psi} \cdot \hat{\Phi} \tag{21a}$$

to rotate eqn. 11 into the correct co-ordinate space. The value of α follows from eqns. 21*a* and 19, viz,

$$\cos\alpha = \csc\nu_i(\cos\theta_1\cos\beta + \sin\theta_1\sin\beta\cos\Phi_0) \tag{21b}$$

Our surface curvature matrix is now given directly from eqns. 4.45 and 11 as

$$\mathbf{C} = \begin{bmatrix} -\left(\dfrac{\cos^2\alpha}{\rho_1^c} + \dfrac{\sin^2\alpha}{\rho_2^c}\right) & -\cos\alpha\sin\alpha\left(\dfrac{1}{\rho_1^c} - \dfrac{1}{\rho_2^c}\right) \\ -\cos\alpha\sin\alpha\left(\dfrac{1}{\rho_1^c} - \dfrac{1}{\rho_2^c}\right) & -\left(\dfrac{\sin^2\alpha}{\rho_1^c} + \dfrac{\cos^2\alpha}{\rho_2^c}\right) \end{bmatrix} \tag{22}$$

$$= \begin{bmatrix} C_{11} & C_{12} \\ C_{12} & C_{22} \end{bmatrix}$$

The incident field curvature matrix is given by eqn. 4, and for plane wave incidence is seen to reduce to zero. Thus from eqn. 4.55a our reflected field curvature matrix is

$$\mathbf{Q}^r(0) = \begin{bmatrix} 2C_{11}\cos\nu_i & 2C_{12} \\ 2C_{12} & 2C_{22}\sec\nu_i \end{bmatrix} \tag{23}$$

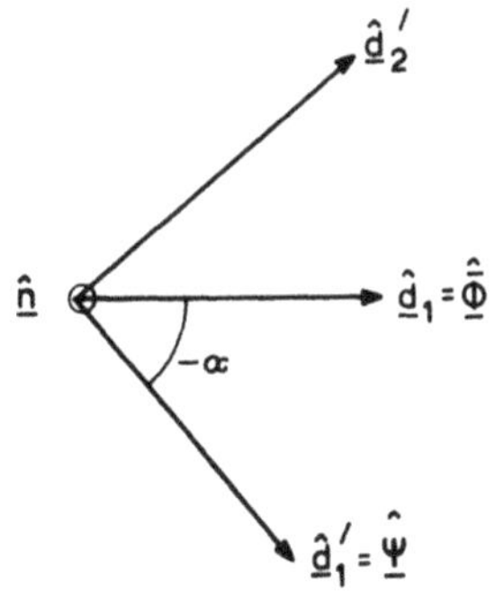

Fig. 7.4 Co-ordinate rotation at the reflection point

From this equation we can derive the principal radii of curvature for the reflected wavefront from eqn. 4.46, viz,

$$\frac{1}{\rho^r_{1,2}} = \tfrac{1}{2}\left(Q^r_{11}(0) + Q^r_{22}(0) \pm \sqrt{[\{Q^r_{11}(0) - Q^r_{22}(0)\}^2 + 4Q^{r2}_{12}(0)]}\right) \tag{24}$$

and, finally, the geometrical optics reflected field can be written as

$$\begin{bmatrix} E^r_\psi(s^r) \\ E^r_\xi(s^r) \end{bmatrix} = \begin{bmatrix} -1 & 0 \\ 0 & 1 \end{bmatrix} \begin{bmatrix} E^i_\psi(P_0) \\ E^i_\xi(P_0) \end{bmatrix} \left[\frac{\rho^r_1 \rho^r_2}{(\rho^r_1 + s^r)(\rho^r_2 + s^r)}\right]^{1/2} \exp(-jks^r) \tag{25}$$

The field components in this equation can be readily obtained from eqns. 17, 19, and 20. To conclude this example we will evaluate the principal radii of curvature ρ^r_1, ρ^r_2 for two well known cases.

Spherical reflector

If the radius of a spherical reflector is given as R then let

$$\tau = R$$

and

$$\tau' = \tau'' = 0$$

so that from eqns. 9 and 10

$$\rho^c_1 = \rho^c_2 = R$$

For an on-axis plane wave $\theta_1 = 0$ and from eqn. 21

$$\cos\alpha = \csc\nu_i \cos\beta$$

and from eqn. 18

$$\sin\nu_i = \cos\beta$$

which gives $\cos\alpha = 1$. The reflection matrix of eqn. 23 now reduces to

$$\mathbf{Q}^r(0) = \begin{bmatrix} -\dfrac{2\cos\nu_i}{R} & 0 \\ 0 & -\dfrac{2\sec\nu_i}{R} \end{bmatrix}$$

from which we derive, via eqn. 24, the well known result

$$\rho_1^r = -\frac{R}{2}\sec\nu_i; \qquad \rho_2^r = -\frac{R}{2}\cos\nu_i \tag{26}$$

Paraboloidal reflector

The profile function of a paraboidal reflector is given as

$$\tau = F\sec^2\frac{\delta}{2}$$

giving

$$\tau' = \tau\tan\frac{\delta}{2}; \qquad \tau'' = \frac{1}{2F}\tau^2(2-\cos\delta)$$

With these values the principal radii of curvature for the reflector as given by eqns. 9 and 10 become

$$\rho_1^c = -2\tau\cos\frac{\delta}{2}; \qquad \rho_2^c = -2\tau\sec\frac{\delta}{2}$$

For an on-axis plane wave we have, as before, $\theta_1 = 1$, $\alpha = 0$, and $\sin\nu_i = \cos\beta$. From this last relationship and eqn. 7 we can deduce that $\nu_i = \dfrac{\delta}{2}$. The reflection matrix of eqn. 23 now becomes

$$\mathbf{Q}^r(0) = \begin{bmatrix} -\tau^{-1} & 0 \\ 0 & -\tau^{-1} \end{bmatrix}$$

which gives the expected result

$$\rho_1^r = \rho_2^r = -\tau \tag{27}$$

7.2 Radiation from a parallel-plate waveguide

One of the simplest problems using the methods of GTD is in evaluating the radiated field from the open end of a parallel-plate waveguide. It has been solved by several methods: aperture field and exact Wiener-Hopf solution (for example see Chapter 15 of Collin and Zucker, 1969); GTD approach, Rudduck and Wu (1969), Lee (1969); and a

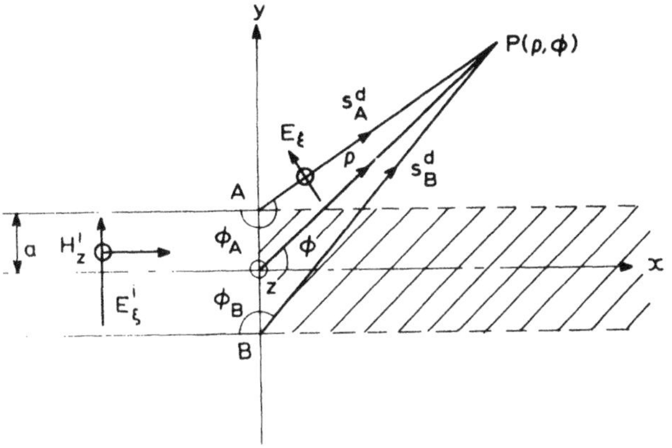

Fig. 7.5 Parallel-plate waveguide supporting the TEM mode

surface integration technique combined with GTD, Wu *et al.* (1969). We will consider this problem in some detail as it allows an exposition of several important features of the GTD method while retaining simplicity. In addition we can compare the approximate methods of GTD and the aperture field approach with the exact solution.

To begin, consider the TEM mode propagating in the parallel-plate waveguide as shown in Fig. 7.5. Initially each plate is considered as an isolated half-plane. Edge diffraction for a general incident field to a half-plane is given from eqns. 5.125 and 5.129 as

$$E^d(s^d) = \mathbf{D}E^i(P_0)\left[\frac{\rho_3}{s^d(\rho_3+s^d)}\right]^{1/2}\exp(-jks^d); \quad \mathbf{D} = \begin{bmatrix} D^e & 0 \\ 0 & D^m \end{bmatrix} \tag{28}$$

where

$$D^{e,m} = -\sigma^{1/2}\{\epsilon^i K_-(v^i) \mp \epsilon^r K_-(v^r)\}$$

$$\sigma = \frac{\rho_1\rho_2(\rho_3+s^d)s^d}{(\rho_1+s^d)(\rho_2+s^d)\rho_3}$$

$$v^{i,r} = \sqrt{(2k\sigma)}\,|\cos\tfrac{1}{2}(\phi \mp \phi_0)|\sin\theta_0$$

Consider the lower half-plane with the edge at B in Fig. 7.5. The incident field at the edge is a plane wave at grazing incidence so that

$$\rho_1 = \rho_2 = \rho_3 = \infty; \qquad \sigma = s_B^d; \qquad \phi_0 = 0$$

Also the field is normal to the edge, hence $\theta_0 = \frac{\pi}{2}$, and is magnetically polarised giving only a ξ component of incident field. With the field at grazing incidence the diffracted field is divided by 2 (see Section

5.11). Edge diffraction from B to a point $P(\rho, \phi)$ in the forward space outside the aperture is now given from eqn. 28 as

$$E_\xi^d = -E_\xi^i(B)\,\epsilon_B^i\,K_-\{\surd(2ks_B^d)|\cos\tfrac{1}{2}\phi_B|\}\exp(-jks_B^d) \quad \text{(from edge } B) \tag{29}$$

A similar diffracted field exists from edge A to the field point at P and is given as

$$E_\xi^d = -E_\xi^i(A)\,\epsilon_A^i\,K_-\{\surd(2ks_A^d)|\cos\tfrac{1}{2}\phi_A|\}\exp(-jks_A^d) \quad \text{(from edge } A) \tag{30}$$

These fields in eqns. 29 and 30 are added to give the total diffracted field on the basis that the two plates comprising the waveguide are behaving as isolated half-planes. This assumption will be modified later to include interaction between them.

If the field point is in the shaded region in Fig. 7.5 then the direct geometrical optics field must also be added to the solution. For example if the field point is on the x-axis at $P(b, 0)$ we have

$$s_A^d = s_B^d = s^d = \surd(a^2 + b^2)$$

$$\phi_A = \phi_B = \frac{\pi}{2} + \chi \quad \text{where} \quad \chi = \sin^{-1}\frac{b}{s^d}$$

$$\epsilon_A^i = \epsilon_B^i = 1$$

and the modified Fresnel integral argument in eqns. 29 and 30 can be written as

$$\{ks^d(1 - \sin\chi)\}^{1/2} = \{k(s^d - b)\}^{1/2}$$

Referring the phase of the incident field to the origin of the (x, y, z) co-ordinate system and taking $|E_\xi^i| = 1$, the electric field $\boldsymbol{E}$ at $P(b, 0)$ along the axis is given from the superposition of eqns. 29 and 30 and the direct optical term. This yields

$$\boldsymbol{E}(b,0) = 0\hat{x} + \hat{y}\left(\exp(-jkb) - 2K_-[\surd(k(s^d - b)\}]\right.$$
$$\left.\exp(-jks^d)\right) + 0\hat{z} \tag{31}$$

If the field point $P(\rho, \phi)$ is now taken to be in the far field, then only the edge diffracted field components will contribute, provided that $\phi \neq 0$. The following relationships are valid if P is in the far field:

$$\hat{\boldsymbol{\xi}} = \hat{\boldsymbol{\varphi}}$$

$$\phi_A = \pi + \phi$$

$$\phi_B = \pi - \phi$$

$$s_A^d = \rho - a \sin \phi$$

$$s_B^d = \rho + a \sin \phi$$

Substituting these quantities into eqns. 29 and 30, and taking the asymptotic form of the modified Fresnel integral gives

$$E_\xi^d \sim E_\xi^i \csc \frac{\phi}{2} \exp(jka \sin \phi) \frac{\exp(-jk\rho)}{\sqrt{(8j\pi k\rho)}} \qquad \text{(from edge } A\text{)} \tag{32a}$$

$$E_\xi^d \sim -E_\xi^i \csc \frac{\phi}{2} \exp(-jka \sin \phi) \frac{\exp(-jk\rho)}{\sqrt{(8j\pi k\rho)}} \qquad \text{(from edge } B\text{)} \tag{32b}$$

Adding these two field components we obtain the electric radiation field, E_ϕ, in the forward region as

$$E_\phi = 4jkaE_\xi^i \cos \frac{\phi}{2} \frac{\sin u}{u} \frac{\exp(-jk\rho)}{\sqrt{(8j\pi k\rho)}}; \qquad u = ka \sin \phi, |\phi| < \frac{\pi}{2} \tag{33}$$

It will be noted that this equation gives a finite non-zero value for the far field on the axis at $\phi = 0$ despite the fact that the individual source terms of eqn. 32 have an infinity for this angle (since it corresponds to the shadow boundary of the half-planes). Also the direct geometrical optics term has not been included in eqn. 33. If we now compare this equation with that of eqn. 31 for the field at all points along the axis, it is seen that this latter equation tends to zero as the far field is approached, i.e. as $b \to s^d$. The paradox of the solution given by eqn. 33 can be resolved by studying the aperture field solution. The equivalent sources, $\boldsymbol{J}_s$, $\boldsymbol{M}_s$ in the aperture of the parallel-plate waveguide are determined from the incident field as

$$\begin{aligned} \boldsymbol{J}_s &= \hat{\boldsymbol{x}} \times \boldsymbol{H}^i = -\hat{\boldsymbol{y}} \sqrt{\left(\frac{\epsilon}{\mu}\right)} E_\xi^i \\ \boldsymbol{M}_s &= \boldsymbol{E}^i \times \hat{\boldsymbol{x}} = -\hat{\boldsymbol{z}} E_\xi^i \end{aligned} \tag{34}$$

and from eqn. 2.21 the three possible solutions for the far field using these currents are

$$\begin{aligned} \boldsymbol{E}_2 &= \hat{\boldsymbol{\varphi}} I \quad \text{where} \quad I = jk2 \frac{\exp(-jk\rho)}{\sqrt{(8j\pi k\rho)}} E_\xi^i \int_{-a}^{a} \exp(jky \sin \phi)\, dy \\ \boldsymbol{E}_1 &= \hat{\boldsymbol{\varphi}} \cos \phi I \\ \boldsymbol{E}_3 &= \tfrac{1}{2}(\boldsymbol{E}_1 + \boldsymbol{E}_2) \end{aligned} \tag{35}$$

If we attempt to solve the integral in eqn. 35 asymptotically using the methods given in Section 2.3, stationary phase points are given by $g'(y) = 0$ where

$$g(y) = y \sin \phi \tag{36}$$

Since all higher order derivatives beyond the first derivative are identically zero a stationary phase evaluation is invalid. This means that the entire aperture distribution is contributing to the field where $g'(y) = 0$, i.e., when $\phi = 0$. We can, however, evaluate the endpoint contributions using eqn. 2.78 for field points removed from the axis (i.e. removed from the stationary phase point) to give

$$I \sim \frac{1}{jk \sin \phi} \{\exp(jka \sin \phi) - \exp(-jka \sin \phi)\} = 2a \frac{\sin u}{u}; \; u = ka \sin \phi \tag{37}$$

This result is equal to a direct evaluation of the integral. As for the GTD analysis the individual endpoint contributions are infinite on the axis. The fact that eqn. 37 equals the exact integration is a consequence of the uniformity and symmetry of the aperture distribution. This is a special case but has been studied in detail here since it is a phenomenon that frequently occurs in GTD analysis of radiation and scattering problems. If the field in the aperture had a symmetrical distribution the endpoint infinities along the axis would still cancel to yield a finite result but we would not, in general, expect it to yield the same value as a direct evaluation of the integral.

Substituting eqn. 37 into eqn. 35 gives

$$\begin{aligned} E_2 &= \hat{\boldsymbol{\varphi}} jk4aE^i_\xi \frac{\sin u}{u} \frac{\exp(-jk\rho)}{\sqrt{(8j\pi k\rho)}} \\ E_1 &= \hat{\boldsymbol{\varphi}} jk4aE^i_\xi \frac{\sin u}{u} \cos\phi \frac{\exp(-jk\rho)}{\sqrt{(8j\pi k\rho)}} \qquad |\phi| < \frac{\pi}{2} \\ E_3 &= \hat{\boldsymbol{\varphi}} jk4aE^i_\xi \frac{\sin u}{u} \cos^2 \frac{\phi}{2} \frac{\exp(-jk\rho)}{\sqrt{(8j\pi k\rho)}} \end{aligned} \tag{38}$$

All these equations and the GTD result of eqn. 33 are identical in the vicinity of the axis. Well removed from the axis all four results differ, with E_3 being the nearest aperture field formulation to the GTD result. It is interesting to note that if the physical optics half-plane diffraction coefficient of eqn. 5.77*b* is used in place of the exact coefficient in eqn. 32, then the field as given by E_3 is obtained. Similarly, for the problem of diffraction through a slit in an infinite screen the field as given by E_2 agrees with the physical optics half-plane diffraction

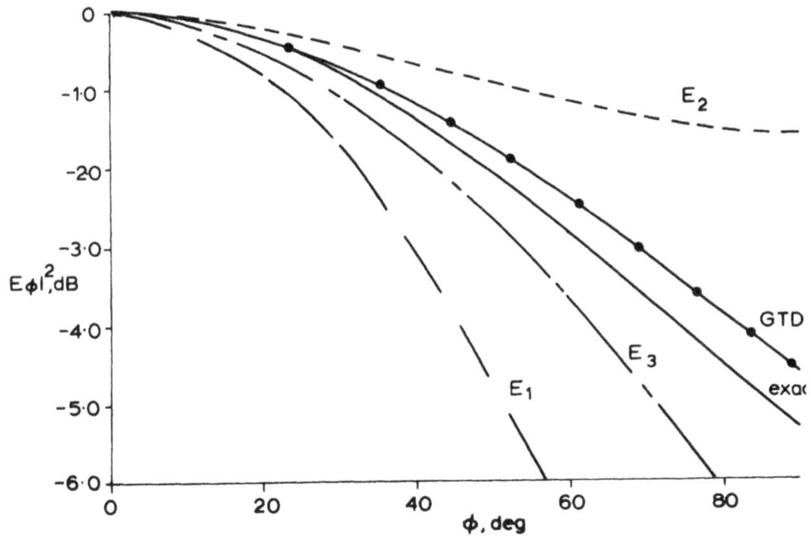

Fig. 7.6 Radiation from a parallel-plate waveguide supporting the TEM mode

coefficient formulation. The third aperture field method result given by E_1 is the same formulation as for physical optics at a perfect conductor.

The above results in the forward region are compared with the exact solution in Fig. 7.6, where the normalised radiation pattern from the exact solution is given by

$$|E_\phi| = \exp\left\{\frac{ka}{2}(\cos\phi - 1)\right\}\left|\frac{\sin u}{u}\right|^{1/2} \quad \text{provided} \quad a < \frac{\lambda}{2} \qquad (39)$$

and we have taken a to be $0{\cdot}17\lambda$. Even with our simple GTD model reasonable agreement is obtained with the exact solution. The aperture field method, however, gives serious errors for angles well removed from the axis.

We turn our attention now to the backward region. For $\frac{\pi}{2} < \phi < \pi$ it is seen from Fig. 7.5 that the line source from the edge at B cannot radiate directly into the far field since it is blocked by the upper half-plane. The field is then determined entirely by diffraction from the edge at A, which for our simple analysis is given by eqn. 32a, viz,

$$E_\phi^d = \csc\frac{\phi}{2}\exp(jka\sin\phi)\frac{\exp(-jk\rho)}{\sqrt{(8j\pi k\rho)}}; \qquad \frac{\pi}{2} < \phi < \pi \qquad (40)$$

where $|E_\xi^i| = 1$. If this equation is used for the backward (upper) region and eqn. 33 for the forward region then a discontinuity in field occurs

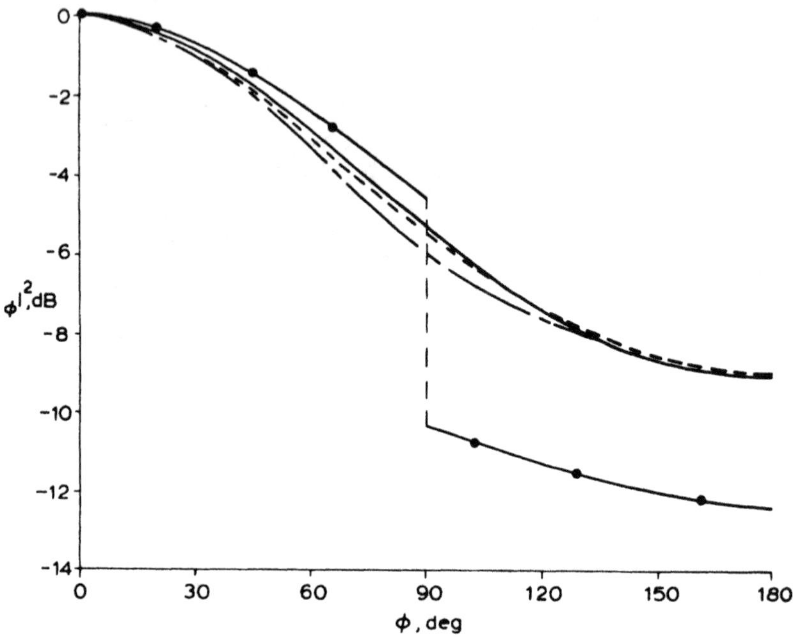

Fig. 7.7 Radiation from a parallel-plate waveguide supporting the TEM mode

———— exact
—●—●— no coupling terms
—·— one coupling term added
—··— addition of slope-diffraction term
} GTD

at $\phi = \frac{\pi}{2}$, as shown in Fig. 7.7. This type of discontinuity in field is a common feature in GTD radiation analysis where an edge diffraction source is blocked over an angular region of space by the geometry of the radiating body. Such results, however, do not constitute a complete first order diffraction analysis. This is because at the discontinuity a shadow boundary of the source being blocked exists. The magnetic line source at B in our example gives rise to diffraction at A which in the total solution represents a higher order diffraction component. Along the shadow boundary at $\phi = \frac{\pi}{2}$, however, this component is of the same order as the original source and must be included in this region if a complete first order solution is to be obtained. If we simply add the diffraction at A due to the source at B to our previous result for $0 < \phi < \pi$, then we will have a complete first order diffraction solution, but an incomplete second order result. (This follows since the higher order diffraction from A to B has been ignored. In the lower half-space this term will be important in the vicinity of $\phi = -\frac{\pi}{2}$).

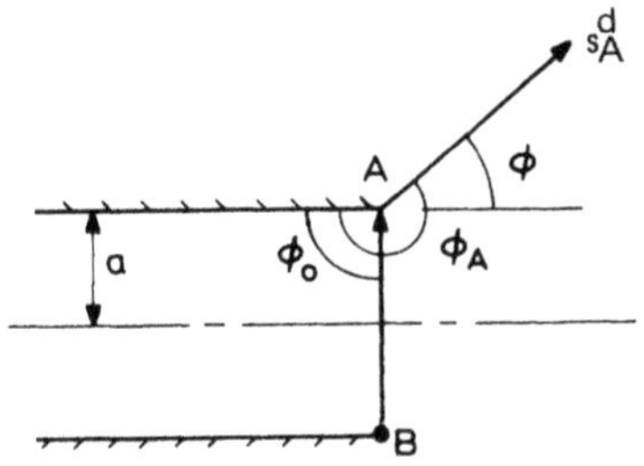

Fig. 7.8 Effective line source at *B* giving coupling effect between the plates

The magnetic line source at B in Fig. 7.8 has an amplitude R_B in the direction of the edge at A determined from eqn. 32b when $\phi = \frac{\pi}{2}$, i.e.,

$$R_B = -\sqrt{(2)} \quad \text{where} \quad |E_\xi^i| = 1$$

Using eqn. 28 we have, as before, only a ξ-component of electric field. The incident field is now a cylindrical wave where

$$\rho_1 = \rho_3 = \infty, \quad \rho_2 = 2a, \quad \theta_0 = \frac{\pi}{2}, \quad \phi_0 = \frac{\pi}{2}$$

giving

$$\sigma = \frac{2as_A^d}{2a + s_A^d}$$

$$\sim 2a \text{ for the far field when } s_A^d \to \infty$$

Our higher order edge diffracted field from A due to the source at B is, from eqn. 28

$$E_\xi^d(s_A^d) = -R_B \frac{\exp(-jk2a)}{\sqrt{(8j\pi k2a)}} \sqrt{\left(\frac{2a}{s_A^d}\right)} \left[\epsilon^i K_-\left\{\sqrt{(4ka)}\left|\cos\tfrac{1}{2}\left(\phi_A - \frac{\pi}{2}\right)\right|\right\}\right.$$

$$\left. + \epsilon^r K_-\left\{\sqrt{(4ka)}\left|\cos\tfrac{1}{2}\left(\phi_A + \frac{\pi}{2}\right)\right|\right\}\right] \exp(-jks_A^d)$$

In the far field we have

$$\phi_A = \pi + \phi, \quad s_A^d = \rho - a\sin\phi, \quad \hat{\boldsymbol{\xi}} = \hat{\boldsymbol{\phi}}$$

and the above equation becomes

$$E_\phi^d = \sqrt{(2)}\exp\{-jka(2 - \sin\phi)\}(\epsilon^i K_-[\sqrt{\{2ka(1 - \sin\phi)\}}]$$

$$+ \epsilon^r K_-[\sqrt{\{2ka(1 + \sin\phi)\}}]) \frac{\exp(-jk\rho)}{\sqrt{(8j\pi k\rho)}} \tag{41}$$

where, in the region $0<\phi<\pi$, $\epsilon^i = 1$ for $\phi<\frac{\pi}{2}$, -1 otherwise, and ϵ^r is always -1. When eqn. 41 is added to the previous GTD result of eqns. 33 and 40 a smooth transition of field occurs across the shadow boundary at $\phi = \frac{\pi}{2}$, as seen from Fig. 7.7. As discussed above this result represents a complete first order solution in the upper half-space but an incomplete second order solution. For a complete second order solution we would need the diffracted field from A to B, plus the third order doubly diffracted term resulting from this latter second order diffraction, to give a smooth transition of field at $\phi = \frac{\pi}{2}$. It will be already noted, however, that the inclusion of eqn. 41 gives a result which is close to the exact solution. While it is perfectly feasible to include higher order diffraction terms, the extra effort is not usually justified for terms beyond second order.

The result obtained so far can be improved further by the inclusion of slope-diffraction in the second order term from the non-uniform line source at B in Fig. 7.8. From eqn. 5.130 the slope-diffraction term $\boldsymbol{E}^{sd}$ is given as

$$\boldsymbol{E}^{sd} = \frac{1}{jk\sin\theta_0}\frac{\partial}{\partial\phi_0}\mathbf{D}\frac{\partial}{\partial n}E^i\left[\frac{\rho_3}{s^d(\rho_3+s^d)}\right]^{1/2}\exp(-jks^d) \quad (42)$$

This equation is only true in general if the direct incident field $\boldsymbol{E}^i$ is zero. Since this is not the case in our example we need to satisfy ourselves that eqn. 42 remains valid in this instance by referring to the derivation of slope-diffraction in Section 5.5.

The slope-diffraction term is determined from the $\zeta_1^{i,r}$ component in eqn. 5.50. In our example $\rho_3 = \infty$ and $\sigma = 2a$ for the far field so that $\zeta_1^{i,r}$ as defined in eqn. 5.51 becomes

$$w_1^{i,r} = E_{\substack{\psi\\ \xi}}^{i,r}\left\{\zeta_1^{i,r} - \sec^3\tfrac{1}{2}\Phi^{i,r}\frac{1}{8a\sqrt{(8j\pi s^d)}}\right\} \quad (43)$$

where, from eqn. 5.52

$$w_1^{i,r} = \frac{\delta_1^{i,r}}{\sqrt{(s^d)}} - \tfrac{1}{2}\int_0^{s^d}\left[\frac{s}{s^d}\right]^{1/2}\nabla^2 w_0^{i,r}(s)\,ds \quad (44)$$

The components $w_0^{i,r}$ and $\delta_1^{i,r}$ in this equation are given from eqns 5.54 and 5.56 as

$$w_0^{i,r}(s^d) = -E_{\substack{\psi\\ \xi}}^{i,r}\frac{\sec\frac{1}{2}\Phi^{i,r}}{\sqrt{(8j\pi s^d)}} \quad (45)$$

$$\delta_1^{i,r} = -\frac{1}{4\sqrt{(8j\pi)}}\sec^3\tfrac{1}{2}\Phi^{i,r}\left(E_{\substack{\psi\\ \xi}}^{i,r}\frac{1}{2a} + \hat{s}^d\cdot\nabla E_{\substack{\psi\\ \xi}}^{i,r}\right) \quad (46)$$

The second term in eqn. 46 yielded the slope-diffraction formulation of eqn. 42. It remains to determine what contribution the first term makes. For this first term only, then

$$\frac{\delta_1^{i,r}}{\sqrt{(s^d)}} = -\frac{1}{8a} E^{i,r}_{\substack{\psi \\ \xi}} \frac{\sec^3 \frac{1}{2}\Phi^{i,r}}{\sqrt{(8j\pi s^d)}} \tag{47}$$

Solving the finite part integral in eqn. 44 via eqn. 5.57 for $\rho_3 = \infty$ we get

$$\tfrac{1}{2}\!\!\!\!\!\!\;\;\;\int_0^{s^d} \left[\frac{s}{s^d}\right]^{1/2} \nabla^2 w_0^{i,r}(s)\, ds = \frac{1}{4} \frac{\sec^3 \frac{1}{2}\Phi^{i,r}}{\sqrt{(8j\pi s^d)}} s^{d^{-1}} \tag{48}$$

which clearly goes to zero in the far field. Thus eqn. 47 gives $w_1^{i,r}$, without the slope-diffraction component, for our problem. Substitution into eqn. 43 shows that $\zeta_1^{i,r} = 0$ for the direct component $E^{i,r}_{\substack{\psi \\ \xi}}$. Eqn. 42 can now be used as the complete slope-diffraction term in the present application.

The radiation pattern of the line source at B in Fig. 7.8 is given by $-\sec\dfrac{\phi_B}{2}$, and the slope of the wavefront incident at A is simply

$$\frac{1}{2a}\frac{\partial}{\partial \phi_B}\left(-\sec\frac{\phi_B}{2}\right) \quad \text{with} \quad \phi_B = \frac{\pi}{2}.$$

This gives $\dfrac{-\sqrt{(2)}}{4a}$ so that in eqn. 42 we have, for the edge at A

$$\frac{\partial}{\partial n}E^i = -\frac{\sqrt{(2)}}{4a}\frac{\exp(-jk2a)}{\sqrt{(8j\pi k2a)}}$$

With $\theta_0 = \frac{\pi}{2}$, $\rho_3 = \infty$, $E_\phi^{sd} = 0$ in eqn. 42 and using $\dfrac{\partial}{\partial \psi_0}D^m$ as given by eqn. 5.131d, the slope-diffraction far field component E_ϕ^{sd} for the non-uniform line source at B diffracting at the edge A in Fig. 7.8 becomes

$$E_\phi^{sd} = \frac{\exp\{-jka(2-\sin\phi)\}}{2j\sqrt{(2ka)}}\left\{\sin\tfrac{1}{2}\left(\frac{\pi}{2}+\phi\right)\left\{2jv^iK_-(v^i) - \sqrt{\left(\frac{j}{\pi}\right)}\right\}\right.$$
$$\left. - \sin\tfrac{1}{2}\left(\frac{3\pi}{2}+\phi\right)\left\{2jv^rK_-(v^r) - \sqrt{\left(\frac{j}{\pi}\right)}\right\}\right\}\frac{\exp(-jk\rho)}{\sqrt{(8j\pi k\rho)}}$$

where

$$v^{i,r} = \sqrt{\{2ka(1 \mp \sin\phi)\}}$$

The addition of this term to our GTD solution yields the remarkable result shown in Fig. 7.7. Note that the plates are only separated by $0{\cdot}34\,\lambda$ and yet with the inclusion of the first coupling ray between these plates, together with its slope-diffraction term, a result is obtained which, from an engineering point of view, is barely distinguishable from the exact solution. This is by no means an isolated example. In many other problems the GTD approach has been shown to yield very good agreement with the exact solution, where it is available, and also with measured data. For small structures, as in the above example, it is the ability to account for higher order diffraction effects not possible with the aperture field method or the physical optics approximation that makes the GTD approach especially useful.

If the parallel-plate waveguide is supporting the TE_1 mode, then this mode can be decomposed into two plane waves propagating between the two plates as shown in Fig. 7.9 where the angle of incidence (and reflection) α, is given by

$$\alpha = \sin^{-1}\left(\frac{\lambda}{4a}\right)$$

As before we consider, initially, each edge in isolation. The incident field is now electrically polarised giving a diffracted far field from the edge at B as

$$E_z^d = -E_z^i\{\sec\tfrac{1}{2}(\phi_B-\alpha)-\sec\tfrac{1}{2}(\phi_B+\alpha)\}\frac{\exp(-jks_B^d)}{\sqrt{(8j\pi ks_B^d)}} \quad \text{(from edge } B\text{)}$$

Similarly for the edge at A

$$E_z^d = -E_z^i\{\sec\tfrac{1}{2}(\phi_A-\alpha)-\sec\tfrac{1}{2}(\phi_A+\alpha)\}\frac{\exp(-jks_A^d)}{\sqrt{(8j\pi ks_A^d)}} \quad \text{(from edge } A\text{)}$$

Adding these effective sources in the far field we obtain the radiation field E_z for $|\phi|<\frac{\pi}{2}$ as

$$E_z = 8E_z^i\,\frac{\cos u\cos\tfrac{1}{2}\phi\sin\tfrac{1}{2}\alpha}{\cos\alpha-\cos\phi}\,\frac{\exp(-jk\rho)}{\sqrt{(8j\pi k\rho)}}; \qquad |\phi|<\frac{\pi}{2} \qquad (49a)$$

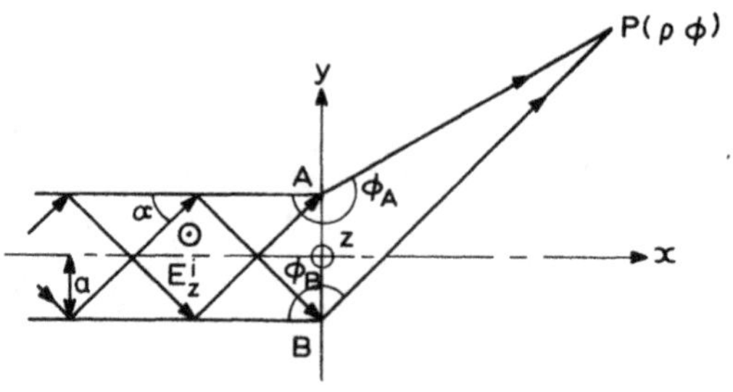

Fig. 7.9 Parallel-plate waveguide supporting the TE_1 mode

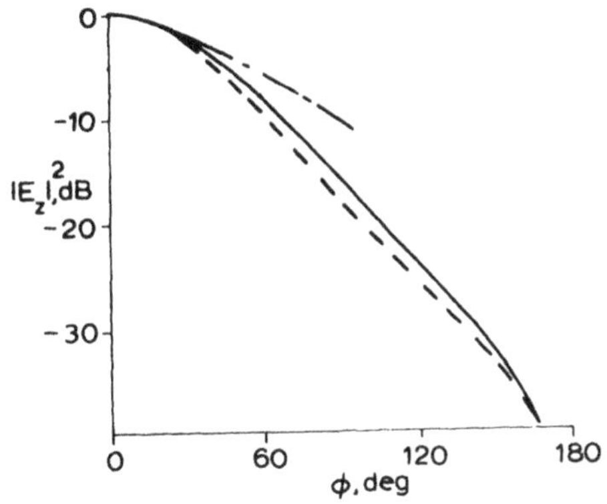

Fig. 7.10 Radiation from a parallel-plate waveguide supporting the TE_1 mode
——— exact
.... GTD with a single coupling term
—·— aperture field

On the shadow boundary when $\phi = \alpha$ this result, as for our previous example, is finite despite the infinities in the individual diffraction terms. The first order coupling term between the plates can be calculated as above. These results are compared in Fig. 7.10 with the normalised exact solution for the TE_1 mode in the parallel-plate waveguide given by

$$|E_z| = \exp\left\{\frac{ka}{2}(\cos\phi - 1)\right\} \cos\tfrac{1}{2}\phi \left| \frac{\cos u}{\left(\frac{\pi}{2}\right)^2 - u^2} \right|^{1/2} \qquad (49b)$$

and with the aperture field method using eqn. 2.21c which yields the normalised result

$$|E_z|_3 = \cos^2\tfrac{1}{2}\phi \left| \frac{\cos u}{\left(\frac{\pi}{2}\right)^2 - u^2} \right| \qquad (49c)$$

It is seen that the trends are similar to the TEM mode case. The aperture field method gives poor agreement except for the region about the axis, while the GTD method gives a result which follows closely the exact solution. The value of a is taken to be 0·42 λ.

7.3 Waveguide with a splash plate

A more complicated problem in GTD is to analyse the effect of a reflecting, or splash, plate situated within a length of 2–4 λ in front

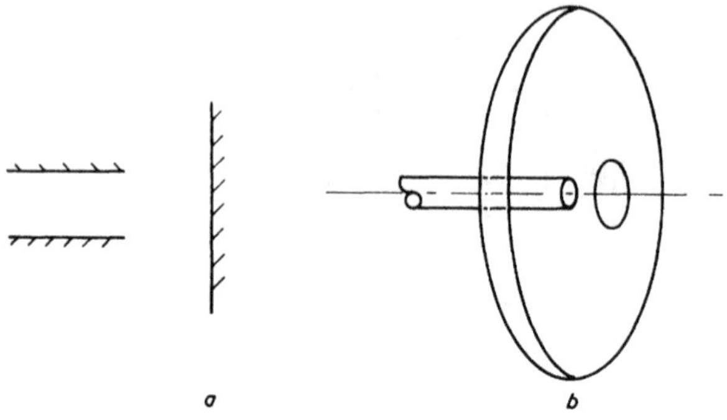

Fig. 7.11 Waveguide with splash plate

of a radiating waveguide aperture, as shown in Fig. 7.11*a*. Initially we will treat this problem as a 2-dimensional one and then extend it to the third dimension. An application of such a system is as a simple low-cost feed for a reflector antenna, as in Fig. 7.11*b*, where the purpose of the plate is to redirect the energy on to the main reflector surface. The splash plate need not be flat and we can shape it in an attempt to improve the efficiency of the overall system. To begin with we shall assume it to be flat and then attempt a shaped plate design. Few mathematical details will be given as they are very similar to the previous example.

The far field radiation pattern from an open ended parallel-plate waveguide was seen in the previous example to be given by a line source radiating from the centre of the waveguide aperture, as at O in Fig. 7.12. With the splash plate placed in front of the aperture a number of ray paths are generated. Provided that the aperture is not too large the splash plate can be considered to be in the far field (even at 2–4 λ from the aperture) of the waveguide. The line source at O generating the far field has an image at O' (see Fig. 7.12) due to reflection from the plate. This line source at O' radiates directly into the far field, but only over a restricted angular sector. Radiation is limited by the physical size of the plate and by blocking of the waveguide. The angular sector over which O' radiates into the far field is given in the upper half-space by $A\hat{O}'C$ in Fig. 7.12. (A similar sector exists in the lower half-space but because of the symmetry of the system we need only consider the upper half-space).

The direct radiation from the image source at O' represents the geometrical optics component of the radiated field. If the waveguide is supporting the TE_1 mode then the polar pattern of this source is

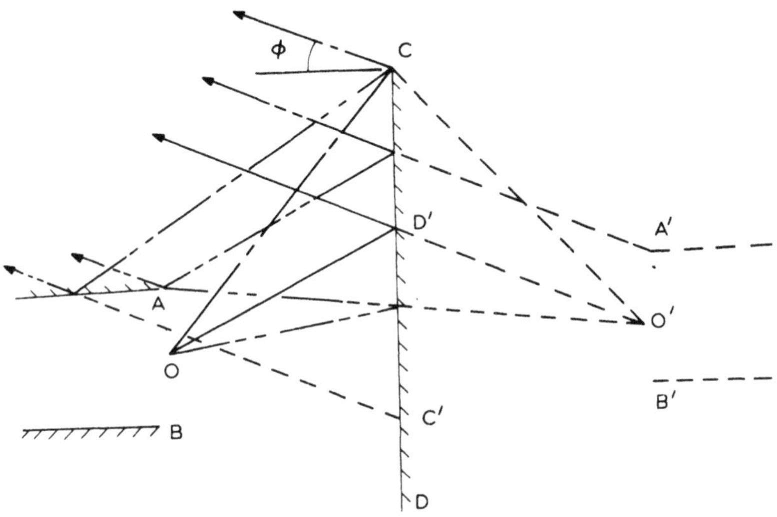

Fig. 7.12 GTD model for splash plate
——— geometrical optics rays
— — — edge rays for splash plate
— - - — coupling rays between splash plate and waveguide;

given by eqn. 49. A specific example is given in Fig. 7.13 for a plate of length 4λ situated a distance 2λ in front of the aperture. The geometrical optics field alone is seen to give a poor approximation to the measured result. Note that our main interest is in the region where $|\phi| < \frac{\pi}{2}$ which would correspond to the direction of the reflector antenna when this system is used as a feed.

To include edge diffraction effects we begin with the plate edges at C and D. These edges are illuminated by a normally incident cylindrical wave from O' so that in eqn 28

$$\rho_1 = \rho_3 = \infty, \quad \rho_2 = O'C, \quad \theta_0 = \frac{\pi}{2}, \quad \phi_0 = O'\hat{C}D$$

and since the TE_1 mode only excites electric polarisation

$$E_\xi^i = 0.$$

The resultant edge diffraction line source at C radiates at all angles over our region of interest, i.e. $0 < \phi < \frac{\pi}{2}$. The source at D, however, is blocked by the waveguide over a considerable part of this sector. Adding these two sources to the geometrical optics field gives the result shown in Fig. 7.13. The edge diffraction field from C has corrected for

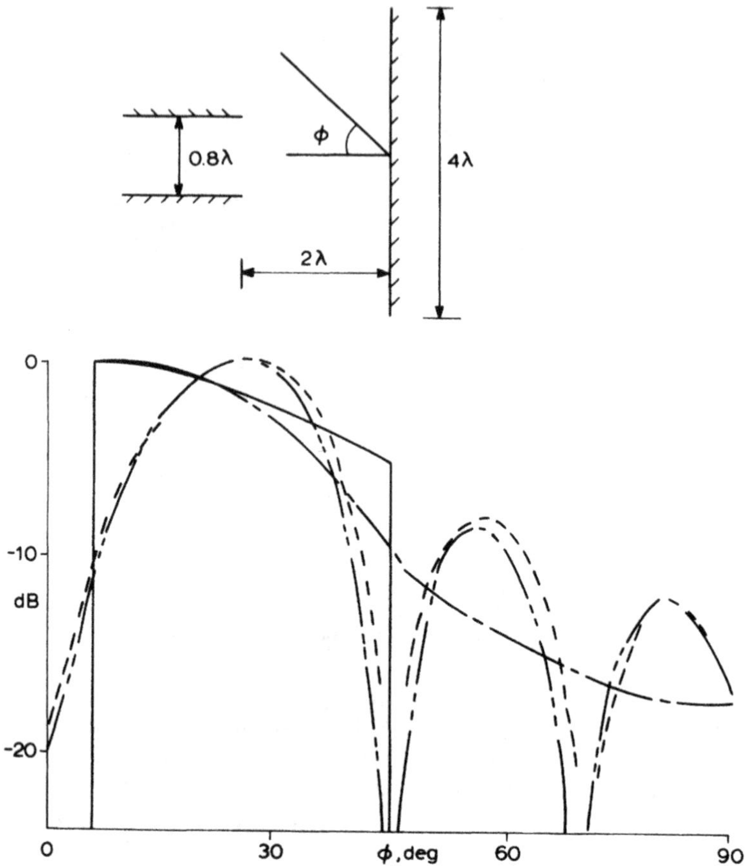

Fig. 7.13 Radiation pattern for a splash plate feed; key as for Fig. 7.12, and ----- measured

the discontinuity in the field at $\phi = 45°$ but a discontinuity still exists at $\phi = 5°$.

The remaining rays for a first order solution are those which account for the coupling between the splash plate and the waveguide. From the line source at C reflection occurs from the upper wall of the waveguide as shown in Fig. 7.12. This can be regarded as coming from the image source at C'. As for the image at O' this source radiates only over a restricted sector of space, in this case from $0<\phi<\frac{\pi}{2}-C\hat{C}'A$ (assuming that the waveguide extends to infinity). The original source at O' diffracts at the waveguide edges at A and B. These sources in turn have images at A' and B' to account for their reflection from

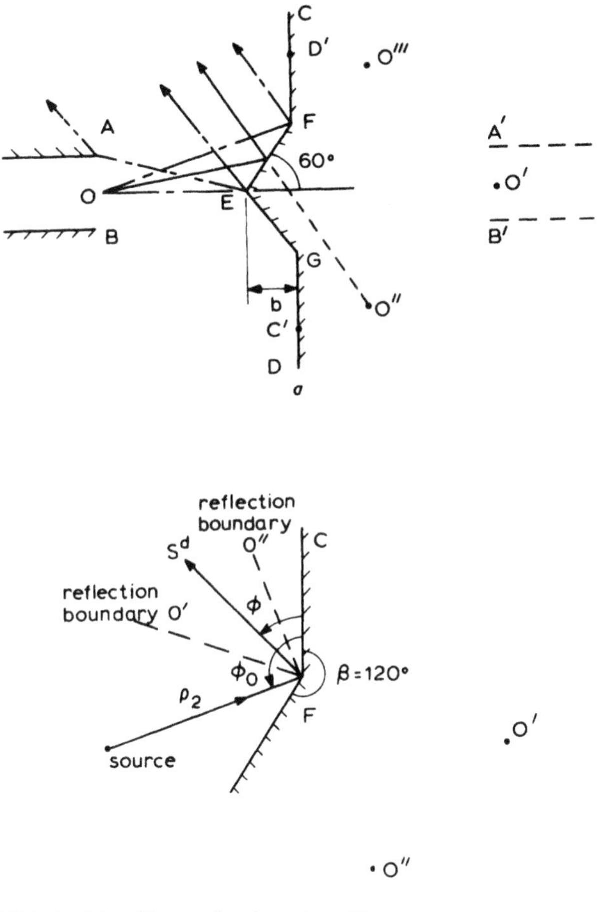

Fig. 7.14 Splash plate with a wedge-shaped profile

the splash plate. Note that the source at A will correct for the discontinuity of field at $\phi = 5^\circ$.

We now have the complete number of sources to give a first order solution to the radiation field in the sector $0 < \phi < \frac{\pi}{2}$. These sources are at O', D, C, C', A, A' and are all half-plane edge diffraction sources except for the one at O'. The total solution is plotted in Fig. 7.13 where very good agreement is obtained with the measured pattern. [To measure the radiation pattern of the 2-dimensional structure we used a parallel-plate system similar to that of Rowe (1953)].

In an attempt to improve the performance of this structure we may place a wedge shape in the centre of the splash plate as in Fig. 7.14*a*. The purpose of this is two-fold. Our first objective is to redirect the reflected rays from the plate away from the waveguide aperture and so reduce the VSWR. Secondly, this redirected energy should be used to increase the radiated field in the vicinity of $\phi = 60°$ so as to make a more effective feed when illuminating a reflector. A semi-angle of 60° is chosen initially for the wedge at E in Fig. 7.14*a* in an attempt to meet this twin objective. The geometrical optics field is now given by the three image sources at O', O'', O'''. These two latter sources are related to the reflected rays from the wedge faces along EF and EG in the figure, and have reduced the range of the source at O' for the upper half-space to the angular sector $C\hat{O}'F$. We also have additional edge diffraction at E, F and G. The behaviour at these edges is similar to the half-plane in giving the diffracted field as radiating from a line source at the edge. More care needs to be taken, however, in calculating the diffraction coefficients for the wedge. Let us concentrate on the interior wedge at F.

Diffraction by a wedge is formulated as in eqn. 28 but with the diffraction coefficients given by

$$
\begin{aligned}
D^{e,m} &= \{h(\Phi^i) + h(-\Phi^i)\} \mp \{h(\Phi^r) + h(-\Phi^r)\}; \quad \Phi^{i,r} = \phi \mp \phi_0 \\
h(\Phi^{i,r}) &= -\epsilon^{i,r}\sqrt{(\sigma)}K_-(v^{i,r})\,\Lambda^{i,r} \\
\Lambda^{i,r} &= \frac{a^{i,r}}{\sqrt{(2)}N}\cot\left(\frac{\pi + \Phi^{i,r}}{2N}\right); \qquad N = \frac{2\pi - \beta}{\pi} \qquad (50) \\
a^{i,r} &= \sqrt{(2)}\cos\tfrac{1}{2}(\Phi^{i,r} + 2n\pi N); \qquad n = 0, \pm 1 \\
\sigma &= \frac{\rho_1\rho_2(\rho_3 + s^d)s^d}{(\rho_1 + s^d)(\rho_2 + s^d)\rho_3}; \qquad v^{i,r} = \sqrt{(k\sigma)}\,|a^{i,r}|\sin\theta_0
\end{aligned}
$$

The problem is to choose the correct value for n in the diffraction term $h(\pm\Phi^{i,r})$. In Fig. 7.14*b* we note that the source illuminates both wedge faces, giving rise to two reflection boundaries and no shadow boundary. Initially we must choose $h(\pm\Phi^r)$ correctly for each reflection boundary to ensure that $|h(\pm\Phi^r)| = \frac{1}{2}$ along these boundaries. This is satisfied if

$$|2n\pi N + \Phi^r| = \pi; \qquad \Lambda^r = 1$$

at the reflection boundaries. If the angle ϕ is measured from the wedge face along CF in Fig. 7.14*b* then to meet these requirements along reflection boundary O' we have $h(-\Phi^r)$ with $n = 0$. For the other boundary at O'' we have $h(\Phi^r)$ with $n = -1$. These results can be obtained directly from eqn. 5.25.

We need now to choose the correct values for n in $h(\pm\Phi^i)$ to satisfy the boundary conditions along the wedge faces. This is important in our example since the reflection boundary along O'' can be very near to the wedge face along CF. The correct values of n are given from eqn. 5.25b and the diffraction coefficient D^e for electric polarisation which we require (assuming that the waveguide is supporting the TE_1 mode) can be written as

$$D^e = h(\Phi^i)_{n=0} + h(-\Phi^i)_{n=-1} - h(\Phi^r)_{n=-1} - h(-\Phi^r)_{n=0} \quad (51)$$

A similar problem exists for the interior wedge at G but we shall not require this edge diffraction source for the field in the upper half-space. The wedge at E is required, however, and its diffraction coefficient is similar to eqn. 51. Since in this case the reflection boundaries are well removed from the wedge faces, the value of v^i will normally be greater than 3, so that we may use the asymptotic value of the modified Fresnel integral. The term $h(\Phi^i)$ then becomes

$$h(\Phi^i) \sim \frac{-\csc\theta_0 \cot\left(\dfrac{\pi + \Phi^i}{2N}\right)}{N\sqrt{(8j\pi k)}}$$

which, as noted previously, is independent of the value of n.

For the radiation field in the sector $0 < \phi < \frac{\pi}{2}$ we have so far considered the geometrical optics sources O', O'' and the edge diffraction sources from the splash plate at C, D, F, and E. As before the source at C will reflect at the upper waveguide wall giving the image at C'. Provided that $b \tan 60^\circ \leqslant a$ then the source at F will not reflect at this wall. To give a smooth transition where the source at E is blocked by the waveguide, we include the second order diffraction term from E to A. Since this source at A is of higher order we do not include its image A'. Our complete first order solution for $0 < \phi < \frac{\pi}{2}$ is therefore given by the sources at O', O'', C, D, C', F, E, and A. Two results are given in Fig. 7.15 where the GTD approach has again been shown to yield good agreement with the measured patterns.

We shall now consider how these results can be extended to a 3-dimensional system where the parallel-plate waveguide becomes a circular waveguide and the reflecting strip becomes a disc. The far field radiated from the waveguide aperture now appears as emanating from a point source situated at the centre of the aperture. To determine the points of edge diffraction along the rim of the disc we can apply Fermat's principle for edge diffraction. For this simple case. however, this is not necessary as we may determine the edge diffraction points from physical considerations in the following way. The illuminating source is situated on the axis of the disc so that at each point along

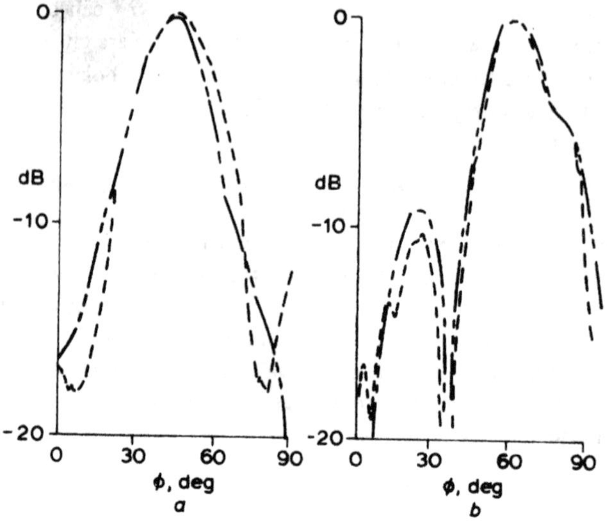

Fig. 7.15 Radiation patterns for wedge-shaped splash plate
(*a*) $b = 0{\cdot}25\lambda$ ———— GTD
(*b*) $b = 0{\cdot}5\lambda$ ········ measured

the edge the field is at normal incidence, i.e. $\theta_0 = \frac{\pi}{2}$. Edge diffracted rays, therefore, remain within the plane of incidence since they do not form a cone. For every radiation plane only two edge diffraction points will contribute to the field, and these points correspond to the intersection of the disc and the chosen radiation plane. Thus our radiation problem essentially reduces to a 2-dimensional one. The difference with the previous analysis is that the incident field will, in general, contain both electric and magnetic polarisation components. In the principal planes, however, only one polarisation will exist, being electric polarisation for the *H*-plane and magnetic polarisation for the *E*-plane. We shall consider only the *H*-plane as it relates to the above analysis.

Consider first the flat disc as in Fig. 7.16*a*. All the edges relate to half-plane diffraction and we can use eqn. 28 directly. The incident field is now a point source where $\rho_1 = \rho_2 = \rho$, and the value of ρ_3 is determined from eqn. 6.63 as

$$\frac{1}{\rho_3} = Q_{11}^i - \frac{1}{\rho_e \sin^2\theta_0}(\hat{s}^i \cdot \hat{n}_e - \hat{s}^d \cdot \hat{n}_e) \tag{52}$$

The component Q_{11}^i of the incident curvature matrix at the point C on the rim is determined from eqn 4.45 as equal to $\frac{1}{\rho}$. At C the cur-

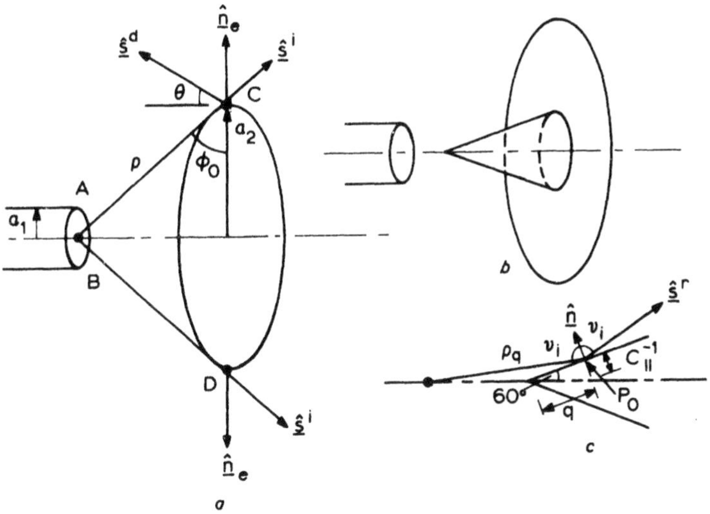

Fig. 7.16 Splash plate analysis for circular waveguide

vature of the edge ρ_e is simply given by a_2. Referring to the angles defined in Fig. 7.16*a* we have for the value of ρ_3 at C

$$\frac{1}{\rho_3} = \frac{1}{\rho} - \frac{1}{a_2}(\cos\phi_0 - \sin\theta) = \frac{\sin\theta}{a_2} \quad \text{(for edge at } C\text{)} \qquad (53a)$$

Similarly for the diffraction points at A, B, and D we get for the far field in the upper half-space

$$\frac{1}{\rho_3} = -\frac{\sin\theta}{a_2} \quad \text{(for edge at } D\text{)} \qquad (53b)$$

$$\frac{1}{\rho_3} = \frac{\sin\theta}{a_1} \quad \text{(for edge at } A\text{)} \qquad (53c)$$

$$\frac{1}{\rho_3} = -\frac{\sin\theta}{a_1} \quad \text{(for edge at } B\text{)} \qquad (53d)$$

From eqn. 28 the value of σ as $s^d \to \infty$ becomes

$$\sigma = \frac{\rho_1\rho_2}{\rho_3} \qquad (54)$$

For our problem $\rho_1 = \rho_2$ at all the edge diffraction points, and it will also be noted from eqn. 53 that at the optical boundaries $\rho_1 = \rho_2 = \rho_3$ giving σ as

$$\sigma = \rho_2 \tag{55}$$

If this quantity is used for σ in place of eqn. 54 then it is found that little error will result in the evaluation of the field. The main advantage of doing this is that the modified Fresnel integral arguments are identical to the 2-dimensional problem solved earlier, and we need only multiply the individual edge source terms in that analysis by

$$\left[\frac{\rho_3}{\rho_3 + s^d}\right]^{1/2} \xrightarrow[s^d \to \infty]{} \sqrt{\left(\frac{\rho_3}{s^d}\right)} \tag{56}$$

to extend it to the 3-dimensional problem. Furthermore, with the incident field given from a point source we normalise the total field by $\exp(-jks^d)/s^d$ so that the multiplication factor of eqn. 56 reduces to $\sqrt{(\rho_3)}$.

For the sources below the axis in Fig. 7.16*a* we note from eqn. 53 that ρ_3 is negative. This is because the axis is a caustic of the edge diffracted rays and the rays in the upper half-space from the bottom edges have passed through this caustic. The correct phase shift through a caustic was determined earlier as $\exp(j\pi/2)$ i.e.,

$$\begin{aligned} \sqrt{(\rho_3)} &= |\rho_3|^{1/2} \qquad \text{for} \quad \rho_3 > 0 \\ \sqrt{(\rho_3)} &= \exp\left(\frac{j\pi}{2}\right) |\rho_3|^{1/2} \qquad \text{for} \quad \rho_3 < 0 \end{aligned} \tag{57}$$

If the geometry of Fig. 7.14 is extended to 3-dimensions we arrive at a cone in the centre of the splash plate as shown in Fig. 7.16*b*. The additional edge sources at F and G are treated as above by multiplying the 2-dimensional result with the appropriate value for $\sqrt{(\rho_3)}$. We must also determine the values of ρ_1^r, ρ_2^r in the diffraction coefficient for the reflection boundary from the cone surface (see eqn. 58 below). At E, however, we have the tip of a cone for which a diffraction coefficient is not yet available. In the present application the tip diffraction has been ignored in the belief that its effect will be small compared to the other sources. That this can be expected can be seen by noting that the factor $\sqrt{(\rho_3)}$ decreases with the radius of curvature of the edge. If extended to the tip of the cone then $\rho_3 = 0$.

The geometrical optics source of O'' in Fig. 7.14 is also modified by the surface of the cone. At a reflection point along the cone as in Fig. 7.16*c* the curvature matrix $\mathbf{C}$ of the surface can be deduced as

$$\mathbf{C} = \begin{bmatrix} (q \tan 60^\circ)^{-1} & 0 \\ 0 & 0 \end{bmatrix}$$

and the reflection matrix from eqn. 4.55*a* becomes

$$\mathbf{Q}^r = \begin{bmatrix} \left(\dfrac{2\cos\nu_i}{q\tan 60^\circ} + \dfrac{1}{\rho_q}\right) & 0 \\ 0 & \dfrac{1}{\rho_q} \end{bmatrix} \tag{58}$$

so that the reflected field from the point P_0 along s^r is modified by

$$\left[\frac{\rho_1^r \rho_2^r}{(\rho_1^r + s^r)(\rho_2^r + s^r)}\right]^{1/2} \xrightarrow[s^r \to \infty]{} \sqrt{(\rho_1^r \rho_2^r)}\,\frac{1}{s^r}$$

where from eqn. 58

$$\rho_1^r = \frac{q\rho_q \tan 60^\circ}{2\rho_q \cos\nu_i + q\tan 60^\circ}, \quad \rho_2^r = \rho_q$$

To compare the theoretical results with measured data we tested a feed system shown in cross-section in Fig. 7.17*a*. Our theoretical model assumed that the waveguide consisted only of a smooth circular pipe. In practice, however, we had waveguide junctions, a waveguide to co-axial transformer, and a co-axial cable, as shown in Fig. 7.17*a*. All of these obstacles will re-radiate to some degree in the backward direction. Since they are well separated compared to the aperture-splash plate dimensions they can be expected to give a high frequency ripple on top of the main radiation from the splash plate. That this is indeed the case is shown in the measured results given in Fig. 7.17*b* and *c*. The GTD result has predicted the correct trends in the radiation pattern, and in fact appears to give the mean change in the pattern. In the immediate vicinity of $\phi = 0^\circ$ there exists a caustic of the edge diffracted rays which has been ignored in this application. Such caustics will be considered in the following example.

7.4 Edge diffracted field from a reflector antenna

In Section 7.1 we derived the geometrical optics field reflected from a reflector antenna under plane wave illumination. The edge diffracted field will now be developed for this problem. We begin by determining the diffraction points on the reflector rim from which emanate the diffracted ray, or rays, through the given field point $P(r_2, \theta_2, \phi_2)$ as in Fig. 7.1. Points of edge diffraction are determined from the ray path $s^i + s^r$ being stationary where one point is on the edge. (Fermat's

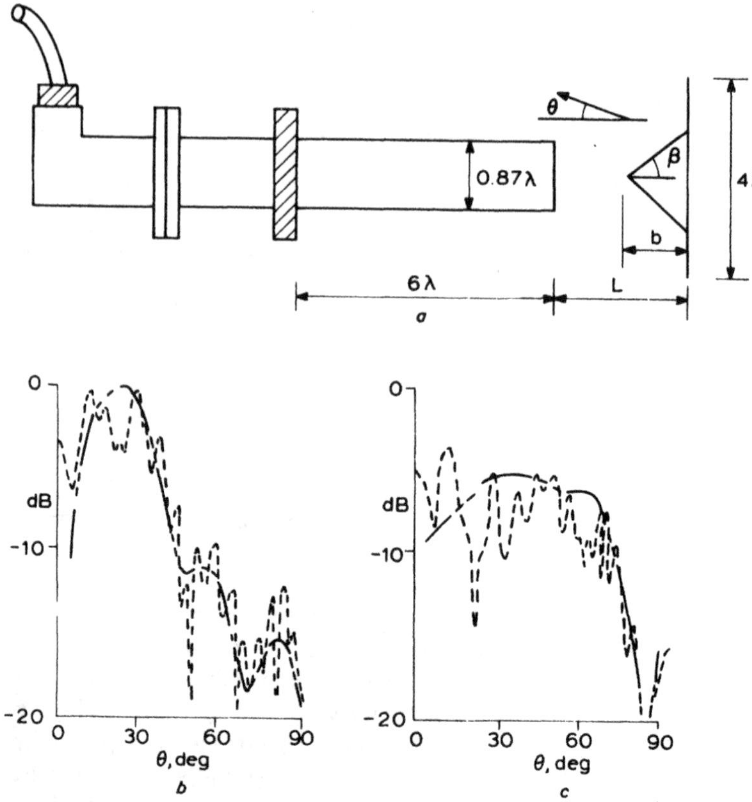

Fig. 7.17 H-plane for circularly symmetric splash plate feeds
(*b*) $L = 2\lambda$, $b = 0$
(*c*) $L = 2{\cdot}5\lambda$, $b = 0{\cdot}5\lambda$, $\beta = 60°$
——— GTD
– – – measured

principle for edge diffraction). If δ_e is the value of the elevation angle in Fig. 7.1 at the reflector rim, then Fermat's principle is satisfied if

$$\frac{\partial s^i}{\partial \Phi} + \frac{\partial s^r}{\partial \Phi} = 0 \quad \text{where} \quad \delta = \delta_e \tag{59}$$

The points of edge diffraction along the reflector rim are at $\Phi = \Phi_e$ where Φ_e is a solution to eqn. 59. With the source at infinity at $s(\infty, \theta_1, 0)$, solving eqn. 59 with $s^{i,r}$ given by eqn. 2, and putting $s^d = s^r$ when $\delta = \delta_e$, as in Fig. 7.18, we get

$$\tan \Phi_e = \frac{r_2 \sin \theta_2 \sin \phi_2}{s^d \sin \theta_1 + r_2 \sin \theta_2 \cos \phi_2} \tag{60}$$

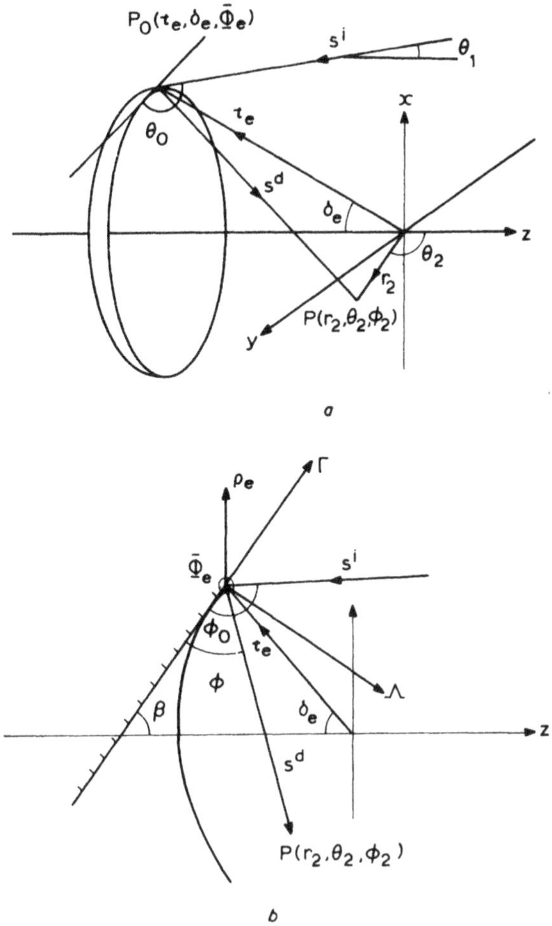

Fig. 7.18 Edge diffraction at a reflector rim

This equation can be readily evaluated using Newton's method. When the positions on the rim have been determined we require, for the solution of the edge diffracted field, $\rho_1^{i,r}$ and $\rho_2^{i,r}$, the principal radii of curvature of the incident and reflected wavefront in the vicinity of the edge diffraction points; ρ_3 the distance between the edge and the caustic created by ray diffraction from the edge; the local incident angles ϕ_0, θ_0 of the incoming field to the edge, and ϕ, the diffraction angle measured from the tangent half-plane to the edge. This last quantity is illustrated in Fig. 7.18. The angle β in the figure was derived previously and is given by eqn. 7.

Referring to Fig. 7.18, the angle ϕ is determined from the relationship

$$\tan\phi = \frac{A_\Lambda}{-A_\Gamma} \tag{61}$$

where

$$A_\Lambda = -\cos\beta(A_x\cos\Phi_e + A_y\sin\Phi_e) + A_z\sin\beta$$

$$A_\Gamma = \sin\beta(A_x\cos\Phi_e + A_y\sin\Phi_e) + A_z\cos\beta$$

and the components A_x, A_y, A_z are obtained from

$$s^d = \mathbf{r}_2 - \boldsymbol{\tau}_e = A_x\hat{x} + A_y\hat{y} + A_z\hat{z} \tag{62}$$

where

$$\mathbf{r}_2 = r_2(\hat{x}\sin\theta_2\cos\phi_2 + \hat{y}\sin\theta_2\sin\phi_2 + \hat{z}\cos\theta_2)$$

$$\boldsymbol{\tau}_e = \tau_e(\hat{x}\sin\delta_e\cos\Phi_e + \hat{y}\sin\delta_e\sin\Phi_e - \hat{z}\cos\delta_e)$$

Finally we get

$$\tan\phi = \frac{r_2\{\cos\alpha\cos\theta_2 - \sin\alpha\sin\theta_2\cos(\phi_2 - \Phi_e)\} + \tau_e\cos(\alpha - \delta_e)}{-r_2\{\sin\alpha\cos\theta_2 + \cos\alpha\sin\theta_2\cos(\phi_2 - \Phi_e)\} - \tau_e\sin(\alpha - \delta_e)} \tag{63}$$

where

$$\alpha = \frac{\pi}{2} - \beta$$

To determine the incident angle ϕ_0 from this equation we make the substitutions

$$r_2 \to r_1 \quad \text{where} \quad r_1 \gg \tau_e$$

$$\theta_2 \to \theta_1$$

$$\phi_2 \to \phi_1 \quad \text{where} \quad \phi_1 = 0$$

so that

$$\tan\phi_0 = \frac{\cos\alpha\cos\theta_1 - \sin\alpha\sin\theta_1\cos\Phi_e}{-\sin\alpha\cos\theta_1 - \cos\alpha\sin\theta_1\cos\Phi_e} \tag{64}$$

In a similar way we determine the angle θ_0 measured from the edge to the incident ray where

$$\tan\theta_0 = \frac{(A_\Gamma^2 + A_\Lambda^2)^{1/2}}{A_{\Phi_e}} = \frac{(1 + \tan^2\theta_1\cos^2\Phi_e)^{1/2}}{-\tan\theta_1\sin\Phi_e} \tag{65}$$

The value of ρ_3 is given by

$$\frac{1}{\rho_3} = Q_{11}^i - \frac{1}{\rho_e \sin^2\theta_0} (\hat{s}^i \cdot \hat{n}_e - \hat{s}^d \cdot \hat{n}_e)$$

Since the incident field is a plane wave then $Q_{11}^i = 0$. The curvature of the edge ρ_e is given by the radius of the reflector a. Along the rim the outward normal $\hat{n}_e$ is

$$\hat{n}_e = \hat{x} \cos \Phi_e + \hat{y} \sin \Phi_e$$

and for the incoming ray to the edge

$$\hat{s}^i = -\hat{x} \sin \theta_1 - \hat{z} \cos \theta_1$$

With these equations and $\hat{s}^d$ determined from eqn. 62 we arrive at

$$\frac{1}{\rho_3} = -\frac{1}{a \sin^2 \theta_0} \left\{ -\sin \theta_1 \cos \Phi_e + \frac{a - r_2 \sin \theta_2 \cos(\phi_2 - \Phi_e)}{s^d} \right\} \tag{66}$$

The values of $\rho_1^{i,r}, \rho_2^{i,r}$ are determined from the analysis given in Section 7.1 at the edge point $(\tau_e, \delta_e, \Phi_e)$.

We have now evaluated ϕ, ϕ_0, θ_0, ρ_3, $\rho_1^{i,r}$, $\rho_2^{i,r}$ and all that remains to be computed for the diffracted field using eqns. 6.62 and 6.68 is to determine the incident field components E_ψ^i, E_ξ^i at the diffraction point. These components can be obtained directly from eqns. 17, 19, and 20 in Section 7.1 at the edge point $(\tau_e, \delta_e, \Phi_e)$.

When the incident field is off-axis (i.e., $\theta_1 \neq 0$) then the diffracted rays will form cones radiating from the edge. The number of edge diffraction points, determined by solving eqn. 60, contributing to a given field point can be as many as four. These rays will generate a caustic surface in the vicinity of which the GTD analysis is invalid. For an on-axis field when $\theta_1 = 0$ this caustic surface will collapse into a line caustic along the axis of the reflector. To simplify this example we shall now restrict ourselves to the on-axis case.

When $\theta_1 = 0$, eqn 60 reduces to

$$\tan \Phi_e = \tan \phi_2; \qquad \theta_1 = 0$$

and the edge diffraction points are given at $\Phi_e = \phi_2, \phi_2 + \pi$. Note that these points are at the intersection of the reflector and the radiation plane for a given ϕ_2. In this plane they are fixed, being independent of r_2, s^d and θ_2. This is not true for an off-axis incident field where the edge diffraction points are, in general, a function of all these parameters. For the edge point at $\Phi_e = \phi_2$ we get from eqns. 63–66

$$\tan \phi = \frac{r_2 \cos(\alpha + \theta_2) + \tau_e \cos(\alpha - \delta_e)}{-r_2 \sin(\alpha + \theta_2) - \tau_e \sin(\alpha - \delta_e)}$$

$$\tan\phi_0 = -\cot\alpha, \quad \therefore \quad \phi_0 = \frac{\pi}{2} + \alpha$$

$$\theta_0 = \frac{\pi}{2}$$

$$\frac{1}{\rho_3} = -\frac{1}{a}\left(\frac{a - r_2 \sin\theta_2}{s^d}\right) = -\frac{1}{a}\cos(\phi - \alpha)$$

and from Section 7.1

$$\rho_1^i = \rho_2^i = \infty$$

$$\rho_1^r = -\frac{a}{\sin 2\alpha}$$

$$\rho_2^r = -\frac{\rho_2^c \cos\alpha}{2}$$

where ρ_2^c is given by eqn. 9. The incident field from eqn. 17 is given as $\hat{x}E_x^i$ so that from eqns. 19 and 20

$$E_\psi^i = -E_x^i \sin\Phi_e = -E_x^i \sin\phi_2$$
$$E_\xi^i = E_x^i \cos\Phi_e = E_x^i \cos\phi_2$$

Substituting into eqn. 6.62 gives, with $|E_x^i| = 1$,

$$\begin{bmatrix} E_\psi^d(s^d) \\ \\ E_\xi^d(s^d) \end{bmatrix} = \begin{bmatrix} D^e & 0 \\ 0 & D^m \end{bmatrix} \begin{bmatrix} -\sin\phi_2 \\ \cos\phi_2 \end{bmatrix} \left[\frac{\rho_3}{s^d(\rho_3 + s^d)}\right]^{1/2} \exp\{-jk(s^d + \tau_e \cos\delta_e)\} \tag{67}$$

and $D^{e,m}$ is given for the curved half-plane by eqn. 6.68. The magnetic diffracted field components are given as

$$H_\psi^d = -\sqrt{\left(\frac{\hat{\epsilon}}{\mu}\right)} E_\xi^d; \qquad H_\xi^d = \sqrt{\left(\frac{\hat{\epsilon}}{\mu}\right)} E_\psi^d \tag{68}$$

Sometimes it is convenient to express the field in terms of the Cartesian co-ordinates x, y, z. From eqns. 19 and 20b we derive

$$\hat{\xi}^d = -\hat{x}\cos\phi_2 \sin(\phi - \alpha) - \hat{y}\sin\phi_2 \sin(\phi - \alpha) - \hat{z}\cos(\phi - \alpha)$$
$$\hat{\psi}^d = -\hat{x}\sin\phi_2 + \hat{y}\cos\phi_2 \tag{69}$$

and the magnetic diffracted field components in the (x, y, z) co-ordinates are

$$\begin{bmatrix} H_x^d \\ \\ H_y^d \\ \\ H_z^d \end{bmatrix} = \sqrt{\left(\frac{\hat{\epsilon}}{\mu}\right)} \begin{bmatrix} \sin\phi_2 \cos\phi_2 \{D^e \sin(\phi-\alpha) + D^m\} \\ \\ \sin^2\phi_2 \sin(\phi-\alpha) D^e - \cos^2\phi_2 D^m \\ \\ \sin\phi_2 \cos(\phi-\alpha) D^e \end{bmatrix}$$

$$\left[\frac{\rho_3}{s^d(\rho_3+s^d)}\right]^{1/2} \exp\{-jk(s^d + \tau_e \cos\delta_e)\} \tag{70}$$

This equation has been evaluated for the edge diffraction point at $\Phi_e = \phi_2$. For the other point at $\Phi_e = \phi_2 + \pi$ the only change in this equation, apart from replacing ϕ_2 with $\phi_2 + \pi$, is in the value of ϕ, which is now given as

$$\tan\phi = \frac{r_2 \cos(\alpha-\theta_2) + \tau_e \cos(\alpha-\delta_e)}{-r_2 \sin(\alpha-\theta_2) - \tau_e \sin(\alpha-\delta_e)}$$

To obtain the scattered field from the reflector for an on-axis incident field, the geometrical optics reflected field as derived in Section 7.1, is added to the contributions from the two edge diffraction points as formulated by eqn. 67. Some examples are given in Fig. 7.20 which we will discuss later.

In eqns. 67 and 70 the quantity $\rho_3/(\rho_3 + s^d)$ gives an infinity for the diffracted field at all points along the axis. This is the axial caustic of the edge diffracted rays mentioned earlier. Since all points along the reflector rim are contributing a diffracted ray to the axis this caustic can only be removed by considering the currents around the entire rim. This can be accomplished by making use of the equivalent current method discussed in Section 6.7. Since the incident field is a plane wave and at normal incidence at each point around the rim, the equivalent currents $\boldsymbol{J}$, $\boldsymbol{M}$ are obtained directly from eqn. 6.60*b*. In this equation the *z*-direction corresponds to the Φ-direction in the present application and the amplitude of the incident field components given by E_z^i, E_ϕ^i in eqn. 6.60*b* becomes $-E_x^i \sin\phi_2$, $E_x^i \cos\phi_2$ respectively. Thus around the edge the equivalent currents are

$$\begin{aligned} J &= \hat{\boldsymbol{\Phi}} \frac{\sin\phi_2}{j\omega\mu} I_{GTD}^e \exp(-jk\tau_e \cos\delta_e) \\ M &= \hat{\boldsymbol{\Phi}} \frac{\cos\phi_2}{jk} I_{GTD}^m \exp(-jk\tau_e \cos\delta_e) \end{aligned} \tag{71}$$

where

$|E_x^i| = 1$, and

$$I_{GTD}^e = \frac{4 \sin\frac{1}{2}\phi \sin\frac{1}{2}\phi_0}{\cos\phi + \cos\phi_0}; \qquad I_{GTD}^m = \frac{4 \cos\frac{1}{2}\phi \cos\frac{1}{2}\phi_0}{\cos\phi + \cos\phi_0} \tag{72}$$

The incident angle ϕ_0 is a constant around the rim being equal to $\alpha + \frac{\pi}{2}$.

The equivalent currents in eqn. 71 are now used in the potential integral solution of eqn. 2.11 which for the magnetic diffracted field radiated by the entire edge is given by

$$H^d = jk\int_0^{2\pi}\left[J \times \hat{s}^d - \sqrt{\left(\frac{\epsilon}{\mu}\right)}\{M - (M\cdot\hat{s}^d)\hat{s}^d\}\right] Gad\Phi ; G = \frac{\exp(-jks^d)}{4\pi s^d} \tag{73}$$

A complication in using the currents given by eqn. 71 is that they are a function of the angle ϕ. This means that in each step of integration we must evaluate ϕ through the relationship given by eqn. 63. The consequent numerical integration is considerably slowed down by this calculation. Since the equivalent current formulation will only be necessary in the region of the axial caustic, we can approximate $I^{e,m}_{GTD}$ by their value on the axis in the region where the field is being evaluated. This means $I^{e,m}_{GTD}$ can be taken outside the integration and the procedure is considerably simplified. Also in the axial region

$$\hat{s}^d \simeq -\hat{P}_e \cos(\phi - \alpha) + \hat{z}\sin(\phi - \alpha)$$

as seen from Fig. 7.18, so that

$$M\cdot\hat{s}^d = 0$$

and

$$J \times \hat{s}^d = \frac{\sin\phi_2}{j\omega\mu}\{\hat{P}_e \sin(\phi - \alpha) + \hat{z}\cos(\phi - \alpha)\} I^e_{GTD} \exp(-jk\tau_e \cos\delta_e)$$

Substituting these equations in eqn. 73 yields

$$H^d = a\exp(-jk\tau_e\cos\delta_e)\sqrt{\left(\frac{\epsilon}{\mu}\right)}\int_0^{2\pi}[\sin\phi_2 I^e_{GTD}\{\hat{P}_e\sin(\phi-\alpha) + \hat{z}\cos(\phi-\alpha)\} - \hat{\Phi}\cos\phi_2 I^m_{GTD}]\, Gd\Phi$$

Expressing this equation in rectangular co-ordinates gives

$$H^d_x = A(I^e_{GTD}\sin(\phi-\alpha) + I^m_{GTD})\int_0^{2\pi}\sin\Phi\cos\Phi\, Gd\Phi$$

$$H^d_y = A(I^e_{GTD}\sin(\phi-\alpha)\int_0^{2\pi}\sin^2\Phi\, Gd\Phi - I^m_{GTD}\int_0^{2\pi}\cos^2\Phi\, Gd\Phi)$$

$$H^d_z = A\, I^e_{GTD}\cos(\phi-\alpha)\int_0^{2\pi}\sin\Phi\, Gd\Phi \tag{74}$$

where $G = \exp(-jks^d)/4\pi s^d$, and $A = a\exp(-jk\tau_e\cos\delta_e)\sqrt{\left(\frac{\hat{\epsilon}}{\mu}\right)}$. The value of ϕ in $I^{e,m}_{GTD}$ is determined along the axis from eqn. 63, i.e., when $\theta_2 = 0$ or π.

Of the three equations in eqn. 74 only H^d_y is non-zero along the axis, therefore we can ignore H^d_x and H^d_z in eqn. 74 for this example. For small angles of off-set (of the incident field) we may still use eqn. 74 by including within the integral the additional phase shift of exp $(jk\tau_e \cos\Phi \sin\delta_e \sin\theta_1)$ of the incident field at the edge. With increasing values of off-set angle θ_1 the axial caustic spreads out into a surface caustic and the equivalent current technique is no longer a suitable method in evaluating the diffracted field at the caustic. This is because of the following two reasons: (i) our approximations leading to eqn. 74 to make the equivalent current method tractable becomes invalid for large θ_1; (ii) at a surface caustic it is not the entire rim that contributes to the field but the interaction between nearby edge diffraction points. The representation would then be in the form of the Airy function as is the case shown in Section 2.3 for a surface caustic of the geometrical field.

It is interesting to compare the GTD approach to the physical optics approximation. The latter method yields a double integral for the scattered magnetic field, H^s, which can be written in the form

$$H^s = \int_0^{2\pi}\int_0^{\delta_e} f(\delta, \Phi)\exp\{jkg(\delta, \Phi)\}d\delta d\Phi$$

A stationary phase evaluation of this integral yields the geometrical optics field as derived in Section 7.1. The next higher order term in the asymptotic expansion of this integral is dependent on the endpoint contribution at $\delta = \delta_e$. Since at δ_e the reflector terminates in an edge, the only difference between this solution and that for GTD given above is in the form of the diffraction coefficients $D^{e,m}$, which for the physical optics approximation are given in Section 5.6. As noted there, the two methods differ only in regions well removed from the optical boundaries.

The axial region field for the endpoint contribution for an on-axis field can be obtained from eqn. 2.101 as

$$\int_0^{2\pi} \frac{f(\delta_e, \Phi)}{-jkg'_\delta(\delta_e, \Phi)} \exp\{jkg(\delta_e, \Phi)\}d\Phi$$

provided that we are well removed from optical boundaries. This formulation contains implicitly the equivalent edge current method and

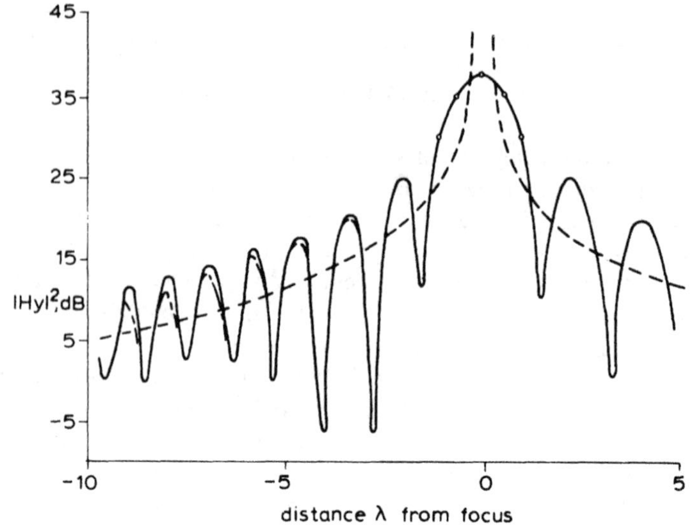

Fig. 7.19 Axial field for a parboloid
$D = 50\lambda$, $F/D = 0{\cdot}35$
- - - - - - - - - geometrical optics
——— equivalent edge-current method using GTD
—·— equivalent edge-current method using asymptotic physical optics
○ ○ ○ physical optics (numerical)

the result is given by eqn. 74 with $I^{e,m}_{GTD}$ replaced by the physical optics terms $I^{e,m}_{PO}$ where from Section 5.6

$$I^e_{PO} = \frac{2\sin\phi_0}{\cos\phi_0 + \cos\phi};\qquad I^m_{PO} = \frac{-2\sin\phi}{\cos\phi_0 + \cos\phi} \tag{75}$$

The advantage of the physical optics formulation is that it can provide an estimate of the field in the vicinity of those caustic regions where the asymptotic methods fail. Fortunately in such regions the phase function is varying relatively slowly and a numerical integration of the double integral usually presents little difficulty.

As an example using GTD and physical optics we give some scattered patterns from a paraboloidal reflector. In Fig. 7.19 the axial field scattered from a paraboloid is shown where both the GTD and physical optics equivalent currents are used in the evaluation of H^d_y, which, from eqn. 74 is given on the axis by

$$H^d_y = a\pi\sqrt{\left(\frac{\hat{\epsilon}}{\mu}\right)}\{I^e\sin(\phi-\alpha) - I^m\}G \tag{76}$$

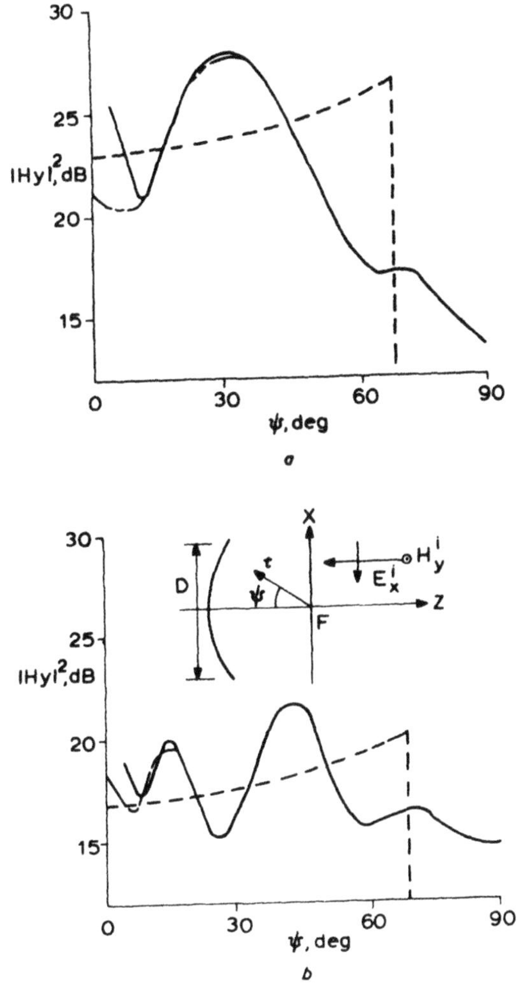

Fig. 7.20 Near field scattered from a paraboloidal reflector illuminated by an incident on-axis plane wave
(a) $t = 1{\cdot}25\lambda$, $D = 50\lambda$, $F/D = 0{\cdot}35$
(b) $t = 2{\cdot}5\lambda$
– – – – geometrical optics
——— GTD and asymptotic physical optics
—–— equivalent edge-current method

As the focus is approached, both the optical term and the equivalent edge current terms, I^e, I^m [given either by eqn. 72 or 75] tend to infinity, but by taking the limit of the combined solution the numerical physical optics result (i.e. solving the double integral) is obtained. A similar phenomenon appeared in the example of Section 7.2. As was the case there, it is a consequence of the symmetry and uniformity of the incident field which produces this result. It will also be noted from Fig. 7.19 that the GTD and asymptotic physical optics solutions differ only at field points well removed from the focus, or in other words, from an optical boundary.

Fig. 7.20 gives two examples of the near field scattered by a paraboloidal reflector for an on-axis incident plane wave of unit amplitude. Such results find use in the evaluation of dual reflector antennas. [For example, see Claydon and James, 1975.] The GTD and asymptotic physical optics solutions are essentially identical since we are never far removed from an optical boundary in these examples. Note also that the equivalent edge current method as given by eqn. 74 has apparently provided a smooth transition for the asymptotic solution in the axial region. However this transition is not a result of the equivalent edge current method going uniformly into the GTD method off-axis but comes about by a judicious choice of the angle ψ as to when to change from one solution to the other. The next example will give one approach where the GTD solution is retrieved uniformly from the equivalent edge current formulation.

7.5 Radiation from a circular aperture with a finite flange

We now give a simple example of a hybrid technique where GTD is used in combination with an integral equation solution. Consider a circular waveguide terminating in an infinite perfectly conducting flange as illustrated in Fig. 7.21. (Although we consider here only

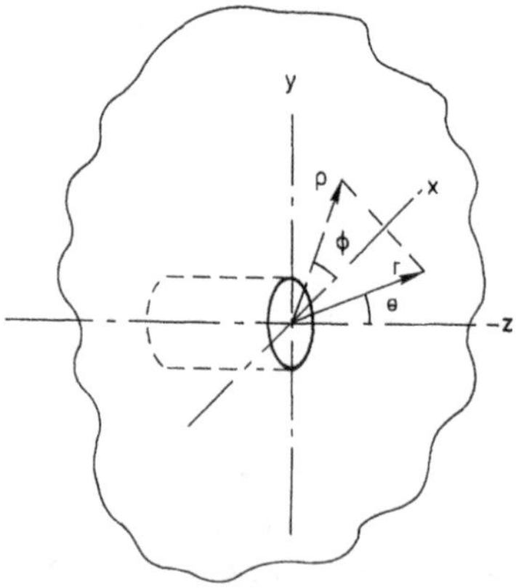

Fig. 7.21 Circular waveguide terminating in an infinite flange

circular waveguides, this restriction does not limit the general application of the method to be described). With the flange extending to infinity, a rigorous solution is possible for both the field in the waveguide aperture and the subsequent radiated field calculated by the aperture field method. An example can be found in Mishustin (1965) where the circular waveguide is excited by the TE_{11} mode. This approach was extended by Hockham (1976) to include the effect of a number of annular slots surrounding the circular waveguide aperture and by Bird (1979) to include coupling of higher-order modes between apertures. For any circularly symmetric system, the electric far-field radiated by the aperture(s) in an infinite flange can be written in the general form

$$\begin{aligned} E_\theta^0 &\sim \sin\phi g(\theta)\frac{\exp(-jkr)}{r} \\ E_\phi^0 &\sim \cos\phi\cos\theta f(\theta)\frac{\exp(-jkr)}{r} \end{aligned} \tag{77}$$

where the complex pattern functions $g(\theta)$, $f(\theta)$ are determined from the rigorous solution. In many practical cases the flange surrounding the aperture will be limited in extent as shown in Fig. 7.22 and the infinite flange model will not be a good representation of the true situation. To account for the effects of the finite flange we use a GTD formulation in conjunction with the integral equation solution for the infinite flange. To begin, the radiated field travelling along the infinite flange is given from eqn. 77 with $\theta = \pi/2$. For a finite flange we make the assumption that this is the incident field at the flange edge. Further, we shall ignore higher-order interaction between the rim and the aperture, consider the rim to be locally a 90° wedge, and assume the parameter τ in Fig. 7.22 to be sufficiently large so as to be able to ignore interaction between the edges created by the finite thickness of the flange. Also since the geometry is symmetrical, we need only consider the field in the half-space $0 \leqq \theta, \phi \leqq \pi$. Fig. 7.22(*b*) shows a cross-sectional view of the flanged aperture at the arbitrary angle of ϕ'. The field incident on the rim at the edges A and B is at grazing incidence. For the magnetically polarized incident field E_θ^i at A, we have from eqn. 77

$$E_\theta^i(A) = \sin\phi g\left(\frac{\pi}{2}\right)\frac{\exp(-jk\rho_{max})}{\rho_{max}} = \sin\phi C^i$$

In the far field $r_A \to \infty$ and the diffracted far-field $E_\theta^d(r_A)$ from the edge at A expressed in the equations given in section 6.8 is given (for grazing incidence) by

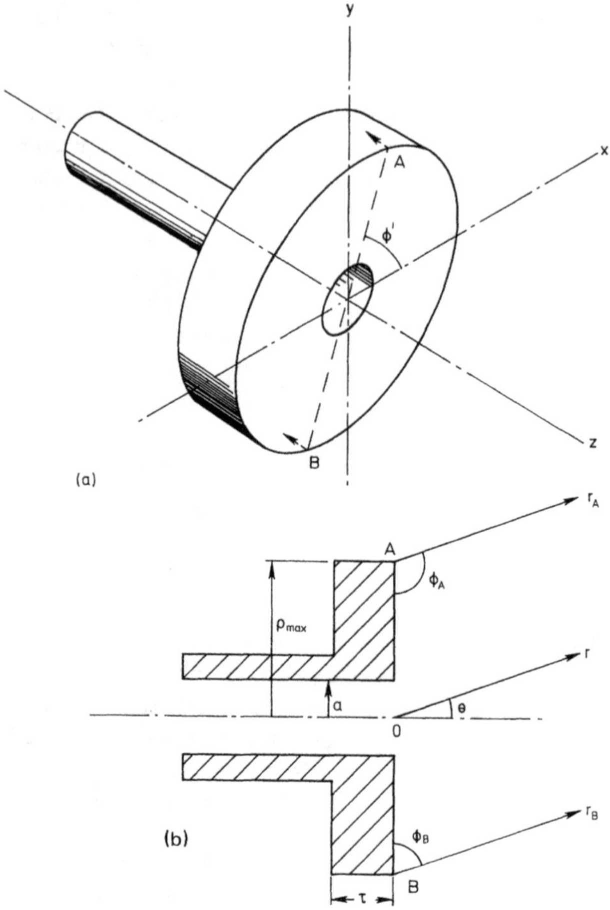

Fig. 7.22 Circular waveguide with a finite flange

$$E_\theta^d(r_A) \sim \sin\phi C^i D_A^m(\theta)\sqrt{\rho_{\max}\csc\theta}\exp\{j(\gamma - kr)\}/r \quad (78)$$

where $\gamma = k\rho_{\max}\sin\theta$

$$D_A^m(\theta) = h(\pi/2+\theta)_{n=-1} + h(-\pi/2-\theta)_{n=0}$$

and the quantity under the square root is the value for ρ_3. Similarly the diffracted far-field from edge B into the upper half-space is given by

$$E_\theta^d(r_B) \sim U(\pi/2-\theta)\sin\phi C^i D_B^m(\theta) j\sqrt{\rho_{\max}\csc\theta}\exp\{j(-\gamma - kr)\}/r \quad (79)$$

where

$$D_B^m(\theta) = h(\pi/2-\theta)_{n=-1} + h(\theta-\pi/2)_{n=0}$$

and the unit step function $U(x)$ accounts for the blocking effect of the flange when $\theta > \pi/2$.

To ensure continuity of the field at $\theta = \pi/2$ we need to include the dominant diffraction component of the interaction between B and A. The diffracted field from B is also at grazing incidence to the edge at A. Further, as a consequence of the axial caustic of the edge diffracted rays, the incident field from B appears as if emanating from a source at the origin. At edge A this incident field is given from eqn. 79 with $\theta = \pi/2$ and $r = \rho_{max}$. To include the interaction between B and A the constant C^i in eqn. 78 is replaced by

$$C^i \to C^i\left(1 + jD_B^m(\pi/2)\frac{\exp(-2jk\rho_{max})}{\sqrt{\rho_{max}}}\right) = C^i C_1^i \tag{80}$$

Adding equations 78–80 to the geometrical optics field, eqn. 77, gives the total far field for the magnetically polarized (to the flange edge) field component as

$$\begin{aligned} E_\theta^t \sim\ & \sin\phi\, \{U(\pi/2-\theta)g(\theta) + C^i\sqrt{j\rho_{max}\csc\theta}\ [C_1^i D_A^m(\theta)\cdot \\ & \cdot \exp\{j(\gamma-\pi/4)\} + U(\pi/2-\theta)D_B^m(\theta) \\ & \cdot \exp\{-j(\gamma-\pi/4)\}]\}\frac{\exp(-jkr)}{r} \end{aligned} \tag{81a}$$

In a similar way we construct the total field for the electrically polarized field component E_ϕ to the flange edge. In this case the direct incident field is zero and the slope-diffraction term becomes the leading term. From Section 5.11 we can write the slope-diffracted field from the edge at A as

$$E_\phi^{sd}(r_A) \sim \cos\phi C^{si} D_A^{e'}(\theta)\sqrt{\rho_{max}\csc\theta}\exp\{j(\gamma-kr)\}/r$$

where

$$\cos\phi C^{si} = \frac{1}{jk}\frac{\partial}{\partial n}E_\phi^i(A),$$

$$D_A^{e'}(\theta) = \frac{\partial}{\partial\phi_{0A}}D_A^e(\theta) = h'(-\pi/2-\theta)_{n=0} - h'(\pi/2+\theta)_{n=-1}.$$

Now we have $(\partial/\partial n) = (-1/\rho_{max})(\partial/\partial\theta)$ so it follows that

$$C^{si} = \frac{1}{jk\rho_{max}^2} f(\pi/2)\exp(-jk\rho_{max})$$

A similar expression exists for the slope-diffracted field from the lower edge at B. In this case the higher-order diffracted field from B to A is zero along the flange, so we write the uniform solution to the total field for the electrically polarized field component as

$$E_\phi^t \sim \cos\phi\,\{U(\pi/2-\theta)\cos\theta f(\theta) + C^{si}\sqrt{j\rho_{\max}\csc\theta}\,[D_A^{e'}(\theta)\cdot$$
$$\cdot\exp\{j(\gamma-\pi/4)\} + U(\pi/2-\theta)D_B^{e'}(\theta)$$
$$\cdot\exp\{-j(\gamma-\pi/4)\}]\}\,\frac{\exp(-jkr)}{r} \qquad (81b)$$

where C^{si}, $D_A^{e'}$ are given above and

$$D_B^{e'}(\theta) = h'(\theta-\pi/2)_{n=0} - h'(\pi/2-\theta)_{n=-1}$$

The above expressions for the diffracted field give infinite values for the field on the axis since this is a caustic of the edge diffracted rays. As in the previous section we can use the equivalent edge current method to overcome this difficulty. In the current example we shall derive uniform equivalent edge currents and present a formulation which goes uniformly from the axial caustic region into the GTD solution of eqn. 81 for field points well-removed from the axial region.

Using the methods of section 6.7, we consider the field diffracted from the edge at A (where $\phi=\phi'$) in Fig. 7.22b as equivalent to the field radiated by elemental currents $J_{\phi'}$, $M_{\phi'}$ situated at the edge. To derive these currents the edge at A is considered to be an element of a line source. This requires the edge diffracted field from a tangential straight edge at A to be equated to the field radiated by electric and magnetic line sources. Thus for magnetic polarization we have

$$E_\theta^d(r_A) = D_A^m C^i \sin\phi'\,\frac{\exp(-jkr_A)}{\sqrt{r_A}}$$

and equating this to the field from a magnetic line current $M_{\phi'}$ at A (see eqn. 6.59) we get

$$M_{\phi'} = -\sqrt{\frac{8\pi}{jk}}\sin\phi' I^m \qquad (82a)$$

Similarly for electric polarisation

$$J_{\phi'} = -\sqrt{\frac{\hat{\epsilon}}{\mu}}\sqrt{\frac{8\pi}{jk}}\cos\phi' I^{e'}. \qquad (82b)$$

With I^m, $I^{e'}$ given by $C^iD_A^m$, $C^{si}D_A^{e'}$, the equivalent currents of eqn. 82 are seen to be uniform. In solving for the diffracted electric far field in the axial region we use these currents in the potential integral solution of eqn. 2.12, viz.

$$E^d(r) \sim -jk\frac{\exp(-jkr)}{4\pi r}\int_0^{2\pi}\left[M\times\hat{r} + \sqrt{\frac{\mu}{\hat{\epsilon}}}(J - J\cdot\hat{r}\hat{r})\right]$$

$$\cdot \exp\{j\gamma\cos(\phi-\phi')\}\cdot\rho_{max}d\phi' \qquad (83)$$

where $\gamma = k\rho_{max}\sin\theta$. On the axis the diffraction coefficients D^m, $D^{e'}$ are independent of the angle ϕ'. As an approximation, we use the on-axis value of these coefficients to simplify the evaluation of eqn. 83 when determining the field in the axial region. With this approximation, and

$$\hat{\Phi}' = \hat{\Phi}\cos(\phi-\phi') + \sin(\phi-\phi')[\hat{r}\sin\theta + \hat{\theta}\cos\theta],$$

eqn. 83 yields the solution

$$\begin{Bmatrix} E^d_\theta \\ E^d_\phi \end{Bmatrix} \sim \sqrt{8j\pi k}\,\rho_{max}\frac{\exp(-jkr)}{4r}\begin{matrix}\sin\phi \\ \cos\phi\end{matrix}\left\{\begin{matrix} I^m \\ I^{e'} \end{matrix}[J_0(\gamma) - J_2(\gamma) + \begin{matrix} I^{e'} \\ I^m \end{matrix}\right.$$

$$\left.\times\cos\theta[J_0(\gamma) + J_2(\gamma)]\right\} \qquad (84)$$

after invoking the relationship

$$\frac{1}{2\pi}\int_0^{2\pi}\begin{matrix}\sin m\phi' \\ \cos m\phi'\end{matrix}\exp\{j\gamma\cos(\phi-\phi')\}d\phi' = j^m J_m(\gamma)\begin{matrix}\sin m\phi \\ \cos m\phi\end{matrix}$$

Equation 83 is seen to give a finite value for the diffracted field on the axis. At field points well-removed from the axis γ becomes large and from the asymptotic expressions for the Bessel functions it follows that

$$J_0(\gamma) + J_2(\gamma) \xrightarrow[\gamma\to\infty]{} 0$$

$$J_0(\gamma) - J_2(\gamma) \xrightarrow[\gamma\to\infty]{} \sqrt{\frac{2}{\pi\gamma}}[\exp\{j(\gamma-\pi/4)\} + \exp\{-j(\gamma-\pi/4)\}] \qquad (85)$$

Substitution of these equations into eqn. 84 does not retrieve the diffracted field as given in eqn. 81 no matter how I^m, $I^{e'}$ are chosen. To provide a uniform solution we use the approach given by Rusch (1981) who includes an additional $J_1(\gamma)$ term to eqn. 84. This does not contribute to the on-axis value of the diffracted field but does permit a uniform transition between the diffracted field component in eqn. 81 to that in eqn. 84. Rusch justifies the additional term by noting that such a term is found in many axially symmetric scattering results. This approach is in the spirit of the UTD formulation since the known field in two regions are matched uniformly through appropriate functions in the transition region, which in this case is the region surrounding the axial caustic.

The uniform solution for the diffracted field in the present example can be written in the following form. First define the quantities $I_{\pm}^{m}$, $I_{\pm}^{e'}$

$$I_{\pm}^{m} = \tfrac{1}{2}C^{i}[C_1^{i}D_A^{m}(\theta) \pm U(\pi/2-\theta)D_B^{m}(\theta)]$$

$$I_{\pm}^{e'} = \tfrac{1}{2}C^{si}[D_A^{e'}(\theta) \pm U(\pi/2-\theta)D_B^{e'}(\theta)]$$

and write the uniform diffracted field as

$$\begin{matrix} E_\theta^d \\ E_\phi^d \end{matrix} \sim \sqrt{8j\pi k}\,\rho_{\max}\frac{\exp(-jkr)}{4r}\begin{matrix}\sin\phi \\ \cos\phi\end{matrix}\left\{\begin{matrix} I_{+}^{m} \\ , \\ I_{+}^{e'}\end{matrix}[J_0(\gamma)-J_2(\gamma)]\right.$$

$$\left. + \begin{matrix} I_{+}^{e'} \\ , \\ I_{+}^{m}\end{matrix}\cos\theta[J_0(\gamma)+J_2(\gamma)] + \begin{matrix} I_{-}^{m} \\ , \\ I_{-}^{e'}\end{matrix} 2jJ_1(\gamma)\right\}; \qquad (86)$$

For large values of γ

$$2jJ_1(\gamma) \xrightarrow[\gamma\to\infty]{} \sqrt{\frac{2}{\pi\gamma}}[\exp\{j(\gamma-\pi/4)\}-\exp\{-j(\gamma-\pi/4)\}]$$

and it is readily shown that eqn. 86 uniformly retrieves the GTD diffracted field component of eqn. 81.

The application of eqn. 86 to a practical problem is demonstrated in Fig. 7.23 where a circular waveguide supporting the TE_{11} mode

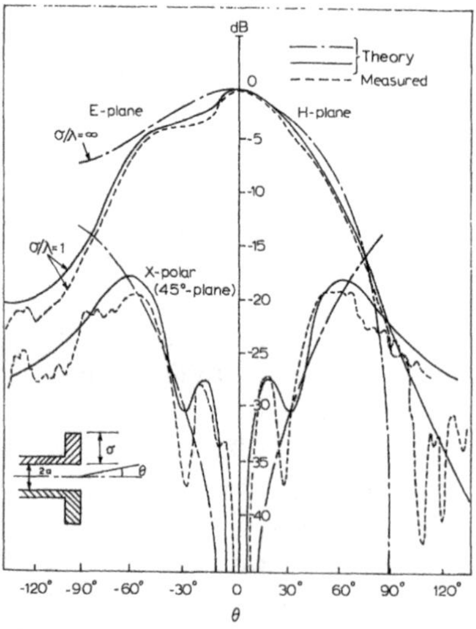

Fig. 7.23 Radiation patterns of circular waveguide ($ka = 2.32$) with a plane flange

terminates in a finite flange. Assuming the dominant mode only in the aperture, the far-field functions $g(\theta), f(\theta)$ are deduced from the aperture field method (assuming an infinite flange) to be

$$g(\theta) = J_1(u)/u, \qquad f(\theta) = J_1'(u)/\{1-(u/1.841)^2\},$$

where $u = ka \sin \theta$ and a is the waveguide radius. To account for a finite flange we write the far field as

$$\begin{aligned} E_\theta &\sim U(\pi/2-\theta)E_\theta^0 + E_\theta^d \\ E_\phi &\sim U(\pi/2-\theta)E_\phi^0 + E_\phi^d \end{aligned} \qquad (87)$$

where E_θ^0, E_ϕ^0 are given in eqn. 77 and E_θ^d, E_ϕ^d given by eqn. 86. The predictions by eqn. 87 for the E, H, and 45°-planes shown in Fig. 7.23 are in close agreement with measured radiation patterns. The importance of including the effect of the finite size of the flange is evident in this example.

References

BIRD, T.S., (1979): 'Mode coupling in a planar circular waveguide array', *IEE Proc. H, Microwave Opt. and Acoustics*, **3**, pp. 172–180.

CLAYDON, B., and JAMES, G.L. (1975): 'Asymptotic methods for the prediction of dual reflector antenna efficiency', *Proc. IEE*, **122**, (12), pp. 1359–1362.

COLLIN, R.E., and ZUCKER, F.J. (1969): 'Antenna theory' (McGraw-Hill).

HOCKHAM, G.A. (1976): 'Investigations of a 90° corrugated horn', *Electron. Lett.*, **12**, pp. 199–201.

LEE, S.W. (1969): 'On edge diffracted rays of an open-ended waveguide', *Proc. IEEE*, **57**, pp. 1445–1446.

MISHUSTIN, B.A. (1965): 'Radiation from the aperture of a circular waveguide with an infinite flange', *Sov. Radiophysics*, **8**, pp. 852–858.

ROWE, R.V. (1953): 'Microwave diffraction measurements in a parallel-plate region', *J. Appl. Phys.*, **24**, pp. 1448–1452.

RUDDUCK, R.C., and WU, D.C.F. (1969): 'Slope diffraction analysis of TEM parallel-plate guide radiation patterns', *IEEE Trans.*, **AP-17**, pp. 797–799.

RUSCH, W.V.T. (1981): 'Modified ring currents to treat axial caustics and slope diffraction in edge-diffraction analysis', *Electron. Lett.*, **17**, pp. 801–803.

WU, D.C.F., RUDDUCK, R.C., and PELTON, E.L. (1969): 'Application of a surface integration technique to parallel-plate waveguide radiation pattern analysis', *IEEE Trans.*, **AP-17**, pp. 280–285.

Index